RANDOM GRAPHS '85

NORTH-HOLLAND MATHEMATICS STUDIES 144

Annals of Discrete Mathematics (33)

General Editor: Peter L. HAMMER
Rutgers University, New Brunswick, NJ, U.S.A.

Advisory Editors

C. BERGE, *Université de Paris, France*
M. A. HARRISON, *University of California, Berkeley, CA, U.S.A.*
V. KLEE, *University of Washington, Seattle, WA, U.S.A.*
J.-H. VAN LINT, *California Institute of Technology, Pasadena, CA, U.S.A.*
G.-C. ROTA, *Massachusetts Institute of Technology, Cambridge, MA, U.S.A.*

NORTH-HOLLAND – AMSTERDAM · NEW YORK · OXFORD

RANDOM GRAPHS '85

Based on lectures presented at the 2nd International Seminar on Random Graphs and Probabilistic Methods in Combinatorics, August 5-9, 1985, organized and sponsored by the Institute of Mathematics, Adam Mickiewicz University, Poznań, Poland.

Edited by

Michał KAROŃSKI

*Adam Mickiewicz University, Poznań, Poland, and
Southern Methodist University, Dallas, TX, U.S.A.*

and

Zbigniew PALKA

Adam Mickiewicz University, Poznań, Poland

1987

NORTH-HOLLAND–AMSTERDAM · NEW YORK · OXFORD

© Elsevier Science Publishers B.V., 1987

All rights reserved. No part of this publication may be reproduced, stored in a retrieval system, or transmitted, in any form or by any means, electronic, mechanical, photocopying, recording or otherwise, without the prior permission of the copyright owner.

ISBN: 0 444 70265 2

Publishers:
ELSEVIER SCIENCE PUBLISHERS B.V.
P.O. Box 1991
1000 BZ Amsterdam
The Netherlands

Sole distributors for the U.S.A. and Canada:
ELSEVIER SCIENCE PUBLISHING COMPANY, INC.
52 Vanderbilt Avenue
New York, N.Y. 10017
U.S.A.

PRINTED IN POLAND

PREFACE

The Second International Seminar on Random Graphs and Probabilistic Methods in Combinatorics was held in Poznań in August, 1985. The conference enjoyed a considerable increase of popularity among combinatorists, probabilists and computer scientists, and gathered 52 participants from 14 countries.

The opening lecture devoted to unsolved problems in the theory of random graphs was delivered by Professor Edgar M. Palmer, while the enumerative results on series-parallel networks were the topic of another special plenary lecture presented by Professor John W. Moon. The participants discussed their recent research results relating to the main theme of the Seminar.

The volume reflects those presentations and covers a wide scope of random graphs subjects such as properties of random subgraphs of the n-cube, random binary and recursive trees, random digraphs, induced subgraphs and spanning trees in random graphs as well as matchings, hamiltonian cycles and closure in such structures. Papers in this collection also illustrate various aspects of percolation theory and its applications, properties of random lattices and random walks on such graphs, random allocation schemes, pseudo-random graphs and reliability of planar networks. Several open problems that were presented during a special session at the Seminar are also included at the end of the volume.

All papers were subject to a refereeing process. We are very grateful to the referees for their outstanding contribution.

The success of the conference was due to its participants, who created a truly open and friendly atmosphere of discussions and cooperation during those rainy days in Poznań. In fact, several new ideas and papers originated at the meeting, which especially encourages us to organize the Seminar biennially in future.

We wish to acknowledge the substantial help provided by the sponsor, Adam Mickiewicz University, and by Professors Julian Musielak and Andrzej Alexiewicz in organizing the meeting. The senior editor expresses his thanks to the Departments of Computer Science, Mathematics and Statistics of the Southern Methodist University, and especially to David Matula, George Reddien, Bill Schucany and David Yun for their kind support while editing the volume. Our sincere thanks are also due to the General Editor and to the Publishers of the Series.

Poznań, September 1986

and Michał KAROŃSKI
Zbigniew PALKA

CONTENTS

PREFACE	V
LIST OF PARTICIPANTS	VII
S. BERG, A variant of Banach's match box problem	1
J. E. COHEN, The sensitivity of expected spanning trees in anisotropic random graphs	9
M. E. DYER, A. M. FRIEZE and L. R. FOULDS, On the strength of connectivity of random subgraphs of the n-cube	17
J. W. ESSAM, Connectedness and connectivity in percolation theory	41
J. GIMBEL, D. KURTZ, L. LESNIAK, E. R. SCHEINERMAN and J. C. WIERMAN, Hamiltonian closure in random graphs	59
G. R. GRIMMETT, Long paths and cycles in a random lattice	69
B. HARRIS, M. MARDEN and C. J. PARK, The distribution of the number of empty cells in a generalized random allocation scheme	77
S. JANSON, Random self-avoiding walks in some one-dimensional lattices	91
J. JAWORSKI and I. H. SMIT, On a random digraph	111
R. KEMP, Additive weights of non-regularly distributed trees	129
P. KIRSCHENHOFER, A tree enumeration problem involving the asymptotics of the "diagonals" of a power series	157
T. ŁUCZAK, On matchings and hamiltonian cycles in subgraphs of random graphs	171
C. MCDIARMID, General percolation and oriented matroids	187
J. W. MOON, Some enumerative results on series-parallel networks	199
E. M. PALMER, Unsolved problems in the theory of random graphs	227
H. PRODINGER, Some recent results on the register function of a binary tree	241
A. RAMESH, M. O. BALL and Ch. J. COLBOURN, Bounds for all-terminal reliability in planar networks	261
A. RUCIŃSKI, Induced subgraphs in a random graph	275
J. SZYMAŃSKI, On a nonuniform random recursive tree	297
A. THOMASON, Pseudo-random graphs	307
K. WEBER, On the independence number of random subgraphs of the n-cube	333
J. C. WIERMAN, Directed site percolation and dual filling models	339
Random Graphs '85: Open problems	353

LIST OF PARTICIPANTS

K. Balińska, *Poznań, Poland*
S. Berg, *Lund, Sweden*
P. Bogacki, *Poznań, Poland*
M. Capobianco, *New York, N.Y., U.S.A.*
J. E. Cohen, *New York, N.Y., U.S.A.*
M. Dondajewski, *Poznań, Poland*
C. C. Y. Dorea, *Brasilia, Brasil*
M. E. Dyer, *Middlesbrough, England*
J. W. Essam, *Egham, England*
W. Fernandez de la Vega, *Orsay, France*
P. Flajolet, *Rocquencourt, France*
A. M. Frieze, *London, England*
T. Gerstenkorn, *Łódź, Poland*
A. Gibbons, *Coventry, England*
J. Gimbel, *Waterville, Maine, U.S.A.*
E. Godehardt, *Cologne, W. Germany*
G. R. Grimmett, *Bristol, England*
J. Gruszka, *Poznań, Poland*
B. Harris, *Madison, Wisconsin, U.S.A.*
S. Janson, *Uppsala, Sweden*
J. Jaworski, *Poznań, Poland*
M. Karoński, *Poznań, Poland*
R. Kemp, *Frankfurt a.M., W. Germany*
J. W. Kennedy, *New York, N.Y., U.S.A.*
P. Kirschenhofer, *Vienna, Austria*
L. Knopik, *Bydgoszcz, Poland*
U. Konieczna, *Bydgoszcz, Poland*
J. Kratochvil, *Prague, Czechoslovakia*
L. Kučera, *Prague, Czechoslovakia*

T. Łuczak, *Poznań, Poland*
N. V. R. Mahadev, *Winnipeg, Canada*
A. Marchetti Spaccamela, *Rome, Italy*
D. W. Matula, *Dallas, Texas, U.S.A.*
C. McDiarmid, *Oxford, England*
J. W. Moon, *Edmonton, Canada*
L. Mutafciev, *Sofia, Bulgaria*
K. Nowicki, *Lund, Sweden*
Z. Palka, *Poznań, Poland*
E. M. Palmer, *East Lansing, Michigan, U.S.A.*
M. Paprzycki, *Poznań, Poland*
H. Prodinger, *Vienna, Austria*
M. Protasi, *L'Aquila, Italy*
L. V. Quintas, *New York, N.Y., U.S.A.*
A. Ramesh, *Waterloo, Ontario, Canada*
Z. Skupień, *Kraków, Poland*
I. H. Smit, *Amsterdam, The Netherlands*
M. Sysło, *Wrocław, Poland*
J. Szymański, *Poznań, Poland*
D. Szynal, *Lublin, Poland*
A. Thomason, *Exeter, England*
K. Weber, *Rostock, G.D.R.*
J. C. Wierman, *Baltimore, Maryland, U.S.A.*

Participants of the 2nd International Seminar on Random Graphs held at Adam Mickiewicz University in Poznań, August 5 - 9, 1985.

A VARIANT OF BANACH'S MATCH BOX PROBLEM

Sven BERG

University of Lund, Sweden

Introduction and summary

Sampling from a certain simple bipartite graph is shown to lead to some interesting combinatorial and statistical problems. A basic probability distribution can be derived from simple arguments without invoking, e.g., the principle of inclusion-exclusion. It is also found that, when describing graph sampling schemes, it might be worthwhile to derive estimators of graph size, even though no specific inference problem is involved.

A picturesque formulation of the graph sampling scheme in terms of pills taken from a medicine bottle is reminiscent of the classical match box problem discussed in texts in probability theory. Certain parallels which can be exploited do exist between the two processes: on the one hand that of taking matches from a match box, on the other that of choosing pills from a medicine bottle. We begin in section 1 with a brief description of Banach's match box problem.

1. Banach's match box problem

This classical problem is given a fairly detailed treatment in Feller's book (1961, p. 157), and we quote:

"A certain statistician always carries one match box in his right pocket and one in his left. When he wants a match, he selects a pocket at random, the successive choices thus constituting Bernoulli trials with $p=1/2$. Suppose that initially each box contained exactly N matches and consider the moment when, for the first time, our mathematician *discovers* that a box is empty. At that moment the other box may contain 0, 1, ..., N matches...".*

Here we will denote the corresponding probability by $u_{x,N}$, which is given by Feller as

$$u_{x,N} = \binom{2N-x}{N} 2^{-2N+x}, \quad x=0, 1, \ldots, N. \tag{1}$$

Feller (pp. 212–213) calculates the expected number of matches — the mean of the discrete distribution (1) — as

$$\mu_x = (2N+1)\binom{2N}{N} \bigg/ 2^{2N} - 1, \text{ or} \tag{2}$$

$$\mu_x \simeq 2\sqrt{N/\pi} - 1 \text{ for large } N.$$

The following recursion formula can be used to calculate moments of all orders

$$u_{x+2v, N+v} = (N-x)^{(v)} u_{x,N}/(N+1)^{[v]}, \quad v=1, 2, \ldots \tag{3}$$

where [ ] denotes ascending factorial. An approximating distribution, when N is large, is obtained via Stirling's formula. If $x/\sqrt{N}$ is set $=y$, then

$$f(y) \simeq \frac{1}{\sqrt{\pi}} \exp(-y^2/4), \quad y>0. \tag{4}$$

To prove that $u_{x,N}$, as defined by (1), sums to unity is not difficult. This is essentially the way Riordan (1958, p. 31) does it.

Lemma. *For $N=1, 2, \ldots$, we have the summation formula*

$$\sum_0^N \binom{N+x}{x} 2^{-x} = 2^N. \tag{5}$$

Proof: First write $S_N = \sum \binom{N+x}{x} 2^{-x} = \sum a_{x,N}$. Then consider the difference $S_{N+1} - S_N$. We find termwise differences $a_{x,N+1} - a_{x,N} = \tfrac{1}{2} a_{x-1, N+1}$, $x=0, 1, \ldots, N$

---

* The match box problem was mentioned to Feller by H. Steinhaus.

and $a_{N, N+1} - a_{N+1, N+1} = 0$. Therefore, $S_{N+1} - S_N = \frac{1}{2}S_{N+1}$. The solution to this difference equation is $S_N = 2^N$. □

If we change the order of summation in (5), then

$$\sum \binom{2N-x}{N} 2^{-x+2N} = 1.$$

Hence the expression (1) does indeed represent a proper discrete probability distribution.

Before leaving the classical match box problem, let us pose just one more question, which is of a statistical nature: How many matches were there to begin with? Or put differently, by observing the variable x having distribution (1), would it be possible to estimate the parameter N? It can be shown that the following is an unbiased estimator of the parameter N:

$$\hat{N}_x = x(x+3)/2. \tag{6}$$

Feller (p. 156) gives a table with match box probabilities for $N=50$ and x ranging from 0 to 29. For the first 10 outcomes, we have included in Table 1 a column showing the unbiased estimate (6).

Table 1.

Match box probabilities (1) for $N=50$ and unbiased estimate (6).

x	$\hat{N}_x$	$u_{x, N}$	$U_{x, N} = \sum_{i \leq x} u_{i, N}$
0	0	.079 589	.079 589
1	2	.079 589	.159 178
2	5	.078 785	.237 963
3	9	.077 177	.315 140
4	14	.074 790	.389 931
5	20	.071 674	.461 605
6	27	.067 902	.529 506
7	35	.063 568	.593 073
8	44	.058 783	.651 855
9	54	.053 671	.705 527
10	65	.048 363	.753 890
.	.	.	.

$$\sum_{0}^{29} \hat{N}_x u_{x, N} = 49.838$$

As is evident from Table 1, it is hard to get a good estimate of the size parameter N on the basis of match box outcomes. The two most likely point estimates, irrespective of parameter value, are $\hat{N}_0=1$ and $\hat{N}_1=2$. The sampling distribution of the estimate (6) is spread out widely along the x-axis. Little is left of the distribution, however, when $x \geqslant 30$.

2. A medicine bottle problem

A medicine bottle contains a total of N double-strength sleeping pills. These pills can easily be divided in half. At bedtime, a certain statistician randomly removes one pill from the bottle and, if it is intact, breaks it into two. The process continues until no intact pills remain in the bottle. We may now pose the following question: How many halves are there left in the bottle?* Another query of a more statistical nature would be the following: Finding no more intact pills in the bottle and observing x halves, our forgetful statistician asks himself: How many double-strength sleeping pills were there initially in the bottle?

The problem is made more precise by introducing a few graph sampling concepts. Initially, we have a bipartite graph, $B_{N,\,N}$ say, consisting of N pairs of nodes. We sample nodes with equal probability and without replacement. After n draws, part of the graph may be reconstructed in the form of a random subgraph of $B_{N,\,N}$ (Cf. fig. 1).

Figure 1. Bipartite graph $B_{N,\,N}$ and random subgraph.

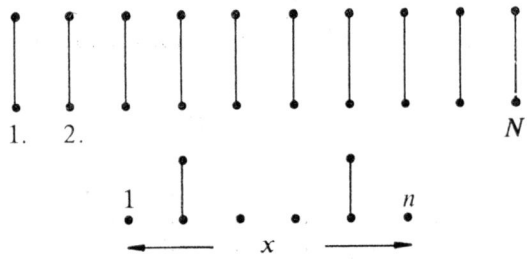

Problem 1. *If n nodes are sampled with equal probability and without replacement from $B_{N,\,N}$, how many different pairs of nodes will occur in the sample?*

We find the probability function (pf):

$$p_{x,\,N} = \binom{N}{x}\binom{x}{n-x}2^{2x-n} \Big/ \binom{2N}{n}, \quad x = \left[\frac{n+1}{2}\right], \ldots, \min(n, N). \qquad (7)$$

---

* I am indebted to Vibeke Horstmann at our department for the idea of this example.

This is a so called factorial series distribution, i.e. it is of the type $N^{(x)}a_x/g(N)$, with series function $g(N) = \binom{2N}{n}$ (Cf. Berg (1974) or *Encyclopedia of Statistical Sciences*, Vol. III). If a random variable x obeys a factorial series law, then the binomial moments of the complementary variable $N-x$ can be written down directly as

$$\beta^{(v)}_{N-x} = E\binom{N-x}{v} = \binom{N}{v} g(N-v)/g(N), \quad v = 1, 2, \ldots \tag{8}$$

where $g(N)$ is the series function.

Let now instead a Bernoulli sampling scheme with inclusion odds θ determine the nodes occurring in the sample. The probability of obtaining n nodes is thus the binomial

$$p(n) = \binom{2N}{n} \theta^n/(1+\theta)^{2N}, \quad n = 0, 1, \ldots, 2N,$$

while the simultaneous probability of having n nodes and x different pairs is given by

$$p(x, n) = \binom{N}{x}\binom{x}{n-x}(2\theta)^{2x-n}(\theta^2)^{n-x}/(1+\theta)^{2N}. \tag{9}$$

Hence the conditional probability of obtaining x different pairs of nodes, given the total n, is

$$p(x/n) = \binom{N}{x}\binom{x}{n-x}2^{2x-n} \Big/ \binom{2N}{n}, \quad x = \left[\frac{n+1}{2}\right], \ldots, \min(n, N),$$

i.e. we recapture the factorial series distribution (7).

The basic pf (7) is best derived just by looking at the graphs in Fig. 1. However, should we invoke, e.g., the principle of inclusion-exclusion here, then at first we obtain a more complex expression than (7). Actually, there is a neat summation formula behind the expression (7):

$$\sum_i \binom{x}{i}(-1)^i \binom{2(x-i)}{n} = \Delta^x \binom{2\theta}{n} = \binom{x}{n-x} 2^{2x-n},$$

where Δ is the forward difference operator.

Let us now look at a practical application of the above. Lanke (1984) reports an epidemiological study, in which there are data on $N=248$ sibships of size 2. Out of the 496 persons involved in the study, $n=108$ were found to suffer from a certain mental illness. In 20 cases both siblings were affected, i.e. a total of $x=88$ sibships. The question now arises whether this concentration is an indication that the illness might be hereditary? The null hypothesis of total randomness, i.e. both homogeneity and independence, gives rise to discrete distributions of the type discussed here. In particular, formula (8) gives the expected number of affected sibship pairs as

$$\mu_x = 248(1 - \binom{494}{108}/\binom{496}{108}) = 96.3,$$

and the standard deviation $\sigma_x = 2.68$. Thus the actual outcome is 3.1 standard deviations below what is expected under randomness, corresponding to a p-value less than 1%. A more detailed statistical analysis of the data could be made using the simultaneous distribution (9).

Problem 2. *If nodes are sampled one at a time without replacement from $B_{N,N}$, how many nodes remain when each pair is included in the sample at least once?*

This is of course the graph sampling version of the sleeping pill problem. Observing the event: "y remaining nodes" is tantamount to observing the outcome: "$2N-y-1$ nodes taken at random result in $N-1$ different pairs being included and the last node sampled comes from the Nth pair". With the aid of formula (7), we can write the required probability as

$$q_{y,N} = \binom{N}{y} 2^y \Big/ \binom{2N}{y} (2N-y)$$

$$= \binom{N-1}{y-1} 2^{y-1} \Big/ \binom{2N-1}{y}, \quad y = 1, 2, \ldots, N. \tag{10}$$

The pill box probability (10) is simply related to the match box probability (1). We have the relation

$$\binom{2N}{N} q_{x,N}(2N-x) = 2^{2N} u_{x,N} x, \quad x = 1, 2, \ldots, N, \tag{11}$$

which can be used to show that (10) represents a *bona fide* discrete probability distribution. Feller (pp. 212–213) calculates the mean of the match box distribution (1). Using his result and the relation (11), we find the expected number of half pills when no intact remains:

$$\mu_y = \sum y q_{y,N} = 2^{2N} \bigg/ \binom{2N}{N} - 1, \tag{12}$$

or, approximately for large N, $\mu_y \simeq \sqrt{\pi N} - 1$.

Finally, using Stirling's formula, one can derive an approximate distribution for the remaining number of nodes for a large bipartite graph. One finds that $y^2/2N$ is approximately distributed as χ^2 with two degrees of freedom.

Problem 3. *A waiting time problem. We continue sampling nodes one at a time and without replacement until the kth completed pair of nodes shows up in the sample. How long does it take?*

Let this happen at the $(y+1)$st step. We then have $y-(k-1)$ distinct pairs of nodes in n draws and one pair repeated at the $(y+1)$st draw. Using again the pf (7), we can write the required probability as

$$p^{(k)}_{y,N} = \binom{N}{y-k+1}\binom{y-k+1}{k-1}(y-2k+2)2^{y-2k+2} \bigg/ \binom{2N}{y}(2N-y),$$

$$y = 2k-1, \ldots, N+k-1. \tag{13}$$

Table 2.

Estimates of size of a bipartite graph and expected values in a waiting time problem.

k	Estimate of size	Expected value	Large N
1	$\hat{N}^{(1)}_y = y(y+3)/4$	$\mu^{(1)}_y = 2^{2N}/\binom{2N}{N} - 1$	$\simeq \sqrt{\pi N} - 1$
2	$\hat{N}^{(2)}_y = y(y+1)/8 - 1/2$	$\mu^{(2)}_y = \dfrac{3}{2} 2^{2N}/\binom{2N}{N} - 1$	$\simeq \dfrac{3}{2}\sqrt{\pi N} - 1$
3		$\mu^{(3)}_y = \dfrac{15}{8} 2^{2N}/\binom{2N}{N} - 1$	$\simeq \dfrac{15}{8}\sqrt{\pi N} - 1$

It is interesting and perhaps somewhat surprising to note that, for $k=1$, (13) reduces to the pf (10) found in the sleeping pill problem.

It is straightforward to establish a recursive relation between the probabilities $p^{(k)}_{y,N}$ and $p^{(k+1)}_{y,N}$. One finds:

$$p^{(k+1)}_{y,N} = \frac{(y-2k)(y-2k+1)}{4k(N-y+k)} p^{(k)}_{y,N}, \quad k = 1, 2, \ldots, y = 2k-1, \ldots, N+k-1. \tag{14}$$

With the aid of unbiased estimates of the parameter N — always quadratic functions of y — the recursion (14) can be used to calculate expected waiting times. A few such results are shown in Table 2 (see p. 7).

References

S. Berg (1974). Factorial series distributions with applications to capture-recapture problems. *Scand. Journ. Statist., 1*.

W. Feller (1961). *An Introduction to Probability Theory and Its Applications* (2nd edition). John Wiley, New York.

J. Lanke (1984). Testing randomness of distribution of cases among sibships. University of Lund Technical Report nr 7.

J. Riordan (1968). *Combinatorial Identities*. John Wiley, New York.

THE SENSITIVITY OF EXPECTED SPANNING TREES IN ANISOTROPIC RANDOM GRAPHS

Joel E. COHEN

The Rockefeller University,
New York, NY 10021-6399, U.S.A.

Consider graphs on the vertex set $V=\{1, 2, ..., n\}$, $1<n<\infty$, in which the edge between vertices i and j occurs with probability $p_{ij}=p_{ji}$, $0 \leq p_{ij} \leq 1$, independently for all edges. Let $P=(p_{ij})$ be the $n \times n$ symmetric matrix of edge probabilities with $p_{ii}=0$, $i=1, ..., n$. Let T be the random number of spanning trees. $E(T|P)$ denotes the redundancy, i.e., the expected number of spanning trees in random graphs with edge probability matrix P. An explicit (determinantal) formula for the sensitivity of the redundancy to changes in any edge probability, namely, $\partial E(T|P)/\partial p_{ij}$, shows that this sensitivity equals the redundancy of random graphs in which vertices i and j have been collapsed to a single vertex or are connected with probability 1. There is an analogous formula for directed graphs.

1. Random-edge graphs

Consider graphs (Tutte 1984) on the set V of vertices, $V=\{1, ..., n\}$, $1<n<\infty$. Suppose that the undirected edge $\{i,j\}$ between i and j occurs with fixed but arbitrary probability $p_{ij}=p_{ji}$, $0 \leq p_{ij} \leq 1$, independently for all distinct pairs i, j that satisfy $1 \leq i < j \leq n$. Define $p_{ii}=0$, $i=1, ..., n$. Let $P=(p_{ij})$ be the $n \times n$ symmetric matrix of edge probabilities. Random graphs or edge probability matrices P with at least two different off-diagonal edge probabilities $p_{ij} \neq p_{gh}$, $i \neq g$ or $j \neq h$, will be called *anisotropic* to distinguish them from *isotropic* random graphs in which, by definition, for some $p \in [0, 1]$, $p_{ij}=p$, for all $i \neq j$. By this definition, anisotropic random graphs necessarily have $n>3$ vertices, since the symmetry of P implies that random graphs with only two vertices are always isotropic: $p_{12}=p_{21}$.

Let T be the (random) number of spanning trees of a random graph, i.e., the number of trees that are incident to every vertex in V. Let $E(.)$ denote expected value or mean. Thus $E(T|P)$ denotes the expected number of spanning trees in random graphs with edge probability matrix P.

Graphs with randomly deleted edges may model a communication network in which communication links are subject to random failure, as in systems of strategic command and control (Ford 1985). The expected number of spanning trees could then be interpreted as a network's redundancy, or mean number of distinct paths of communication among all verices. The number of spanning trees in a fixed graph is sometimes called the complexity of the graph, so the expected number of spanning trees in random graphs could be called the mean complexity of the random graphs.

In efforts to alter the redundancy of a network, it is obviously useful to know how the redundancy changes with small changes in the probability of survival of each edge, namely, $\partial E(T|P)/\partial p_{ij}$, $i, j = 1, \ldots, n$. Buzacott (1980, p. 323) calls $\partial E(T|P)/\partial p_{ij}$ the "importance factor" of edge $\{i,j\}$ for the redundancy $E(T|P)$. The first purpose of this note (sections 2, 3) is to compute the importance factors, or sensitivities, for anisotropic random graphs (see eq. (2)). I shall also define anisotropic random directed graphs and compute (section 4) the importance factors for them (see eq. (4)).

The second purpose of this note is to prove (section 5) that the sum of the importance factors for anisotropic random graphs is larger, the more evenly the edge probabilities are distributed, in the following sense.

For any $n \times n$ real matrix A with zero diagonal ($a_{ii} = 0$, $i = 1, \ldots, n$), let $p(A) = \sum_{i,j} a_{ij}/[n(n-1)]$ be the average of the off-diagonal elements of A. Define $\bar{A}$ to be the equisummed matrix of such a (real, zero diagonal) matrix A if $\bar{A}$ has zero diagonal elements and all off-diagonal elements $\bar{a}_{ij}$, $i \neq j$, equal to $p(A)$. By construction, $\sum_{i,j} a_{ij} = \sum_{i,j} \bar{a}_{ij}$.

For an anisotropic matrix P of edge probabilities, $\bar{P}$ gives the edge probabilities of isotropic random graphs with the same total of edge probabilities. For $0 \leq \alpha \leq 1$, define $P_\alpha = (1-\alpha)P + \alpha\bar{P}$. As α increases from 0 to 1, P_α becomes closer to the edge probabilities of isotropic random graphs, while the total of edge probabilities remains constant.

I showed previously that if P is anisotropic, then the expected number $E_\alpha(T) \equiv E(T|P_\alpha)$ of spanning trees of random graphs with edge probabilities P_α increases strictly with α in $[0, 1]$ (Cohen 1986, Cor. 3.3). Thus, for a given sum of edge probabilities, the redundancy $E_\alpha(T)$ increases, the closer the matrix of edge probabilities is to being isotropic (as measured by increasing α in $P_\alpha = (1-\alpha)P + \alpha\bar{P}$).

Here I show that for anisotropic random graphs,

$$S(\alpha) \equiv \sum_{i<j} \partial E(T|P_\alpha)/\partial p_{ij}$$

increases strictly with α in $[0, 1]$. Thus for a given sum of edge probabilities, the total sensitivity increases, the closer the matrix of edge probabilities is to being isotropic as measured by α.

Finally, I conjecture (section 6) that the closer edge probabilities are to being isotropic, as measured by α in P_a, the more nearly uniform are the sensitivities after normalization by their total $S(\alpha)$. "Nearness to uniform" is made precise by a claim about majorization (Marshall and Olkin 1979), which I am able to prove only for $n=3$ vertices. Because I lack a proof for general n or a counter-example, this note is only a progress report. I would welcome a resolution of the conjecture.

2. Direct computation of importance factors

A formula (1) for $E(T|P)$ from Cohen (1986) generalizes the matrix-tree theorem attributed to Kirchhoff.

For any $n \times n$ real matrix $A=(a_{ij})$, following the terminology of Tutte (1984, p. 138), define the $n \times n$ Kirchhoff matrix $K(A)$ of A by

$$K_{ii} = \sum_{j \neq i} a_{ij}, \quad i=1, \ldots, n,$$

$$K_{ij} = -a_{ij}, \quad 1 \leq i \neq j < n.$$

Let det A be the determinant of A and let $A(i_1, \ldots, i_q)$ be the $(n-q) \times (n-q)$ principal submatrix of A left after deleting rows and columns $i_1, \ldots, i_q$, for $1 \leq q \leq n$. Thus det $[A(i)]$ is the determinant of the $(n-1) \times (n-1)$ matrix that remains after row and column i are deleted from A. Then (Cohen 1986)

$$E(T|P) = \det\{[K(P)](h)\}, \quad h=1, \ldots, n. \tag{1}$$

Then, by the chain rule,

$$\partial E(T|P)/\partial p_{ij} = (\partial \det\{[K(P)](h)\}/\partial\{[K(P)](h)\}_{ii})(\partial\{[K(P)](h)\}_{ii}/\partial p_{ij})$$

$$+ (\partial \det\{[K(P)](h)\}/\partial\{[K(P)](h)\}_{ij})(\partial\{[K(P)](h)\}_{ij}/\partial p_{ij})$$

$$+ (\partial \det\{[K(P)](h)\}/\partial\{[K(P)](h)\}_{ji})(\partial\{[K(P)](h)\}_{ji}/\partial p_{ij})$$

$$+ (\partial \det\{[K(P)](h)\}/\partial\{[K(P)](h)\}_{jj})(\partial\{[K(P)](h)\}_{jj}/\partial p_{ij}).$$

Now take $h=i$. Then row and column i of $K(P)$ are absent from $\{[K(P)(i)\}$, so the first three terms on the right vanish. Then $\partial\{[K(P)](i)\}_{jj}/\partial p_{ij}=1$ implies

$$\partial E(T|P)/\partial p_{ij} = \det\{[K(P)](i,j)\}. \tag{2}$$

Comparison of (2) with Corollary 2.1 of Cohen (1986) establishes an unanticipated identity: the sensitivity $\partial E(T|P)/\partial p_{ij}$ of the redundancy $E(T|P)$ to p_{ij} equals the redundancy of random graphs in which vertices i and j have been collapsed to a single vertex (call it i^*). (In the random graph where i and j have been collapsed to i^*, an edge $\{g, h\}$, $g \neq i^*$ and $h \neq i^*$, is assumed to occur with the probability that $\{g, h\}$ occurs in the original graph, and an edge $\{h, i^*\}$, $h \neq i^*$, occurs in the collapsed graph with the probability that either edge $\{h, i\}$ or $\{h, j\}$ occurs in the original graph; and all edges in the collapsed graph occur independently.)

3. Computation of importance factors using Buzacott's formula

Buzacott (1980, p. 323) established that, for any measure Q of reliability in which an element p_{ij} of P appears linearly, the importance factor of that element satisfies

$$p_{ij}(\partial Q/\partial p_{ij}) = Q - Q_0$$

where Q_0 is the value of the reliability measure for random graphs in which p_{ij} is replaced by 0 (and, necessarily for graphs, p_{ji} is also replaced by 0) and all other elements of P remain unchanged. Essam (these Proceedings, p. 51) observed that the above formula holds if Q is the expectation of any function defined on all possible subsets of edges.

For fixed i and j, let P_0 have all elements equal to the corresponding elements of P except that $(P_0)_{ij} = (P_0)_{ji} = 0$. Then, in Buzacott's formula, take

$$Q = E(T|P) = \det\{[K(P)](i)\},$$

$$Q_0 = E(T|P_0) = \det\{[K(P_0)](i)\}.$$

If e_j denotes the column n-vector with all elements 0 except a 1 as the jth element, then

$$j\text{th column of } P_0 = (j\text{th column of } P) - p_{ij} e_i.$$

Therefore

$$\det\{[K(P_0)](i)\} = \det\{[K(P)](i)\} - p_{ij} \det\{[K(P)](i, j)\}.$$

Hence, from Buzacott's formula

$$p_{ij}(\partial E(T|P)/\partial p_{ij}) = \det\{[K(P)](i)\} - \det\{[K(P_0)](i)\}$$
$$= p_{ij} \det\{[K(P)](i,j)\},$$

as found directly in (2).

4. Importance factors for expected spanning intrees of digraphs

Let T_1 denote the number of spanning intrees (Tutte 1984, Cohen 1986) to vertex 1 in random digraphs in which p_{ij} is the probability of a dart (directed edge) from vertex i to vertex j, $i \neq j$ (contrary to the direction of darts in Tutte 1984), and all darts occur independently. If P is the (possibly asymmetric) matrix of dart probabilities, with 0 diagonal elements, then (Cohen 1986)

$$E(T_1|P) = \det\{[K(P)](1)\}. \qquad (3)$$

Fix any i and j with $i \neq j$ and let $\delta_{gh} = 1$ if $g = h$, $\delta_{gh} = 0$ if $g \neq h$. Then elementary calculations similar to those above lead to

$$\partial E(T_1|P)/\partial p_{ij}$$
$$= (1 - \delta_{1i})\{\det\{[K(P)](1,i)\} \qquad (4)$$
$$+ (\delta_{1j} - 1)(-1)^{i+j} \det\{[K(P)](1,i;1,j)\}\},$$

where $[K(P)](1,i;1,j)$ denotes the $(n-2) \times (n-2)$ matrix that remains after deleting rows 1 and i and columns 1 and j from $K(P)$. This formula shows that $E(T_1|P)$ is completely insensitive to changes in the probabilities p_{1h} of darts from vertex 1 to any other vertex, as expected.

To illustrate, (4) gives for $n = 3$

$$\partial E(T_1|P)/\partial p_{21} = p_{31} + p_{32},$$
$$\partial E(T_1|P)/\partial p_{23} = p_{31},$$

both of which follow by differentiating (3), namely,

$$E(T_1|P) = p_{21}p_{31} + p_{23}p_{31} + p_{21}p_{32}.$$

5. Total sensitivity of random graphs increases with evenness of probabilities

As in sections 1, 2 and 3, let P be a symmetric matrix of edge probabilities for anisotropic random graphs (with three or more vertices) and let $S(\alpha)$ be the sum of the sensitivities when the edge probability matrix is P_α.

If x and y are two real n-vectors, $x=(x_1, ..., x_n)^T$, $y=(y_1, ..., y_n)^T$, let $x_{[1]} \geqslant ... \geqslant x_{[n]}$ denote the elements of x in decreasing order, and similarly for y. Following Marshall and Olkin (1979), say that x is majorized by y and write

$$x \prec y \text{ if and only if } \sum_{i=1}^{k} x_{[i]} < \sum_{i=1}^{k} y_{[i]},$$

$$k=1, ..., n-1,$$

and

$$\sum_{i=1}^{n} x_{[i]} = \sum_{i=1}^{n} y_{[i]}.$$

Theorem. *For $n=3$, $S(\alpha)$ is constant in α. For $n>3$, $S(\alpha)$ is strictly increasing in $\alpha \in [0, 1]$ to a maximum of*

$$S(1) = (n-1)[p(P)n]^{n-2}.$$

Proof. By (2), $S(\alpha)$ is the sum of all principal minors of order $n-2$ of $K(P_\alpha)$, which is identical to the $(n-2)$nd elementary symmetric function (ESF) of the eigenvalues of $K(P_\alpha)$ (e.g., Marshall and Olkin 1979, p. 504). The $(n-1)$-vector $\underset{\sim}{\mu}(\alpha) = (\mu_1(\alpha), ..., \mu_{n-1}(\alpha))^T$ of the $n-1$ largest eigenvalues of $K(P_\alpha)$, under the labelling $\mu_1(\alpha) \geqslant \mu_2(\alpha) \geqslant ... \geqslant \mu_{n-1}(\alpha) \geqslant \mu_n(\alpha) = 0$, has strictly positive elements when $\alpha > 0$ (Cohen 1986). If $n=3$, the first ESF is just the trace of $K(P_\alpha)$, which is a constant, equal to the trace of $K(P)$ by construction. If $n>3$, then the $(n-2)$nd ESF is a strictly Schur-concave function of $\underset{\sim}{\mu}(\alpha)$ (a fact attributed to I. Schur by Marshall and Olkin 1979, p. 78). All terms in the ESF that contain $\mu_n(\alpha) = 0$ as a factor vanish. Combining Lemmas 3.1 and 3.3 of Cohen (1986) shows that, if μ_i is the ith biggest eigenvalue of $K(P)$, then $\mu_i(\alpha) = (1-\alpha)\mu_i + \alpha p(P)n$, for $i=1, ..., n-1$, and if $0 < \alpha_1 < \alpha_2 \leqslant 1$, then $\underset{\sim}{\mu}(\alpha_2) \prec \underset{\sim}{\mu}(\alpha_1)$ but $\underset{\sim}{\mu}(\alpha_2)$ is not a permutation of $\underset{\sim}{\mu}(\alpha_1)$. By strict Schur-concavity of the $(n-2)$nd ESF, $S(\alpha_1) < S(\alpha_2)$.

In the isotropic case, all $n-1$ elements of the vector $\underset{\sim}{\mu}(1)$ are equal to $p(P)n$, so the $(n-2)$nd ESF is $[p(P)n]^{n-2} \binom{n-1}{n-2} = (n-1)[p(P)n]^{n-2}$. □

Since, for every $i=1,\ldots,n$, $\det\{[K(P_\alpha)](i)\}$ increases strictly with $\alpha \in [0, 1]$ when P is anisotropic (Cohen 1986, Cor. 3.3), and since the Theorem just proved shows that $S(\alpha)$ also increases with α in this case, it is natural to ask whether, for all $1 \leq i < j \leq n$, $\det\{[K(P_\alpha)](i,j)\}$ also increases with $\alpha \in [0, 1]$. It is easy to construct numerical examples that show that, for some $i<j$, $\det\{[K(P_\alpha)](i,j)\}$ may decrease strictly as α increases.

6. Relative sensitivity of random graphs: a conjecture about majorization

Let $r_{ij}(\alpha)$ be the relative sensitivity of $E(T|P)$ to the probability of edge $\{i,j\}$, that is,

$$r_{ij}(\alpha) = (\partial E(T|P_\alpha)/\partial p_{ij})/S(\alpha).$$

Let $\underset{\sim}{r}(\alpha)$ be the $\binom{n}{2}$-vector of relative sensitivities $r_{ij}(\alpha)$ in some order, such as lexicographically by subscript.

Conjecture. *If the matrix P of edge probabilities is anisotropic, then*

$$0 \leq \alpha_1 < \alpha_2 \leq 1 \text{ implies } \underset{\sim}{r}(\alpha_2) \prec \underset{\sim}{r}(\alpha_1)$$

and $\underset{\sim}{r}(\alpha_1)$ is not a permutation of $\underset{\sim}{r}(\alpha_2)$.

The conjecture asserts that the nearer P_α is to an isotropic matrix, the more nearly equal are the importance factors of the different edges.

Proof of conjecture for $n=3$. By direct computation from (2),

$$\partial E(T|P_\alpha)/\partial p_{12} = (1-\alpha)(p_{31}+p_{32}) + 2\alpha p(P),$$
$$\partial E(T|P_\alpha)/\partial p_{13} = (1-\alpha)(p_{21}+p_{23}) + 2\alpha p(P),$$
$$\partial E(T|P_\alpha)/\partial p_{23} = (1-\alpha)(p_{12}+p_{13}) + 2\alpha p(P).$$

The sum of these sensitivities, $S(\alpha)$, equals the sum of the elements of P or trace $K(P_\alpha)$, which is independent of α and is positive. Call the sum c. Thus $c\underset{\sim}{r}(\alpha)$ is the vector of sensitivities and has the form $(1-\alpha)x + \alpha\bar{x}$, where x is the non-zero non-negative vector $(p_{31}+p_{32}, p_{21}+p_{23}, p_{12}+p_{13})$ and $\bar{x}$ is the vector with all elements equal to the mean $2p(P)$ of the elements of x. It follows that $\underset{\sim}{r}(\alpha_2) \prec \underset{\sim}{r}(\alpha_1)$

by a very slight generalization (from positive vectors to non-negative vectors) of Lemma 3.1 of Cohen (1986). Since not all elements of x can be equal, $\underset{\sim}{r}(\alpha_2)$ cannot be a permutation of $\underset{\sim}{r}(\alpha_1)$. □

The conjecture for general n is buttressed by at least one hundred examples with randomly chosen P and $\alpha = 0, 1/4, 1/2, 3/4, 1$ for each of $n = 4, 5$ and 6.

ACKNOWLEDGEMENT

This work was supported in part by U.S. National Science Foundation grant BSR 84-07461, a Fellowship from the John D. and Catherine T. MacArthur Foundation, and the hospitality of Mr. and Mrs. William T. Golden. My participation in "Random Graphs '85" in Poznań was made possible by the exchange program of the U.S. National Academy of Sciences and the Polish Academy of Sciences.

References

Cohen, J. E. 1986 Connectivity of finite anisotropic random graphs and directed graphs. Math. Proc. Camb. Phil. Soc. 99 (2): 315–330, March.

Ford, D. 1985 A reporter at large: The Button. The New Yorker 43–91, 1 April 1985; 49–92, 8 April 1985.

Marshall, A. W. and Olkin, I. 1979 Inequalities: Theory of Majorization and Its Applications. New York: Academic.

Tutte, W. T. 1984 Graph Theory. Reading, Mass.: Addison-Wesley.

ON THE STRENGTH OF CONNECTIVITY OF RANDOM SUBGRAPHS OF THE n-CUBE

M. E. DYER*, A. M. FRIEZE** and
L. R. FOULDS***

* *Teesside Polytechnic,*
  Middlesbrough, Cleveland TS1 3BA, England
** *Queen Mary College, University of London,*
  London E1 4NS, England
*** *University of Waikato,*
  Hamilton, New Zealand

Suppose V_n is randomly sampled from the vertex set C^n of the n-cube so that $\Pr(x \in V_n) = p_v$ independently for each $x \in C^n$. Let $E(V_n)$ denote the edges of the subgraph of C^n induced by V_n under the usual adjacency relation in C^n. Suppose that A_n is now randomly sampled from $E(V_n)$ so that $\Pr(a \in A_n) = p_e$ independently for each $a \in E(V_n)$. Let $\Gamma_n = (V_n, A_n)$ be the random graph so produced. We show that for $s \geq 0$ integer, c constant and

$$p_e p_v = \tfrac{1}{2} + (\tfrac{1}{2} s \ln n + c)/n$$

that

$$\lim_{n \to \infty} \Pr(\Gamma_n \text{ is } s+1\text{-connected})$$

$$= 1 - \lim_{n \to \infty} \Pr(\Gamma_n \text{ is } s\text{-connected})$$

$$= e^{-\lambda}$$

where $\lambda = (\lim_{n \to \infty} p_v) \times e^{-2c}/s!$.

1. Introduction

We consider the set $C^n = \{0, 1\}^n$, the vertex set of the unit hypercube. For any two points $x, y \in C^n$ we call x, y *adjacent* (or *neighbours*) if they differ in precisely one coordinate. This relation endows C^n with a graph structure. For any set

$V \subseteq C^n$, let

$$E(V) = \{\{x, y\} : x, y \in V \text{ and } x, y \text{ are adjacent}\}.$$

Let p_e, p_v, $0 \leq p_e, p_v \leq 1$ satisfy

$$p = p_e p_v = \tfrac{1}{2} + (\tfrac{1}{2} s \ln n + c)/n$$

where $s \geq 0$ is an integer.

We construct a random subgraph $\Gamma_n = (V_n, A_n)$ of C^n in the following manner:

(a) V_n is randomly sampled from C^n so that $\Pr(x \in V_n) = p_v$ independently for each $x \in C^n$;

(b) A_n is randomly sampled from $E(V_n)$ so that $\Pr(a \in A_n) = p_e$ independently for each $a \in A_n$.

The following theorem is the main result of this paper. It establishes the asymptotic probable connectivity of Γ_n.

Theorem 1.1.

$$\lim_{n \to \infty} \Pr(\Gamma_n \text{ is } s\text{-connected}) = 1 - e^{-\lambda}$$

$$\lim_{n \to \infty} \Pr(\Gamma_n \text{ is } s+1\text{-connected}) = e^{-\lambda}$$

where

$$\lambda = \bar{p}_v e^{-2c}/s! \tag{1.1}$$

with $\bar{p}_v = \lim_{n \to \infty} p_v$. (Note that $\tfrac{1}{2} \leq \bar{p}_v \leq 1$.)

The first results on this problem are due to Saposhenko [10] and Burtin [3]. They considered the case $s = 0$, $p_v = 1$ and p_e constant. Burtin considered the case $p_e \neq \tfrac{1}{2}$, and showed that if $p_e < \tfrac{1}{2}$ then Γ_n is almost surely not connected, but is almost surely connected when $p_e > \tfrac{1}{2}$. Saposhenko considered the case $p_e = \tfrac{1}{2}$, and gives various properties of Γ_n. (Note that our use of "almost surely" (a.s.) simply means with probability tending to 1 as $n \to \infty$.)

Erdös and Spencer [5] tightened this result and considered the case $s = 0$, $p_v = 1$ and $p_e = \tfrac{1}{2} + c/n$ for constant c. See also Bollobás [2]. The case $s = 0$, $p_e = 1$ and $p_v = \tfrac{1}{2}$ was also considered by Saposhenko [9], where the random subset

V_n is considered to define a random boolean function of n variables. (See also the paper [11] by Weber for a short review.) Theorem 1.1 generalises the above results in two ways. First we consider a model in which both random edge and vertex deletions occur, and secondly we examine the *strength* of connectivity of the resulting graph.

2. Notation and preliminaries

We will write $N=2^n$, ln denotes natural logarithms and lg logs to base 2.

For $x, y \in C^n$, the (Hamming) distance between x and y is the minimal number of edges in a path connecting x and y in $G(C^n) = G_n$. The graph G_n has N vertices and $\tfrac{1}{2}nN$ edges. More generally we may consider the number $H_k(n)$ of connected subgraphs of size k which are vertex-induced subgraphs of G_n. Since any component of size $(k-1)$ has at most $(k-1)n$ incident edges, we have

$$H_k(n) \leqslant N.n.2n. \ldots (k-1)n = (k-1)!\, n^{k-1} N. \tag{2.1}$$

We will have recourse to the following simple inequalities:
For any $0 < r \leqslant n$,

$$\left(\frac{ne}{r}\right)^r \geqslant \binom{n}{r} \geqslant \left(\frac{n}{r}\right)^r. \tag{2.2}$$

For any $S \subseteq C^n$, let $B(S) = \{y \in C^n - S : y \text{ is adjacent to some } x \in S\}$. Thus $B(S)$ is the set of neighbours of S. More generally we can consider the set $B^i(S)$ of points which are within a (shortest) distance i from S, defined by

$$B^0(S) = S, \quad B^i(S) = B\Big(\bigcup_{j=0}^{i-1} B^j(S)\Big).$$

Clearly $B(S) = B^1(S)$. Now, for $1 \leqslant k \leqslant 2^n$, let $b(k) = \min\{|B(S)| : |S| = k\}$.

We first establish the following facts about $b(k)$, which will be required in the analysis of the main problem. The following theorem describes sets which attain $b(k)$.

Theorem 2.1. (Katona [8]). *For each $i = 0, 1, \ldots, n$ let $Z_i = \{x \in C^n : d(x, 0) = i\}$. Order the $x \in C^n$ so that all $x \in Z_i$ occur before all $x \in Z_j$ if $i < j$, and within each Z_i order the vertices lexicographically. Then $b(k)$ is attained by the set $S = M(k)$ which consists of the first k vertices in this ordering.*

Note that $|Z_i| = \binom{n}{i}$ for $i = 0, 1, \ldots, n$. Write

$$T(i) = \sum_{j=0}^{i} \binom{n}{j} = \left|\bigcup_{j=0}^{i} Z_j\right|.$$

Then, for any k, let $q(k)$ be defined by

$$T(q-1) < k \leq T(q). \tag{2.3}$$

Thus $M(k)$ consists of $\bigcup_{j=0}^{q-1} Z_j$ together with the (lexicographic) first $(k - T(q-1))$ vertices in Z_q.

We deduce the following facts. The first deals with "small" sets.

Lemma 2.2. For $1 \leq k \leq n+1$, $b(k) = kn - \tfrac{1}{2}(k-1)(k+2)$.

Proof. If $k = 1$, the formula gives $b(k) = n$. Thus we may assume $k \geq 2$ and hence $q(k) = 1$. Hence, letting e_j be the jth unit vector, and $e_0 = 0$, $M(k) = \{e_0, e_1, e_2, \ldots, e_{k-1}\}$. Now $M(k)$ has precisely the following neighbours:

(i) All vertices of the form $e_i + e_j$ ($0 \leq i \leq k-1$, $k \leq j \leq n$)

(ii) All vertices of the form $e_i + e_j$ ($1 \leq i \leq j \leq k-1$)

and all these are distinct. There are $k(n-k+1)$ of type (i) and $\binom{k-1}{2}$ of type (ii). Thus

$$b(k) = k(n-k+1) + \binom{k-1}{2} = kn - \tfrac{1}{2}(k-1)(k+2). \quad \square$$

Lemma 2.3. For all k, $b(k) > \left(\dfrac{n-2q-1}{q+1}\right)k$ where $q = q(k)$.

Proof. Let $t = k - T(q-1)$. Now $M(k)$ has the following neighbours:

(i) All vertices of Z_q except for the first t.

(ii) All vertices in Z_{q+1} adjacent to the first t in Z_q.

There are clearly $\binom{n}{q} - t$ of type (i). To count those of type (ii), consider $G(Z_q \cup Z_{q+1})$. This is bipartite with each vertex in Z_q having degree $(n-q)$, and each vertex in Z_{q+1} degree $(q+1)$. Thus by a simple count of edges, any t vertices in Z_q must have at least $(n-q)t/(q+1)$ neighbours in Z_{q+1}. Thus there is at least this number of type (ii) vertices. Therefore

$$b(k) \geq \binom{n}{q} - t + (n-q)t/(q+1), \quad \text{i.e.}$$

$$b(k) \geq \binom{n}{q} + (n-2q-1)t/(q+1) \tag{2.4}$$

$$= (n-2q-1)k/(q+1) + \left\{\binom{n}{q} - \binom{n-2q-1}{q+1}T(q-1)\right\}.$$

Thus we need only show that $T(q-1) < \dfrac{q+1}{n-2q-1}\binom{n}{q}$ to complete the proof. This follows since

$$T(q-1) = \sum_{i=1}^{q} \binom{n}{q-i}$$

$$= \binom{n}{q} \sum_{i=1}^{q} \frac{q(q-1)\dots(q-i+1)}{(n-q+1)\dots(n-q+i)}$$

$$\leq \binom{n}{q} \sum_{i=1}^{q} \left(\frac{q}{n-q+1}\right)^i$$

$$< \binom{n}{q} \frac{q/(n-q+1)}{1-q/(n-q+1)} \quad \text{provided } q < \tfrac{1}{2}(n+1)$$

$$= \binom{n}{q} \frac{q}{n-2q+1}$$

$$< \frac{q+1}{n-2q-1}\binom{n}{q} \quad \text{as required.} \quad \square$$

Note that for $q > \tfrac{1}{2}(n-1)$ the right hand side of the inequality in Lemma 2.3 is negative, and hence of no use. We will therefore need the following result, which states roughly that sets which are large, but not too large, have many neighbours. Its exact statement conforms with later requirements.

Lemma 2.4. *Let* $K = K(n) = [2^{\frac{1}{4}\sqrt{n \lg n}}]$. *Then, for large* n, *if* $K \leq k \leq N - K$ *then* $b(k) > 2^{\frac{1}{4}\sqrt{n \lg n}}$.

Proof. First note that $0 \leq t \leq \binom{n}{q}$ in (2.4) implies

$$b(k) \geq \min\left\{\binom{n}{q}, \binom{n}{q+1}\right\}. \tag{2.5}$$

Let $m = \lfloor \frac{1}{2}\sqrt{n/\lg n} \rfloor$. A simple calculation yields

$$T(m) = (1+o(1))\binom{n}{m}$$

$$\leq (1+o(1))(ne/m)^m < K$$

for large n.

Using $T(m-1) + T(n-m) = N$, we obtain $T(n-m) > N - K$ also. Thus $K \leq k \leq N - K$ implies $m < q(k) < n - m$. The result now follows from (2.5) and (2.2).

□

3. Minimum degree of Γ_n

As one might expect from previous work on the strength of connectivity of random graphs, e.g. Bollobás [1], Erdös and Rényi [4] or Fenner and Frieze [6], the connectivity of Γ_n is essentially determined by its minimum degree.

Let v_t denote the number of vertices of degree t in Γ_n. We have the following result.

Theorem 3.1.

(a) *For* $s \geq 1$, $c_n \not\to -\infty$, $\lim\limits_{n \to \infty} \Pr\left(\sum\limits_{t=0}^{s-1} v_t > 0\right) = 0$;

(b) *For* $s \geq 0$, $c_n \to c$, $\lim\limits_{n \to \infty} \Pr(v_s = i) = e^{-\lambda}\lambda^i/i!$ $(i \geq 0)$;

(c) *For* $c_n \to +\infty$, $\lim\limits_{n \to \infty} \Pr(v_s > 0) = 0$;

(d) *For* $c_n \to -\infty$, $\lim\limits_{n \to \infty} \Pr\left(\sum\limits_{t=0}^{s} v_t > 0\right) = 1$;

where λ is given by (1.1). Thus v_s is asymptotically Poisson with mean λ.

Proof.

(a) For $x \in C^n$ and $t < s$, we have that

$$\Pr(x \in V_n \text{ and degree of } x = t)$$

$$= p_v \binom{n}{t} p^t (1-p)^{n-t}$$

$$< n^t 2^{-n} \left(1 - \frac{s \ln n + 2c}{n}\right)^n (1 + o(1))$$

$$< n^t 2^{-n} n^{-s} e^{-2c} (1 + o(1)).$$

Thus, for $t < s$, the expected number of vertices of degree t is $O(n^{-1})$, and hence part (a) of the theorem follows.

(b) We use inclusion-exclusion. Let E_x be the event that $x \in C^n$ is a vertex of degree s in Γ_n. For any $S \subseteq C^n$, write $E_S = \bigcap_{x \in S} E_x$ and

$$\theta_t(n) = \sum_{|S| = t} \Pr(E_S).$$

Then the inclusion-exclusion formula gives

$$\Pr(v_s = i) = \sum_{t=i}^{N} (-1)^{t-1} \binom{t}{i} \theta_t(n) \tag{3.1}$$

and the sum on the right hand side of (3.1) alternates in value about the left hand side.

Let $\alpha_t = \lim_{n \to \infty} \theta_t(n)$, then using the alternation property of the sum in (3.1) it may be shown that

$$\lim_{n \to \infty} \Pr(v_s = i) = \sum_{t=i}^{\infty} (-1)^{t-1} \binom{t}{i} \alpha_t \tag{3.2}$$

provided the sum converges. Thus it remains to estimate α_t. We may obviously assume $n > t$. Let S be any set with $|S| = t$. Note that

$$\Pr(E_S) = \left(p_v \binom{n}{s} p^s (1-p)^{n-s}\right)^t$$

$$= \mu_t(n), \quad \text{say},$$

unless two vertices in S are either adjacent or within a distance 2 of each other (i.e. they have a common neighbour). We show that the contribution to $\sum_{|S|=t} \Pr(E_S)$ from such sets S is asymptotically negligible. Let this contribution be $\beta_t(n)$. Now there are at most

$$N\left(n+\binom{n}{2}\right)\binom{N-2}{t-2}<n^2 N^{t-1}$$

sets S which either have two adjacent vertices or two vertices with a common neighbour. For such sets S we have

$$nt - \tfrac{1}{2}(t-1)(t+2) \leq M = |B(S)| \leq nt.$$

For each $x \in B(S)$, let y_x be a particular neighbour of x in S. Let $T = \{x \in B(S) : x \in V_n \text{ and } \{x, y_x\} \in A_n\}$. If E_S occurs then $|T| \leq ts$. But

$$\Pr(|T| \leq ts) \leq \sum_{r=0}^{ts} \binom{M}{r} p^r (1-p)^{M-r}$$

$$= (1+o(1))\binom{M}{ts} p^{ts}(1-p)^{M-ts}$$

$$\leq (1+o(1)) M^{ts} 2^{-M}.$$

Thus

$$0 \leq \beta_t(n) \leq (1+o(1))(tn)^{ts} 2^{t^2} N^{-t} n^2 N^{t-1}.$$

Hence, as $n \to \infty$, $\beta_t(n) \to 0$, as claimed.

It remains to bound the number of sets S which have no common neighbour. There are clearly at most $\binom{N}{t} \leq N^t/t!$ such sets, and at least

$$N\left(N-\left(1+n+\binom{n}{2}\right)\right)\cdots\left(N-(t-1)\left(1+n+\binom{n}{2}\right)\right)/t!$$

$$> (N-tn^2)/t!$$

Therefore

$$\frac{(N-tn^2)^t}{t!} \mu_t(n) \leq \theta_t(n) - \beta_t(n) \leq \frac{N^t}{t!} \mu_t(n).$$

Letting $n \to \infty$, we obtain

$$\lambda^t/t! \leq \alpha_t \leq \lambda^t/t!$$

Hence the right hand side of (3.2) is

$$\sum_{t=i}^{\infty} (-1)^{t-1} \binom{t}{i} \lambda^t/t! = \lambda^i e^{-\lambda}/i! \qquad (3.3)$$

(c) Proceeding as in (a), we find that the expected number of vertices of degree s tends to zero, since in this case $e^{-2c_n} \to 0$.

(d) In this case, the expected number of vertices of degree at most s tends to infinity. A routine use of the Chebyshev inequality yields the result. One, of course, must compute the variance, but this is a simple exercise whose details are left to the reader. □

Corollary 3.2. *If $c_n \to c$, then*

$$\lim_{n \to \infty} \Pr(\delta(\Gamma_n) = s) = 1 - e^{-\lambda} \quad \text{and}$$

$$\lim_{n \to \infty} \Pr(\delta(\Gamma_n) = s + 1) = e^{-\lambda}$$

where δ denotes the minimum degree of the graph, as usual.
For $c_n \to +\infty$, we have

$$\lim_{n \to \infty} \Pr(\delta(\Gamma_n) \leq s) = 0,$$

and for $c_n \to -\infty$,

$$\lim_{n \to \infty} \Pr(\delta(\Gamma_n) \leq s) = 1.$$

Corollary 3.3. *If $p_v p_e = 1 + c_n/n$ where $c_n \to -\infty$, then*

$$\lim_{n \to \infty} \Pr(\Gamma_n \text{ is connected}) = 0.$$

4. Threshold for connectivity

We start our proof of Theorem 1.1 with the case $s = 0$. In this case we have to establish the limiting probability that Γ_n is connected. In view of Corollary 3.3 we may assume that $c_n \not\to -\infty$. We shall only treat the case $c_n \to c$ in detail. All the calculations go through *a fortiori* for $c_n \to \infty$.

Since c_n plays only a minor role in the subsequent analysis, we shall assume for convenience that $c_n = c$.

Let

$$\Pi(n, k) = \Pr(\Gamma_n \text{ has a component of size } k)$$

and

$$\Pi(n, k_1, k_2) = \Pr(\Gamma_n \text{ has a component of size } k, k_1 \leq k \leq k_2).$$

Clearly

$$\Pi(n, k_1, k_2) \leq \sum_{k=k_1}^{k_2} \Pi(n, k).$$

Corollary 3.2 can be re-expressed here as

$$\lim_{n \to \infty} \Pi(n, 1) = 1 - e^{-\lambda} \tag{4.1}$$

where $\lambda = \bar{p}_v e^{-2c}$.

We can therefore prove our theorem for the case $s = 0$ by showing

$$\lim_{n \to \infty} \Pi(n, 2, \tfrac{1}{2}N) = 0. \tag{4.2}$$

5. Small components

We show here that the probability that Γ_n has a component of size k, $2 \leq k \leq 2^{\frac{1}{2}\sqrt{n \lg n}}$, is very small. Our estimates of $\Pi(n, k)$ in this section are based on the bound

$$\Pi(n, k) \leq H_k(n) \, p_v^k (1-p)^{b(k)}. \tag{5.1}$$

The right hand side of (5.1) is an upper bound to the expected number of components of size k in Γ_n. To see this, let $S \subseteq C^n$ and $|S| = k$. For each $y \in B(S)$, choose an edge $e(y)$ from y to a vertex in S. If S is a component of Γ_n, then $V_n \supseteq S$ and $e(y) \notin A_n$ for $y \in B(S)$. Thus

$$\Pr(S \text{ is a component of } \Gamma_n) \leq p_v^{|S|} (1-p)^{|B(S)|}$$

and (5.1) follows.

We shall actually use a weakening of (5.1) to

$$\Pi(n,k) \leq H_k(n) 2^{-b(k)} \left(1 - \frac{2c}{n}\right)^{b(k)}$$

$$\leq H_k(n) 2^{-b(k)} e^{2k|c|}. \quad \text{(For } c_n \to \infty, \tag{5.2}$$

we would obviously omit the modulus).

Lemma 5.1. For $2 \leq k \leq n^2/2$ and large enough n, $\Pi(n,k) \leq 2^{-\frac{1}{4}kn}$.

Proof.

(i) First consider $k=2$, i.e. an isolated edge. Now C^n has only $n2^{n-1}$ edges altogether. Thus, using (5.2) and Lemma 2.2, we obtain

$$\Pi(n,2) \leq n2^{n-1} 2^{-2(n-1)} e^{4|c|} \leq 2^{-\frac{1}{2}n}$$

for large n.

(ii) $3 \leq k \leq \frac{1}{2}n$.

$$H_k(n) \leq 2^n n^{k-1}(k-1)! \leq 2^{n+2k \lg n}. \tag{5.3}$$

Thus, using (5.2) and Lemma 2.2,

$$\Pi(n,k) \leq 2^{n+2k \lg n - (kn - \frac{1}{2}(k-1)(k+2))} e^{2k|c|}$$

$$\leq 2^{-(1-o(1))kn} \quad \text{for } k \geq 3.$$

(iii) $\frac{1}{2}n \leq k \leq \frac{1}{2}n^2$.

In this range of k, $q(k) \leq 2$ and so $b(k) \geq k(n-5)/3$, using Lemma 2.3. Using (5.2) and (5.3) we obtain

$$\Pi(n,k) \leq 2^{n+2k\lg n - k(n-5)/3} e^{2k|c|}$$

$$\leq 2^{-\frac{1}{4}kn} \text{ for large enough } n. \quad \square$$

Lemma 5.2. For $\frac{1}{2}n^2 \leq k \leq K(n) = \lceil 2^{\frac{1}{2}\sqrt{n \lg n}} \rceil$ and large n,

$$\Pi(n,k) \leq 2^{-k\sqrt{n \lg n}/3}.$$

Proof. Now $H_k(n) \leq (2kn)^k$ and $b(k) > k(n-2q-1)/(q+1)$ by Lemma 2.3, where $q = q(k)$. Thus, using (5.2),

$$\Pi(n,k) \leq (2kn/2^{(n-2q-1)/(q+1)})^k e^{2k|c|}. \tag{5.4}$$

Let $m = \lfloor\sqrt{n/\lg n}\rfloor$. It is easy to see from (2.2) that $T(m-1) > \binom{n}{m-1} > K$. Thus, within our range of k, $q(k) \leq m$. Consequently, for large n, $(n-2q-1)/(q+1) > 9\sqrt{n \lg n}/10$. Hence, from (5.4),

$$\Pi(n,k) \leq (2n2^{\frac{1}{2}\sqrt{n\lg n}}/2^{9\sqrt{n\lg n}/10})^k e^{2k|c|}$$

$$\leq 2^{-k\sqrt{n\lg n}/3} \text{ for large } n. \quad \square$$

Corollary 5.3.

$$\Pi(n, 2, 2^{\frac{1}{2}\sqrt{n\lg n}}) \leq 2^{-n/3} \quad \text{and}$$

$$\Pi(n, n, 2^{\frac{1}{2}\sqrt{n\lg n}}) \leq 2^{-n^2/5},$$

in both cases for large enough n. $\quad \square$

We conclude this section by showing that there are few vertices in "very small" components. We prove only a fairly crude bound.

Lemma 5.4. *The probability that Γ_n has a total of more than n^5 vertices in components of size at most n is less than $2^{-n^2/5}$, for large enough n.*

Proof. The condition of the Lemma implies that for some $1 \leq k \leq n$, there must be at least n^4 vertices in components of size k. Call two components S, S' of size k independent if $B(S) \cap B(S') = \emptyset$. Consider an associated graph with vertices S, and an edge SS' whenever S and S' are not independent. Now, for any vertex S, $|B(S)| \leq kn$ and each vertex in $B(S)$ is adjacent to at most $(n-1)$ other components S'. Hence the associated graph has degree bound $k\binom{n}{2}$. It follows easily that it possesses an independent set of size $r = (n^4/k)/(kn(n-1)+1)$, since it must have at least n^4/k vertices. Thus $r \geq n^2/k^2$. Consider first $k=1$. The probability that Γ_n contains n^2 independent isolated vertices is no more than $\binom{N}{n^2}(1-p)^{n^3}$ $= o(2^{-n^2})$. For $k \geq 2$, following the lines of the proof of Lemma 5.1, the probability of at least r independent components of size k is less than $(H_k(n)2^{-b(k)}e^{2k|c|})^r$ $\leq 2^{-n^2/4}$. Thus the probability that for any $1 \leq k \leq n$, there are at least n^4 vertices in components of this size is less than $n2^{-n^2/4} < 2^{-n^2/5}$ for large enough n. $\quad \square$

Remark 5.5. We say that $x \in C^n$ is an *isolated non-vertex* if $(\{x\} \cup B(\{x\})) \cap V_n = \emptyset$. Arguing as in Lemma 5.4, we may show that

$$\Pr(C^n \text{ has at least } n^4 \text{ isolated non-vertices}) = o(2^{-n^2}).$$

6. Large components

We show that there are no disconnected large components by "bootstrapping" the results of §5. The approach is to dissect C^{n+1} into two copies of C^n in an obvious way. We require the following result, which follows immediately from results of Hoeffding [7].

Lemma 6.1. *For large n,*

$$\Pr((21\bar{p}_v - 2)N/20 \leqslant |V_n| \leqslant (1+\bar{p}_v)N/2)$$

$$\geqslant 1 - e^{-2^{\frac{1}{2}n}}.$$

Proof. From Hoeffding's results, noting that $|V_n|$ is a sum of N independent zero-one random variables, each with expectation p_v.

$$\Pr(|V_n| \leqslant (1-\varepsilon)Np_v) \leqslant e^{-\frac{1}{2}\varepsilon^2 Np_v} \tag{6.1}$$

$$\Pr(|V_n| \geqslant (1+\varepsilon)Np_v) \leqslant e^{-\frac{1}{3}\varepsilon^2 Np_v} \tag{6.2}$$

for any ε, $0 \leqslant \varepsilon \leqslant 1$.

Putting $\varepsilon = 1/n$ and using some obvious estimates in (6.1) and (6.2) gives the conclusion. □

In view of Lemma 6.1, it is sufficient to show that, with high probability, Γ_n possesses no connected component of size at most $(1+\bar{p}_v)N/4$, i.e. that

$$\lim_{n \to \infty} \Pi(n, 2, (1+\bar{p}_v)N/4) = 0.$$

Definition. For any $0 \leqslant f \leqslant n$, an *$f$-face* of C^n is a maximal subset for which the values of $(n-f)$ coordinates are constant.

We now consider C^{n+1} in order to derive a relationship between $\Pi(n+1, n+1, 4k)$ and $\Pi(n, n, k)$ for large (but not too large) values of k.

We will denote the n-face of C^{n+1} on which $x_j = i$ ($i = 0, 1$) by C_{ij}^{n+1}. Let $X_{ij} = V_{n+1} \cap C_{ij}^{n+1}$. Then we observe that X_{ij} is a random subset of the n-cube

C_{ij}^{n+1}, and that for $i=0, 1$ these are independent of each other. This simple observation provides the main tool.

Consider a fixed value of j, and for convenience drop this subscript. We will say a subset S_0 of X_0 is *incident* with a subset S_1 of X_1 if there exist $x_0 \in X_0$, $x_1 \in X_1$ with $\{x_0, x_1\} \in A_{n+1}$. Clearly if S_0, S_1 are incident they lie in the same connected component of Γ_{n+1}.

We define a component of X_{ij} to be a component of the subgraph of Γ_{n+1} induced by X_{ij}. Let us call such a component of any X_{ij} *trivial* if it has at most n vertices, and *large* if it has at least $K = \lceil 2^{\frac{1}{2}\sqrt{n \lg n}} \rceil$ vertices.

Lemma 6.2. *The probability that, for any i, j there exists a large component in X_{ij} which is not incident with any non-trivial component of $X_{1-i,j}$ is less than $2^{-n^2/6}$.*

Proof. We condition on X_{ij}, i.e. we consider it fixed and use the inequality

$$\Pr(A) = \sum_X \Pr(A|X_{ij}=X) \Pr(X_{ij}=X)$$

$$\leqslant \max_X \Pr(A|X_{ij}=X)$$

for any event A. Again we will drop the suffix j temporarily and assume without loss that $i=0$. Suppose that X_0 has a component S of size at least K. Since X_1 is independent of X_0, it is unconditioned. If S is incident with s vertices of X_1, then it follows that

$$\Pr(s \leqslant n^5) \leqslant \binom{K}{n^5}(1-p)^{K-n^5}$$

which, substituting for K and noting that $\binom{K}{n^5} \leqslant K^{n^5}$, clearly implies that $s > n^5$ with probability $1 - o(2^{-\frac{1}{2}K})$, say. However, from Lemma 5.4, with probability at least $1 - 2^{-n^2/5}$, this implies that S is incident with a non-trivial component of X_1. Thus the probability that the Lemma fails is at most

$$2^{-n^2/5} + o(2^{-\frac{1}{2}K}) \text{ for this } S, i, j. \tag{6.3}$$

There are obviously less than 2^n large components in X_0, and C^{n+1} has only $2(n+1)$ n-faces. Thus (6.3) needs to be inflated by a factor of less than $2^n \times 2(n+1)$. For large enough n this gives the conclusion. □

Now, if S_0, S_0' are both components of X_0, we will say S_0 is *bridged* to S_0' if they are both incident to a connected component S_1 of X_1, for any such S_1. Clearly if S_0, S_0' are bridged, they lie in the same component of Γ_{n+1}.

Lemma 6.3. *The probability that any large component (of any X_{ij}) with less than $(1+\bar{p}_v)N/4$ vertices is not bridged to a distinct non-trivial component is less than $2^{-n^2/6}$ for large n.*

Proof. Let S be any such component of X_0, as in Lemma 6.2. Then $K<|S| <(1+\bar{p}_v)N/4$. Thus, from Lemma 6.1, with probability at least $1-e^{-2^{\frac{1}{4}n}}$,

$$|X_0-S|>(21\bar{p}_v-2)N/20-(1+\bar{p}_v)N/4 \geqslant N/20.$$

Hence $|S \cup B_0(S)| \leqslant N-|X_0-S| < N-N/20 < N-K$, where $B_0(S)$ is the set of neighbours of S in C_{0j}^{n+1}. Thus, from Lemma 2.4,

$$|B_0^2(S)|=|B_0(S \cup B_0(S))|>2^{\frac{1}{4}\sqrt{n \lg n}}=K_1, \text{ say}.$$

Now every vertex of $B_0^2(S)$ is clearly either

(i) a vertex of some other component of X_0, or

(ii) adjacent to a vertex of some other component of X_0, or

(iii) an isolated non-vertex.

However, from Remark 5.5, with probability at least $1-2^{-n^2}$, there are less than n^4 vertices of type (iii). Thus $B_0^2(S)$ has at least K_1-n^4 vertices of types (i) and (ii). Now, since there are only $1+n+\binom{n}{2}<n^2$ vertices within a distance at most 2 from any given vertex, there are at least $(K_1-n^4)/n^2$ vertex-disjoint paths of length 2 edges (or 3 vertices) to vertices of types (i) or (ii) in $B_0^2(S)$ from S. Since a vertex of another component can be reached in at most one further step from the terminal vertex of each such path, and each vertex has degree n in C^n, there are thus at least $(K_1-n^4)/n^3=K_2$ vertex-disjoint paths of length at most 4 vertices into a distinct component of X_0. Now the corresponding path exist in X_1 with probability at least $p_v^4 p_e^3 > 1/20$, say, for large n. Thus the probability that no more than n^5 of these paths exist in X_1 is at most

$$\binom{K_2}{n^5}(19/20)^{K_2-n^5}=o(2^{-n^3}).$$

Thus, using Lemma 5.4, the probability that the Lemma fails for S is at most

$$e^{-2^{\frac{1}{2}n}}+2^{-n^2}+o(2^{-n^3})+2^{-n^2/5}. \tag{6.4}$$

Again, as in Lemma 6.2, (6.4) needs to be inflated by a factor of at most $(n+1)2^{n+1}$ to give the conclusion. □

Before proving the main result of this section, we need one further Lemma.

Lemma 6.4. *Any component of Γ_{n+1} with at most $n+1$ vertices is contained entirely in some C_{ij}^{n+1}.*

Proof. We prove a stronger statement which clearly implies the Lemma.

Claim. *Any connected subgraph Y of Γ_{n+1} with at most $k \leq n+1$ vertices is contained entirely in some $(k-1)$-face of C^{n+1}.*

Proof of Claim. By induction on k. It is obvious for $k=1$. For $k>1$, fix any connected subgraph Y' of Y with $(k-1)$ vertices. By hypothesis this lies entirely in some $(k-2)$-face F. All vertices of F agree on $(n-k+3)$ "constant" coordinates. We now add the excluded vertex, which is connected by an edge to Y', and hence is either in F or is joined to it by an edge of C_{n+1}. Thus its coordinates agree with those of F on all but at most one of the constant coordinates. Thus the coordinates of all vertices in Y agree on at least $(n-k+2)$ coordinates, which gives the result. □

We now prove the main result of this section. In order to avoid confusing the flow of the argument with too many probability statements, we will write σ_1 implies $(\varepsilon) \sigma_2$ for two events σ_1, σ_2 if the implication holds on an event of probability at least $1-\varepsilon$. The whole argument then holds on an event of probability at least $(1-\Sigma\varepsilon)$, where the sum is over the values of ε in all such implications used. We will have to amend our notation slightly so that

$$\Pi(n, k_1, k_2, p)$$

$$= \Pr(\Gamma_n \text{ contains a connected component of size } k,$$

$$k_1 \leq k \leq k_2, \text{ assuming } p_v p_e = p).$$

The need for this amendment comes from

Lemma 6.5. *If $K(n) \leq k \leq (1+\bar{p}_v)N/8$, then*

$$\Pi(n+1, n+1, 4k, \tfrac{1}{2}+c/(n+1))$$
$$\leq 2(n+1)\Pi(n, n, k, \tfrac{1}{2}+c/(n+1)) + 2^{-n^2/7}.$$

Proof. Suppose Γ_{n+1} has a component S of size $n+1 \leq |S| \leq 4k$. Choose now any sub-component (i.e. connected subset) of S with $n+1$ vertices, then by Lemma 6.4 there exists i, j such that C_{ij}^{n+1} contains a connected sub-component of S with at least $n+1$ vertices. We will assume that $i=0$. We wish to show that, with high probability, either X_0 or X_1 contains a sub-component of S with size in the range n to k inclusive. Let us call such a component of X_0 or X_1 a *bad* component. Let $S_0 = S \cap X_0$. We have $(n+1) \leq |S_0|$. If $|S_0| \leq k$, then S_0 is a bad component. Therefore assume $|S_0| > k$. If $|S_0| \leq 2k \leq (1+\bar{p}_v)N/4$, then by Lemma 6.3 this implies $(2^{-n^2/6})$ that S_0 is bridged to a non-trivial component $S_0' \subseteq S$ of X_0. Thus $|S_0'| \geq n+1$. If $|S_0'| \leq k$, then it is a bad component. Otherwise $|S_0'| > k$, so $|S_0| + |S_0'| > 2k$. Thus $|X_1 \cap S| < 2k$. However, Lemma 6.2 implies $(2^{-n^2/6})$ that S_0 is incident with a non-trivial component $S_1 \subseteq S$ of X_1. Thus $|S_1| \geq (n+1)$. If $|S_1| \leq k$, then it is a bad component. Otherwise $|S_1| > k$, and Lemma 6.3 implies $(2^{-n^2/6})$ that it is bridged to a component $S_1' \subseteq S$ with $|S_1'| \geq (n+1)$. Clearly $|S_1'| < 2k - |S_1| < k$, and hence S_1' is a bad component.

Otherwise, we must assume $|S_0| > 2k$, then this again implies $(2^{-n^2/6})$ that either S_1 or S_1' is a bad component.

The above proof clearly holds on an event of probability at least $1 - 5.2^{-n^2/6} \geq 1 - 2^{-n^2/7}$ for large n. On this event the existence of S implies the existence of a bad component in some C_{ij}^{n+1}, which has probability at most $2(n+1)\Pi(n, n, k, \tfrac{1}{2}+c/(n+1))$, since there are $2(n+1)$ such n-faces. □

We can iterate the formula of Lemma 6.5. There is a technical point to check, that for large n, $k \geq K(n) = 2^{\frac{1}{2}\sqrt{n \lg n}}$ implies $4k \geq K(n+1)$. This follows easily from the fact that

$$\sqrt{(n+1)\lg(n+1)} - \sqrt{n \lg n} \to 0 \text{ as } n \to \infty.$$

We only iterate as long as $4k \leq (1+\bar{p}_v)2^{n-1}$, which is the maximum component size we wish to consider in C^{n+1}.

Thus, iterating r times (provided r is not too large), it follows that, using fairly crude estimates,

$$\Pi(n+r, n+r, 4^r k, c/(n+r))$$
$$\leqslant 2^r (n+r)^r \Pi(n, n, k, c/(n+r)) + 2^{-n^2/8}. \tag{6.5}$$

Putting $k = K(n)$ in (6.5) and noting from Corollary 5.3 that for large n we have $\Pi(n, n, K(n), c/(n+r)) \leqslant 2^{-n^2/5}$, it follows that $\Pi(n, n, n, c/(n+r)) + 2^{-n^2/8} < 2^{-n^2/9}$, say. Also we may assume $r < n$, since we must have $4^r K(n) \leqslant 2^{n+r} = |C^{n+r}|$.

Thus the right side of (6.5) is bounded by $(4n)^n 2^{-n^2/9} < 2^{-n^2/10}$, say. Therefore we have

$$\Pi(n+r, n+r, 4^r K(n), \tfrac{1}{2} + c/(n+r)) < 2^{-n^2/10}. \tag{6.6}$$

Now put $r = r(n) = \lfloor n - 2 + \lg(1 + \bar{p}_v) - \lg K(n) \rfloor$ to show that

$$\Pi(n+r, n+r, (1+\bar{p}_v) 2^{n+r-3}, \tfrac{1}{2} + c/(n+r)) \tag{6.7}$$
$$< 2^{-n^2/10}.$$

We would like to replace $(n+r)$ by m in (6.7) to obtain an inequality which is valid for all large m. We are hampered by the fact that, for some m, there may be no n such that $m = n + r(n)$. Therefore let

$$\varphi(m) = \max\{n + r(n) : n + r(n) \leqslant m\}.$$

Since $(n+1) + r(n+1) - (n + r(n)) \leqslant 2$, we deduce that

$$\varphi(m) = m - \alpha(m) \text{ where } \alpha(m) = 0 \text{ or } 1.$$

Define $\psi(m)$ by $\varphi(m) = \psi(m) + r(\psi(m))$, and note that $\psi(m) \geqslant \varphi(m)/2$. We thus have from (6.7) that

$$\Pi(\varphi(m), \varphi(m), (1+\bar{p}_v) 2^{\varphi(m) - 3\alpha(m)}, \tfrac{1}{2} + c/\varphi(m))$$
$$\leqslant 2^{-m^2/50}$$

for large m.

Applying Lemma 6.5 $\alpha(m) + 1$ times yields

$$\Pi(m+1, m+1, (1+\bar{p}_v) 2^{m-1}, \tfrac{1}{2} + c/(m+1))$$
$$\leqslant 2^{-m^2/60}$$

for large m.

Finally, putting $n=m+1$ and dropping the fourth parameter

$$\Pi(n, n, (1+\bar{p}_v)2^{n-2}) \leq 2^{-n^2/70} \text{ for large } n.$$

This, combined with Corollary 5.3 and Lemma 6.1, gives the theorem for the case $s=0$. □

7. General case: $s > 0$

We shall use an induction on s based on the partition of C^{n+1} into two copies of C^n, somewhat similarly to §6. We first note, however, that

Lemma 7.1. *If $s \geq 1$, then*

$$\Pr(\Gamma_n \text{ has a vertex of degree at most } s-1) = O(1/n) \tag{7.1a}$$

$$\Pr(\Gamma_n \text{ is not connected}) = O(1/n). \tag{7.1b}$$

Proof. The calculation in Theorem 3.1(a) gives (7.1a) — see the paragraph following (3.1).

$$\Pi(n, 2, N/2) \leq 2^{-n/3} + 2^{-n^2/70}$$

follows from the calculations done for the case $s=0$, and so (7.1b) follows also. □

Definition. A set S is a *proper disconnector* (PD) of a connected graph G if the subgraph of G induced by $V(G) - S$ is (a) not connected, and (b) contains no isolated vertices.

Let now $\Delta(n, s) = \Pr(\Gamma_n$ has a PD S with $|S| = s)$. In view of Theorem 3.1(b) we have only to prove

$$\lim_{n \to \infty} \Delta(n, s) = 0 \tag{7.2}$$

in order to prove our theorem.

For a set $S \subseteq V_n$, let Γ_n/S denote the subgraph of Γ_n induced by $V_n - S$.

Lemma 7.2.

$$\Pr(\text{There exists a PD } S, |S| = s, \text{ of } \Gamma_n \text{ such that} \\ \Gamma_n/S \text{ has a component } Z, 2 \leq |Z| \leq K(n)) \\ \leq 2^{-\frac{1}{2}n}.$$

Proof. If we put

$$\alpha(s, k) = \Pr(\text{There exists a pair } S, Z \text{ as above}, |Z| = k)$$

$$\leq H_k(n)\binom{N}{s}(1-p)^{b(k)-s}$$

$$\leq H_k(n) 2^{-b(k)+s(n+1)} e^{2k|c|}. \tag{7.3}$$

The right hand side of (7.3) bounds the expected number of such pairs — $H_k(n)$ counts the number of Z's, $\binom{N}{s}$ bounds the number of S's and $(1-p)^{b(k)-s}$ bounds the probability that Z is a component of Γ_n/S.

For $k \geq s+2$, we proceed exactly as in Lemmas 5.1 and 5.2 to estimate $H_k(n) 2^{-b(k)}$. This yields

$$\alpha(s, k) = O(2^{-2n/3}) \text{ for } k \geq s+2.$$

For $k \leq s+1$, we see that if Z, S exist, then Z contains a pair of adjacent vertices x, y for which

$$|N(\{x, y\})| \leq s + k - 2 < 2s$$

where for $T \subseteq V_n$, $N(T) = \{v \in V_n - T : v \text{ is adjacent in } C^n \text{ to some } w \in T\}$.

But

$$\Pr(\text{There exist adjacent } x, y \text{ such that } |N(\{x, y\})| < 2s)$$

$$\leq Nn\binom{2n-2}{2s-1}(1-p)^{2n-2s}$$

$$= O(n^{2s} 2^{-n}).$$

We shall now use the relationship between C^{n+1} and C^n similarly to the proof of Lemma 6.4. Let now

$$\Delta(n, s, p)$$

$$= \Pr(\Gamma_n \text{ has a PD } S \text{ with } |S| = s, \text{ and } p_e p_v = p).$$

Lemma 7.3.

$$\Delta(n+1, s, p) \leq \sum_{t=1}^{s-1} \Delta(n, t, p) + O(1/n) \qquad (7.4)$$

where $p = \frac{1}{2} + (\frac{1}{2} s \ln n + c)/n$ and $s \geq 1$.

Proof. We shall, somewhat loosely, refer to the subgraph induced by a subset Y of V_n by Y itself. This should not lead to any confusion, given the context.
Let $X_i = X_{i1}$ for $i = 0$ or 1 as in §6.
Let D denote the event that Γ_{n+1} contains a PD S with $|S| = s$. We note that

$$D \subseteq E_\alpha \cup E_{\beta 0} \cup E_{\beta 1} \cup E_\gamma \cup E_\delta \qquad (7.5)$$

where

E_α is the event that X_i has minimum degree at most $s-1$ for some $i = 0$ or 1.

$E_{\beta i}$ is the event: $S \subseteq X_i$ but not E_α, for $i = 0, 1$, and D occurs.

E_γ is the event $S_i = S \cap X_i$ is a PD of X_i, $0 < |S_i| < s$, for some $i = 0$ or 1, then the event D occurs but not E_α or $E_{\beta 0}$ or $E_{\beta 1}$. (Note that $|S_i| < s$ because of $\bar{E}_{\beta 0}$ and $\bar{E}_{\beta 1}$. Because of $\bar{E}_\alpha$, $S_i \neq N(x, i) = \{y \in X_i : \{x, y\} \in A_n\}$.)

E_δ is the event that D occurs and $X_i - S$ is connected for $i = 0, 1$ and there are no more than s vertices $x \in X_0$ such that $v(x) \in X_1$ and $\{x, v(x)\} \in A_{n+1}$. Here $v(x)$ is obtained from x by changing its first coordinate.

We note first that Lemma 3.1(a) shows

$$\Pr(E_\alpha) = O(1/n) \qquad (7.6)$$

and that

$$\Pr(E_\gamma) \leq 2 \sum_{t=1}^{s-1} \Delta(n, t, p), \quad (E_\gamma = \emptyset \text{ for } s = 1). \qquad (7.7)$$

Furthermore

$$\Pr(E_\delta) \leq \Pr(|\{x \in X_0 : v(x) \in X_1, \{x, v(x)\} \in A_{n+1}\}| \leq s)$$

$$\leq e^{-2^{\frac{1}{2}n}} + \binom{N}{s}(1-p)^{(21\bar{p}_v - 2)N/20 - s}, \qquad (7.8)$$

using Lemma 6.1 and the fact that, given X_0, X_1, the edges joining X_0, X_1 are unconditioned.

Let us now consider the event $E_{\beta 0}$ (and hence by symmetry $E_{\beta 1}$). We have

$$E_{\beta 0} \subseteq E_1^* \cup E_2^* \cup E_3^* \cup E_4^* \tag{7.9}$$

where

E_1^* is the event that X_1 is not connected.

E_2^* is the event that $E_{\beta 0}$ occurs, X_1 is connected and $S = N(x, 0)$ and $\{x, v(x)\} \in A_{n+1}$, some $x \in X_0$. (If $S = N(x, 0)$ and $\{x, v(x)\} \notin A_{n+1}$, then $S = N(\{x\})$ and hence S is not a PD.)

E_3^* is the event that $E_{\beta 0}$ occurs, X_1 is connected and $X_0 - S$ contains a component of size k, $2 \leq k \leq K(n)$.

E_4^* is the event that $E_{\beta 0}$ occurs, X_1 is connected and $X_0 - S$ contains a component T of size $k > K(n)$ such that $\{x, v(x)\} \notin A_{n+1}$ for all $x \in T$.

Now Lemma 7.1 gives

$$\Pr(E_1^*) = O(1/n). \tag{7.10}$$

Lemma 7.2 gives

$$\Pr(E_3^*) \leq 2^{-\frac{1}{2}n} \tag{7.11}$$

and clearly

$$\Pr(E_4^*) \leq (N/K)(1-p)^K \leq 2^{-2\sqrt{n \lg n}/3} \tag{7.12}$$

for n large.

Let us now consider the event E_2^*. We note that

$$E_2^* \subseteq E_3^* \cup E_4^* \cup E_5^* \tag{7.13}$$

where

E_5^* is the event $\bar{E}_\alpha$, D and $X_0 - S$ contains an isolated vertex x' different from the x defining E_2^*.

Now $|B(\{x\}) \cap B(\{x'\})| \leq 2$ for any $x, x' \in C^n$, and thus, given $\bar{E}_\alpha$, we have

$$|N(x, 0) \cup N(x', 0)| \geq 2s - 2 \text{ for } x, x' \in X_0.$$

Thus

$$E_5^* = \emptyset \text{ for all } s \geq 3. \tag{7.14}$$

We have now only to consider

Case 1: $s=1$

Now E_5^* implies that X_0 contains two vertices x, x' of degree 1 with $d(x, x')=2$. Thus

$$\Pr(E_5^*) \leq Nn^4(1-p)^{2n-4} < 2^{-\frac{1}{2}n} \text{ for large } n. \tag{7.15}$$

Case 2: $s=2$

In this case, E_5^* implies E_b^*, the event that X_0 contains 2 vertices x, x' of degree 2 with $d(x, x')=2$.

Now,

$$\Pr(E_b^*) \leq Nn^6(1-p)^{2n-6} < 2^{-\frac{1}{2}n} \tag{7.16}$$

for n large. The Lemma now follows from (7.5) to (7.16). □

Theorem 1.1 now follows easily from Lemma 7.3, the case $s=0$ and Theorem 3.1.

ACKNOWLEDGMENT

We are grateful to the referees for correcting numerous minor errors, and for supplying reference [10].

References

[1] B. Bollobás, "Random graphs", in *Combinatorics* (H. N. V. Temperley, Ed.) London Mathematical Society Lecture Notes Series 52, Cambridge University Press, 1981, pp. 80–102.

[2] B. Bollobás, "The evolution of the cube", in *Combinatorial Mathematics* (C. Berge et al., Eds.), North-Holland, 1983, pp. 91–97.

[3] Yu. D. Burtin, "On the probability of connectedness of a random subgraph of the n-cube" (in Russian), Problemy Peredači Informacii 13 (1977) 90–95.

[4] P. Erdös and A. Rényi, "On the strength of connectedness of a random graph", Acta Math. Acad. Sci. Hungar. 12 (1961) 261–267.

[5] P. Erdös and J. Spencer, "Evolution of the n-cube", Computers and Mathematics with Applications 5 (1979) 33–39.

[6] T. I. Fenner and A. M. Frieze, "On the connectivity of random m-orientable graphs and digraphs", Combinatorica 2 (1982) 347–359.

[7] W. Hoeffding, "Probability inequalities for sums of bounded random variables", Journal of the American Statistical Association 58 (1963) 13–30.

[8] G. O. H. Katona, "The Hamming-sphere has minimum boundary", Studia Sci. Math. Hungar. 10 (1975) 131–140.

[9] A. A. Saposhenko, "Geometric structure of almost all boolean functions" (in Russian), Problemy Kibernet. 30 (1975) 227–261.

[10] A. A. Saposhenko, "Metric properties of almost all boolean functions", Diskretny Analiz 10 (1967) 91–119 (in Russian).

[11] K. Weber, "Random graphs – a survey", Rostock Math. Kolloq. 21 (1982) 83–98.

CONNECTEDNESS AND CONNECTIVITY IN PERCOLATION THEORY

J. W. ESSAM

Royal Holloway and Bedford New College,
University of London,
Egham Hill, Egham, Surrey TW20 OEX, U.K.

We consider percolation on finite graphs and infinite crystal lattice graphs. Results are obtained which relate to the pair m-connectedness function $C_{uv}^{(m)}(p)$ which is the probability of finding m edge-disjoint paths which are open from vertex u to vertex v. Most of the results which were previously derived for $m=1$ extend to general m and counter-examples are provided in other cases.

The first sixteen terms in the power expansion of $S_u^{(2)}(p)$, the expected number of sites which are bi-connected to a given lattice site, is obtained for the plane square lattice. This is used to obtain an estimate of $\gamma^{(2)}$, the critical exponent describing the divergence of $S_u^{(2)}(p)$ at p_c.

1. Introduction

Percolation theory was introduced as a mathematical problem in graph theory by Broadbent and Hammersley (1957). They consider a locally finite graph G, in which only finitely many vertices are adjacent to a given vertex, and assign two possible states, open and closed, to each edge. The probability p that each edge is open, independently of all other edges, is a given parameter of the problem. The percolation probability $P_u(p)$ is the probability that vertex u is connected to infinitely many other vertices by paths which use only open edges. The problem as originally posed was to discuss the conditions on G for the existence of a critical probability defined by

$$p_c = \sup\{p \in [0, 1] \mid P_u(p) = 0\}. \tag{1.1}$$

The value of p_c is the same for any two vertices which are a finite distance apart.

Domb (1959) noticed the similarity between percolation and models of phase transitions which occur in solid state physics. For example magnetic systems which have local ferromagnetic interactions go into an ordered state having a spontaneous magnetic moment below a certain critical temperature T_c analogous to the critical probability. Also transitions which are specifically of a percolative nature can occur in solids, thus an aggregate of conducting and non-conducting material will only conduct electricity if the proportion of conducting material is above a critical threshold. Physicists have therefore become interested in percolation on lattice graphs where the vertices of G are the atoms and the edges are the nearest-neighbour bonds of the lattice (see Essam 1972 and 1980 and Stauffer 1979 for reviews).

More recently mathematicians have become interested in percolation on lattice graphs for which it is known that p_c exists and it has been obtained exactly for a number of two-dimensional lattices (see Kesten 1982 for details and references).

Another percolation function which arises naturally in the context of the susceptibility of a dilute ferromagnet is the mean size, $S_u(p)$, of the cluster containing vertex u (Domb 1959). This is defined to be the expected number of vertices which may be reached by an open path from vertex u. Here we shall be concerned with its generalisation $S_u^{(m)}(p)$ which is the expected number of vertices which may be reached from u by at least m edge disjoint open paths (i.e. the expected number of vertices which are m-connected from u). Almost all of the discussion also applies to connection by vertex-disjoint paths.

Further percolation thresholds $p_c^{(m)}$ may be defined by

$$p_c^{(m)} = \sup\{p \in [0, 1] \,|\, S_u^{(m)}(p) < \infty\}. \tag{1.2}$$

For a number of two-dimensional percolation models it has been shown that $p_c^{(1)} = p_c$ (Kesten 1982). Stimulated by the numerical results reported here, Grimmett (1985) has shown for the square lattice that with $P_u^{(m)}(p)$ defined as the probability that infinitely many vertices are m-connected from u, and

$$\pi_c^{(m)} = \sup\{p \in [0, 1] \,|\, P_u^{(m)}(p) = 0\} \tag{1.3}$$

then $\pi_c^{(1)} = \pi_c^{(2)} = \pi_c^{(3)} = \pi_c^{(4)}$. Since clearly $p_c^{(m)} \leq \pi_c^{(m)}$ and $p_c^{(m)} \leq p_c^{(m+1)}$, $m = 1, 2, 3$, then because for this lattice $\pi_c^{(1)} \equiv p_c = p_c^{(1)}$ (Kesten 1982) it follows that $p_c^{(m)} = p_c$. This is not a general result as may be seen by considering a regular infinite Cayley tree of degree three with each edge replaced by a pair of parallel edges. For this graph $p_c = p_c^{(1)} = 1 - 1/\sqrt{2}$ and $p_c^{(2)} = 1/\sqrt{2} > p_c^{(1)}$. A safe conjecture is that $p_c^{(m)} = p_c$ for percolation problems on all regular lattice graphs in two-dimensions with single edges connecting nearest-neighbour vertices and probably higher dimensions too, although as yet our only data is for the square lattice.

On the basis of excellent numerical evidence (see for example Essam et al. 1986) and renormalisation group arguments (Young and Stinchcombe 1975) physicists believe that the approach to the percolation threshold of the above functions on a lattice graph is a power law:

$$P^{(m)}(p) \sim (p - \pi_c^{(m)})^{\beta^{(m)}}, \quad p \downarrow \pi_c^{(m)} \tag{1.4}$$

and

$$S^{(m)}(p) \sim (p_c^{(m)} - p)^{-\gamma^{(m)}}, \quad p \uparrow p_c^{(m)}. \tag{1.5}$$

There is no proof of these statements as yet but see Kesten (1982, Ch. 8) for progress in this direction. The critical exponents $\beta^{(m)}$ and $\gamma^{(m)}$ are believed to be the same for all lattice graphs with a given space dimension and for this reason are of more fundamental significance than the percolation thresholds which depend on the local lattice structure. A widely believed conjecture (Den Nijs 1979) is that in two dimensions $\beta^{(1)} = 5/36$ and $\gamma^{(1)} = 2\frac{7}{18}$. For $m > 1$ there are no such results and it is one of the objectives of this paper to obtain numerical estimates for $m = 2$ on the square lattice. We shall also review and extend some exact results for the derivatives of percolation functions with respect to p.

2. Low density series expansion for $S_u^{(m)}(p)$

2.1. *The pair m-connectedness*

The function $S_u^{(m)}(p)$ may be expanded as a power series in p and the method we use to obtain the terms up to order p^N for a lattice graph has been discussed in the case $m = 1$ by Essam (1972). However in this case a number of simplifying rules were discovered (Arrowsmith and Essam 1977) and here we consider to what extent these may be used for $m > 1$. The method is to write

$$S_u^{(m)}(p) = \sum_v C_{uv}^{(m)}(p) \tag{2.1}$$

where the pair m-connectedness function $C_{uv}^{(m)}(p)$ is the probability that vertex v is m-connected from u. (A field-theoretic analysis of the pair bi-connectedness ($m = 2$) in the context of critical phenomena theory has been given by Harris and Lubensky (1983).) If $n_{uv}^{(m)}$ is the least number of edges required to m-connect v then $C_{uv}^{(m)}(p)$ is of order $p^{n_{uv}^{(m)}}$ and to determine the power expansion to order N we need consider only vertices for which $n_{uv}^{(m)} \leq N$. Furthermore if G is the subgraph of the lattice graph induced by these vertices then

$$C_{uv}^{(m)}(p) = C_{uv}^{(m)}(p, G) + \mathrm{O}(p^{N+1}), \tag{2.2}$$

where $C_{uv}^{(m)}(p, G)$ is the pair m-connectedness for the finite graph G. (For convenience we suppress the dependence of G on m since it is a fixed parameter of the expansion.) Since therefore $S_u^{(m)}(p)$ and $S_u^{(m)}(p, G)$ are the same to order p^N a direct approach would be to compute the expected number of vertices which are m-connected from u in G by inspection of all possible configurations of open and closed edges of G. Unfortunately to obtain the coefficients up to p^{16} for $m=2$, which we give at the end of this section, the number of edges in G is 252 and examination of all 2^{252} configurations would require years of computing time. A different approach to the calculation of $C_{uv}^{(m)}(p, G)$ is now considered.

2.2. The subgraph expansion

Let $\mathcal{M}$ be the collection of all subsets of the edges E of G which minimally m-connect u to v. In the case $m=1$, $\mathcal{M}$ is the collection $\mathcal{P}$ of edge sets of all simple paths from u to v. For $m>1$, $\mathcal{M}$ may be found by considering all combinations of m edge-disjoint paths from $\mathcal{P}$ and then selecting from their edge sets only the ones which are minimal. By inclusion and exclusion

$$C_{uv}^{(m)}(p, G) = \sum_{\mathcal{M}' \subseteq \mathcal{M}} (-1)^{|\mathcal{M}'|+1} p^{e'} \tag{2.3}$$

where e' is the number of edges in the union U' of the subsets $\mathcal{M}'$. Grouping together all subsets for which $U'=E'$ gives

$$C_{uv}^{(m)}(p, G) = \sum_{E' \subseteq E} d^{(m)}(G') p^{|E'|} \tag{2.4}$$

where G' is the subgraph of G with edge set E' together with its incident vertices and

$$d^{(m)}(G') = \sum_{\substack{\mathcal{M}' \subseteq \mathcal{M}: \\ U'=E'}} (-1)^{|\mathcal{M}'|+1}. \tag{2.5}$$

The weight factor $d^{(m)}(G')$ associated with the subgraph G', usually called its d-weight, depends only on G' and not on G since in (2.5) we can clearly replace $\mathcal{M}$ by $\mathcal{M}(G')$ the subset of $\mathcal{M}$ the members of which only use edges of G'. Subsets $\mathcal{M}'$ for which $U'=E'$ are said to cover G'.

The computational advantages of (2.4) are:

(i) only subsets E' need be considered for which G' has at least one cover and $|E'| \leq N$,

(ii) two subgraphs may be considered equivalent if they are isomorphic (i.e. they differ only by a relabelling of the vertices other than u and v); $d^{(m)}(G')$ need therefore only be calculated once for each class,

(iii) $d^{(m)}(G')$ is a topological invariant (i.e. vertices of degree two, other than u and v, may be suppressed without changing its value); for $m=1$ it is related to the β-invariant of Crapo (1967) (see Essam 1972).

The calculation of the series coefficients thus falls into three parts. (a) The generation of a list of non-isomorphic two-rooted coverable subgraphs with $\leq N$ edges. For each of these: (b) the calculation of $d^{(m)}$ and (c) the determination of the number G' which are isomorphic. Parts (a) and (c) occur in many problems of statistical mechanics on lattices and may be dealt with by general methods (Martin 1974). Only (b) will be discussed here.

In fact (2.5) may be used directly but at order sixteen the computing time on a PRIME 750 is of the order of days. The following proposition, the proof of which we consider below, saves time for the larger graphs.

Proposition 2.1.

$$d^{(m)}(G') = \sum_{H \in \mathcal{D}(G')} \vec{d}^{(m)}(H) \tag{2.6}$$

where $\mathcal{D}(G')$ is the set of directed graphs obtained by directing the edges of G' in all possible ways. The "directed d-weight", $\vec{d}^{(m)}(H)$, is defined by (2.5) where $\mathcal{M}$ is now the collection of subsets of edges of G' which minimally m-connect u to v by paths which follow the direction of the edges.

The directed d-weights arise in their own right as the weights in the power expansion of $C_{uv}^{(m)}(p)$ for directed percolation. The latter occurs in many physical contexts (Kinzel 1983) and was also considered in the original work of Broadbent and Hammersley (1957). It was shown by Arrowsmith and Essam (1977) that $\vec{d}^{(1)}(H)$ is zero unless H is coverable by directed paths from u to v and is also acyclic. If H satisfies these conditions it was also shown that

$$\vec{d}^{(1)}(H) = (-1)^{c(H)} \tag{2.7}$$

where $c(H)$ is the cycle rank of H. Hence, up to a sign, $d^{(1)}(G')$ is the number of acyclic directings of G' which give rise to coverable directed graphs. No simple formula such as (2.7) has been found for $m > 1$.

The proof of Proposition 2.1 will now be pursued. First define an extended set of directed graphs $\mathcal{D}^+(G')$ obtained from G' by replacing each subset (possibly empty) of the edges of G' by an anti-parallel pair of directed edges (arcs) and then directing the remaining edges in all possible ways.

Lemma 2.1. *With $d^{(m)}(H)$ defined as in* Proposition 2.1

$$d^{(m)}(G') = \sum_{H \in \mathcal{D}^+(G')} \vec{d}^{(m)}(H). \qquad (2.8)$$

Proof. In (2.3) replace $\mathcal{M}$ by the collection $\vec{\mathcal{M}}$ of arc sets defined as follows. Consider the collection of all combinations of m edge-disjoint paths from $\mathcal{P}$. Each path of a combination may be considered to have an arc set obtained by directing its edges parallel to the path. $\vec{\mathcal{M}}$ is constructed by taking each combination, forming the union of the arcs generated by its paths and then selecting only the minimal arc sets so formed. This construction was used by Arrowsmith (1979) in his proof of (2.6) in the case $m=1$. U' is now the union of the directed edges in $\mathcal{M}'$ and e' is the number of edges in U' but counting a pair of anti-parallel edges only once. Equation (2.8) follows by assigning each subset $\mathcal{M}'$ of $\vec{\mathcal{M}}$ to the directed graph H having arc-set U'.

Lemma 2.2. *For $H \in \mathcal{D}^+(G')$, $\vec{d}^{(m)}(H)$ is the coefficient of $p^{a(H)}$ in $C_{uv}^{(m)}(p, H)$, the pair m-connectedness for the directed percolation problem on H in which each arc has probability p of being open. Here $a(H)$ is the number of arcs in H.*

Proof. This follows by replacing G by H and d by $\vec{d}$ in (2.4) and noting that H is the only subgraph with $a(H)$ arcs.

Lemma 2.3. *For $H \in \mathcal{D}^+(G')$,*

$$C_{uv}^{(m)}(p, \tilde{H}) = C_{uv}^{(m)}(p, H) \qquad (2.9)$$

where $\tilde{H}$ is obtained from H by replacing each anti-parallel pair of arcs by an undirected edge. For this partially directed graph $C_{uv}^{(m)}(p, H)$ is defined as usual but allowing the connecting paths to pass either way along the undirected edges.

Proof. In the case $m=1$ and $H=D(G')$, the graph obtained by replacing all edges of G' by anti-parallel arc-pairs, the result has been proved by McDiarmid (1980, 1981) using his "clutter percolation theorem". The extension to general m and any $H \in \mathcal{D}^+(G')$ is tedious but straightforward and will not be given here. It is only necessary to establish that the arc-clutter $\mathcal{M}(H)$ used in calculating $C_{uv}^{(m)}(p, H)$, as in (2.3), has properties (C) and (C*) of McDiarmid (1980). This may be done using an argument based on McDiarmid's Lemma (McDiarmid 1980). The clutter $\vec{\mathcal{M}}(H)$ of the clutter percolation theorem is that used in calculating $C_{uv}^{(m)}(p, \tilde{H})$.

Proof of Proposition 2.1. Notice that $a(\tilde{H})<a(H)$ unless H has no anti-parallel arc pairs and hence, using Lemmas 2.2 and 2.3, $\vec{d}^{(m)}(H)=0$ for $H \notin \mathscr{D}(G)$. The result follows by restricting the sum in (2.8) to $\mathscr{D}(G)$.

It has already been pointed out that in the case $m=1$ many of the terms in the sum (2.6) are zero and this is also true for $m>1$ although the characterisation of which terms are zero is not so simple.

It follows immediately from (2.5) applied to $\vec{d}^{(m)}(H)$ that $\vec{d}^{(m)}(H)=0$ unless H is coverable by some subset of $\mathscr{M}$. Also if H is not coverable for $m=1$ it cannot be coverable for $m>1$, however the converse is not true. For example, Figure 1 shows two directings both of which are coverable by directed paths ($m=1$) but only (b) is coverable by the union of a subset of pairs of edge-disjoint paths ($m=2$). In case (a) there is only one such pair which does not cover. In case (b) $\vec{d}^{(2)}=1$ and in both cases $d^{(1)}=1$.

It was also pointed out that in the case $m=1$ the only coverable graphs for which $\vec{d}^{(1)}(H)=0$ are cyclic. However the graph of Figure 2 is coverable by pairs

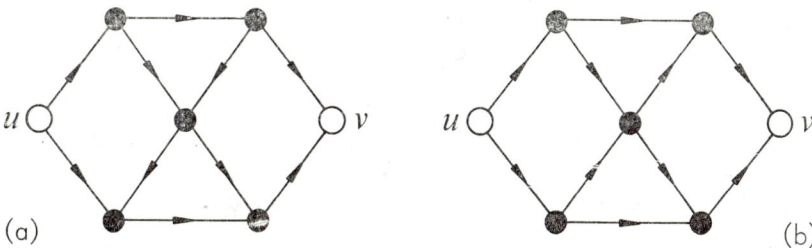

Figure 1.

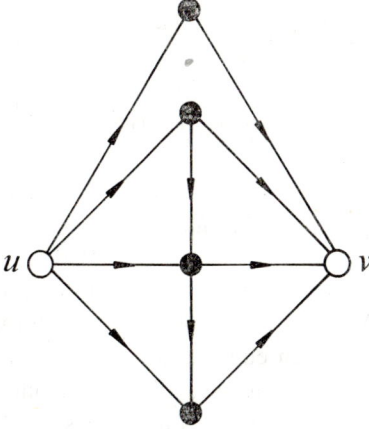

Figure 2.

of edge-disjoint paths and is not cyclic but nevertheless $\vec{d}^{(2)}=0$. On the other hand no counter-example to the following conjecture was found among the graphs required to calculate $S_u^{(2)}(p)$ to order 16 on the square lattice.

Conjecture. $\vec{d}^{(m)}(H)=0$ whenever H is cyclic.

Finally we turn to the result $|\vec{d}^{(1)}(H)| \leq 1$. A counter-example to the extension of this to $m>1$ is the two vertex graph with n parallel arcs connecting u and v for which it may be shown that $\vec{d}^{(2)}=(-1)^n(n-1)$. The latter result follows easily from the general result for parallel graphs which we now consider along with series graphs.

The following results hold for G undirected, directed or partially directed. If G is the series combination of G_A and G_B then for any m

$$C_{uv}^{(m)}(p,G) = C_{uw}^{(m)}(p,G_A) C_{wv}^{(m)}(p,G_B). \tag{2.10}$$

In the case that G is the parallel combination it follows by an inclusion and exclusion argument that

$$C_{uv}^{(1)}(p,G) = C_A^{(1)} + C_B^{(1)} - C_A^{(1)} C_B^{(1)} \tag{2.11}$$

where $C_i^{(m)} = C_{uv}^{(m)}(p, G_i)$. This is not easily extended to $m>1$ but for $m=2$

$$C_{uv}^{(2)}(p,G) = C_A^{(2)} + C_B^{(2)} - C_A^{(1)} C_B^{(2)} - C_A^{(2)} C_B^{(1)} + C_A^{(1)} C_B^{(1)}. \tag{2.12}$$

Equating coefficients of $p^{e(G)}$ in these three equations, where $e(G)$ is the number of edges (directed and undirected) in G gives

$$d_{uv}^{(m)}(G) = d_{uw}^{(m)}(G_A) d_{wv}^{(m)}(G_B) \quad \text{(series)} \tag{2.13}$$

$$d_{uv}^{(1)}(G) = -d_A^{(1)} d_B^{(1)} \tag{2.14}$$

$$d_{uv}^{(2)}(G) = -d_A^{(1)} d_B^{(2)} - d_A^{(2)} d_B^{(1)} + d_A^{(1)} d_B^{(1)}. \quad \text{(parallel)} \tag{2.15}$$

2.3. Expansion of $S_u^{(2)}(p)$ for the square lattice

The graph list required to obtain the expansion of $S_u^{(2)}(p)$ to order sixteen was constructed as follows. A list of non-isomorphic subgraphs of the square lattice having between four and sixteen edges and no articulation point was compiled. There were 241 of these. For each one, all distinct one and two-rootings were determined together with the $d^{(2)}$-weights in the case of two-rootings. The resulting information was used to make a list of all non-isomorphic series combinations with ⩽16 edges, each complete with coverable two-rootings and weights. For each of the above two-rooted graphs the number of occurrences on the square lattice was computed to yield the following expansion.

$$S_u^{(2)}(p) = 1 + 12p^4 + 60p^6 - 48p^7 + 428p^8 - 504p^9$$
$$+ 2608p^{10} - 3944p^{11} + 15980p^{12} - 28516p^{13}$$
$$+ 101064p^{14} - 21215p^{15} + 68630^{16} + \ldots.$$

If the alternating signs were to continue the expansion would certainly not converge as far as p_c. However the Padé approximant method of analysis we have used (Baker 1961, Adler et al. 1983) does not require such convergence in order to estimate p_c and the exponent $\gamma^{(2)}$. The results are consistent with $p_c = \frac{1}{2}$ and using this value gives the biased estimate

$$\gamma^{(2)} \simeq 1.44 \pm 0.03. \tag{2.16}$$

Details of the calculations will be published elsewhere together with results from other similar expansions.

Hyperscaling theory applied to the pair m-connectedness (Harris and Lubensky 1983, Harris 1983) yields

$$\beta^{(m)} = \tfrac{1}{2}(d\nu - \gamma^{(m)}) \tag{2.17}$$

where ν is the exponent which characterises the divergence of the connectedness length at p_c (Essam 1980). A widely believed conjecture (Den Nijs 1979) is that $\nu = \frac{4}{3}$ for two-dimensional lattices ($d=2$). Our estimate of $\gamma^{(2)}$ together with hyperscaling yields

$$\beta^{(2)} \simeq 0.61 \pm 0.02. \tag{2.18}$$

The above estimates are consistent with the inequality $\gamma^{(2)} \leqslant \gamma^{(1)}$ which follows from $S_u^{(2)}(p) \leqslant S_u^{(1)}(p)$ together with the assumed form (1.5). The corresponding inequality $\beta^{(2)} \geqslant \beta^{(1)}$ is also satisfied and both appear to be strict for this lattice.

3. Sensitivity and some weight factor identities

In this section it will be shown that a number of results in the literature may be viewed as cases of a general sensitivity formula. A general parallelism between

3.1. General formulae

Consider the probability space in which the events are the $2^{e(G)}$ configurations of open and closed edges of the graph G with edge set E. If the configurations are labelled by the subset E' of open edges

$$\text{pr}(E') = \prod_{e \in E'} p_e \prod_{e \in E \setminus E'} (1 - p_e) \tag{3.1}$$

where we have allowed the probability p_e that the edge e is open to vary from edge to edge. If Y is a random variable then its expectation value is defined by

$$\mathscr{E}(Y) = \sum_{E' \subseteq E} Y(E') \text{pr}(E') \tag{3.2}$$

and expanding the second product in (3.1) we obtain:

Proposition 3.1. *If $X(E'')$ is defined by*

$$X(E'') = \sum_{E' \subseteq E''} (-1)^{|E'' \setminus E'|} Y(E') \tag{3.3}$$

then inversely

$$Y(E') = \sum_{E'' \subseteq E'} X(E'') \tag{3.4}$$

and

$$\mathscr{E}(Y) = \sum_{E'' \subseteq E'} X(E'') p^{E''} \tag{3.5}$$

where $p^{E''} = \prod_{e \in E''} p_e$.

The variable X will be called the weight corresponding to Y. Further Cardy (1973) has shown that:

Proposition 3.2. *With X defined as in* Proposition 3.1 *and $\delta_e Y(E') \equiv Y(E') - Y(E' \setminus e)$ where $e \in E$*

$$\delta_e Y(E') = \sum_{E'' \subseteq E'} \varepsilon_e(E'') X(E'') \tag{3.6}$$

and ε_e is the indicator that e is open. Note that if $e \notin E'$ then $Y(E'\backslash e) = Y(E')$ and $\delta_e Y(E') = 0$.

Definition 3.1. An increasing (decreasing) event is one the indicator of which can only change from 0 to 1 (1 to 0) on opening a further edge.

Definition 3.2. In any configuration E', the edge e is critical for a given event if the indicator of that event changes on deletion of e (i.e. changing its state from open to closed).

Proposition 3.3. *If Y is the indicator for an increasing event then $\delta_e Y$ is the indicator that e is critical for that event. For a decreasing event $-\delta_e Y$ is the indicator that e is critical.*

Corollary 3.1. *If Y is the indicator for the increasing event $\mathcal{J}$ then with X defined as in* Proposition 3.1

$$\sum_{E'' \subseteq E'} |E''| X(E'') = \text{number of edges in } E' \text{ which are critical for } \mathcal{J}. \tag{3.7}$$

Proof. Combining Prop. 3.2 and Prop. 3.3

$$\sum_{E'' \subseteq E'} \varepsilon_e(E'') X(E'') = \begin{cases} 1 & \text{if in } E', \, e \text{ is critical for } \mathcal{J} \\ 0 & \text{if not}. \end{cases} \tag{3.8}$$

The result follows on summing over $e \in E$.

Cardy's formula (Prop. 3.2) has its probabilistic counterpart in the following relation between difference and derivative:

Proposition 3.4. *For any random variable Y*

$$\mathcal{E}(\delta_e Y) = p_e \frac{\partial}{\partial p_e} \mathcal{E}(Y). \tag{3.9}$$

Proof. Replacing Y by $\delta_e Y$ and X by $\varepsilon_e X$ in Proposition 3.1

$$\mathcal{E}(\delta_e Y) = \sum_{E'' \subseteq E'} \varepsilon_e(E'') X(E'') \tag{3.10}$$

and the result follows since the RHS is the logarithmic derivative of the RHS of (3.5).

Results which are equivalent or similar to Proposition 3.4 have already appeared in the literature (Buzacott 1980, Russo 1981). The result also holds for site percolation (Kesten 1982, Ch. 4).

Corollary 3.2. *If $\mathscr{I}$ is an increasing event then*

$$p_e \frac{\partial}{\partial p_e} \mathrm{pr}(\mathscr{I}) = \mathrm{pr}(e \text{ is critical for } \mathscr{I}). \tag{3.11}$$

For a decreasing event only the sign of the RHS need be changed.

Proof. This is an immediate consequence of Props. 3.3 and 3.4.

Corollary 3.3. *If $\mathscr{I}$ is an increasing event and $p_e = p$ for all $e \in E$ then*

$$p \frac{d}{dp} \mathrm{pr}(\mathscr{I}) = \mathscr{E}(\# \text{ edges which are critical for } \mathscr{I}). \tag{3.12}$$

3.2. Applications

(a) If $\gamma_{uv}^{(m)}$ is the indicator for the event that v is m-connected from u then

$$C_{uv}^{(m)}(p, G) = \mathscr{E}(\gamma_{uv}^{(m)}). \tag{3.13}$$

If $Y = \gamma_{uv}^{(m)}$ then comparing (2.4) and (3.5) shows that $X = d^{(m)}$. An edge will be said to be m-critical if it is critical for the m-connectedness of v from u. Since m-connectedness is an increasing event we have, from Corollary 3.1, that

$$\sum_{E'' \subseteq E'} |E''| d^{(m)}(G'') = \text{number of } m\text{-critical edges in } G'. \tag{3.14}$$

Corollary 3.3 yields the corresponding probabilistic statement

$$p \frac{d}{dp} C_{uv}^{(m)}(p, G) = \mathscr{E}(\lambda_{uv}^{(m)}) \tag{3.15}$$

where $\lambda_{uv}^{(m)}$ is the number of m-critical edges. This result is due to Coniglio (1982) who presented a different proof in the case $m = 1$. An edge which is 1-critical is also called nodal or u-v separating.

The above results apply also to directed and partially directed percolation models.

(b) For undirected percolation the number of clusters $\omega(E')$ in the configuration E' is the number of connected components in the corresponding subgraph G'. By convention G' includes all of the vertices in G and a vertex which does not belong to any edge counts as a cluster. We note that if $Y=\omega$ then $-\delta_e Y(E')$ is the indicator for the event that e is an isthmus of some component of G'. From Proposition 3.4

$$p_e \frac{\partial}{\partial p_e} \mathscr{E}(\omega) = -\text{pr}(e \text{ is an isthmus}) \qquad (3.16)$$

and if $p_e = p$ for all $e \in E$

$$p \frac{d}{dp} \mathscr{E}(\omega) = -\mathscr{E}(\# \text{ isthmuses}). \qquad (3.17)$$

Now if $e = (v_1, v_2)$

pr (e is an isthmus)

$$= p_e \times \text{pr (no open path from } v_1 \text{ to } v_2 \text{ except via } e). \qquad (3.18)$$

With $q_e = 1 - p_e$, (3.16) and (3.18) may be combined to give the result of Fortuin and Kasteleyn (1972)

$$q_e \frac{\partial}{\partial q_e} \mathscr{E}(\omega) = \text{pr(no open path from } v_1 \text{ to } v_2)$$

$$= 1 - C^{(1)}_{v_1 v_2}(p, G). \qquad (3.19)$$

(c) Second derivatives

To determine the second derivative of $\mathscr{E}(\omega)$ we require $(\partial/\partial p_f)$ pr (e is an isthmus). If $f \neq e$, the event that e is an isthmus is decreasing and in configuration E', f is critical for this event if and only if (e, f) is a minimal articulation pair for some component of G'. Using Corollary 3.2 and (3.16) gives

$$\left(p_f \frac{\partial}{\partial p_f}\right)\left(p_e \frac{\partial}{\partial p_e}\right) \mathscr{E}(\omega)$$

$$= \text{pr } (e, f \text{ is a minimal articulation pair}) \qquad (3.20)$$

and, in the case $p_e = p$ for all $e \in E$, since $\mathscr{E}(\omega)$ is linear in p_e,

$$p^2 \frac{d^2}{dp^2} \mathscr{E}(\omega) \tag{3.21}$$

$$= 2\mathscr{E}(\# \text{ minimal articulation pairs}).$$

The extension to higher derivatives is not so straightforward. At order three it is true that minimal articulation sets of three edges are relevant. However, an articulation set (e, f) which is minimal for E' may no longer be minimal for $E' \backslash g$ and this is a compensating factor which must also be allowed for. The same difficulty arises even at order two in the case of derivatives of the pair connectedness and we now consider this problem.

With the notation of (a) it follows from Corollary 3.2 that

$$p_e \frac{\partial}{\partial p_e} C_{uv}^{(1)}(p, G)$$

$$= \text{pr}(e \text{ is a } u\text{-}v \text{ separating edge}). \tag{3.22}$$

A u-v separating edge, or nodal edge, is one belonging to all paths from u to v. In configuration E', $f(\neq e)$ is critical for the event that e is nodal either if (i) (e, f) is a minimal u-v separating pair or (ii) if both e and f are nodal edges. In case (ii) e is no longer a nodal edge of $E' \backslash f$. Using Proposition 3.5 with Y the indicator that e is nodal gives

$$\left(p_f \frac{\partial}{\partial p_f} \right) \left(p_e \frac{\partial}{\partial p_e} \right) C_{uv}^{(1)}(p, G)$$

$$= \text{pr}(e, f \text{ are both nodal})$$

$$- \text{pr}(e, f \text{ form a minimal } u\text{-}v \text{ separating pair}) \tag{3.23}$$

and hence

$$p^2 \frac{d^2}{dp^2} C_{uv}^{(1)}(p, G)$$

$$= 2\mathscr{E}(\# \text{ pairs of edges both of which are nodal}$$

$$- \# \text{ minimal } u\text{-}v \text{ separating pairs}). \tag{3.24}$$

(d) Higher derivatives and a relation between d-weights and connectivity

If $F_l \subseteq E$, where $|F_l| = l$, then from (3.5)

$$\prod_{e \in F_l} \left(p_e \frac{\partial}{\partial p_e} \right) \mathscr{E}(Y)$$

$$= \sum_{E'' \subseteq E} X(E'') p^{E''} \prod_{e \in F_l} \mathscr{E}_e(E'') \tag{3.25}$$

and replacing X by $X \prod_{e \in F_l} \mathscr{E}_e$ in (3.4) and using (3.5) gives

$$\prod_{e \in F_l} \left(p_e \frac{\partial}{\partial p_e} \right) \mathscr{E}(Y) = \mathscr{E}(Z) \tag{3.26}$$

where

$$Z(E') = \sum_{E'' \subseteq E'} X(E'') \prod_{e \in F_l} \mathscr{E}_e(E''). \tag{3.27}$$

The following theorem goes some way towards evaluating Z in the case that $Y = \gamma_{uv}^{(1)}$.

Theorem 3.1. *If G has local edge-connectivity* (Harary 1969) $\kappa > 0$ *between vertices u and v and $F_l = \{e_1, \ldots, e_l\}$ is a subset of its edges then for $1 \leq l \leq \kappa$*

$$\sum_{E' \subseteq E} d^{(1)}(G') \prod_{e \in F_l} \mathscr{E}_e(E') = \begin{cases} (-1)^{\kappa+1} & \text{if } l = \kappa \text{ and } F_l \text{ is a} \\ & u\text{-}v \text{ separating set} \\ 0 & \text{otherwise} \end{cases} \tag{3.28}$$

where $\mathscr{E}_e(E')$ is defined by (3.5).

Notice that this determines Z in the case $Y = \gamma_{uv}^{(1)}$ for E' such that $\kappa' \geq l$. The following corollary is an extension of theorem 7.1 of Cardy (1973).

Corollary 3.4.

$$\sum_{E' \subseteq E} d^{(1)}(G') |E'|^l = \begin{cases} (-1)^{\kappa+1} \kappa! \sigma & \text{for } l = \kappa \\ 0 & \text{for } 1 \leq l < \kappa \end{cases} \tag{3.29}$$

where σ is the number of u-v separating sets of order κ.

We first prove the following lemma.

Lemma. Let

$$Y_l(E) \equiv \sum_{E' \subseteq E} d^{(1)}(G') \prod_{e \in F_l} \mathscr{E}_e(E') \tag{3.30}$$

then if $e_1, e_2, ..., e_n$ is not a u-v separating set then, for $1 \leq l \leq n$, $Y_l(E) = 0$.

Proof. True for $l=1$ using (3.8) with $X=d^{(1)}$. Assume that it is true for $l=m$ for some m in the range $1 \leq m < n$. Now by (3.6) with $Y = Y_m$,

$$Y_{m+1}(E) = Y_m(E) - Y_m(E \backslash e_{m+1}). \tag{3.31}$$

But $Y_m(E) = 0$ by assumption and, since from the given condition $e_1, e_2, ..., e_m$ is not an articulation set of $E \backslash e_{m+1}$, $Y_m(E \backslash e_{m+1}) = 0$ and hence $Y_{m+1}(E) = 0$. The lemma follows by induction.

Proof of Theorem. For $l < \kappa$, the theorem follows from the lemma since, by Menger's theorem (Harary 1969), because there are κ edge-disjoint paths connecting u and v there can be no u-v separating set with less than κ edges.

In the case $l = \kappa$ we need to further prove that if F_κ is a u-v separating set then

$$Y_\kappa(E) = (-1)^{\kappa+1}. \tag{3.32}$$

Setting $m = \kappa - 1$ in (3.31) and using $Y_{\kappa-1}(E) = 0$ gives

$$Y_\kappa(E) = -Y_{\kappa-1}(E \backslash e_\kappa). \tag{3.33}$$

But the graph with edge set $E \backslash e_\kappa$ has connectivity $\kappa - 1$ and the result follows by induction since for $\kappa = 1$, if e_1 is a separating edge then $Y_1(E) = 1$ by (3.8) with $X = d^{(1)}$.

ACKNOWLEDGEMENTS

I would like to thank Professor Joel E. Cohen for his correspondence and drawing my attention to the paper of Buzacott. I am also grateful to the referee for his useful comments and for the references to Russo's work which I had overlooked in the original version of this manuscript.

References

Adler, J., Moshe, M. and Privman, V., (1983), Corrections to scaling for percolation, *Annals of the Israel Phys. Soc.* **5**, 397–423.

Arrowsmith, D. K., (1979), Percolation theory on multirooted directed graphs, *J. Math. Phys.* **20**, 101–3.

Arrowsmith, D. K. and Essam, J. W. (1977), Percolation theory on directed graphs, *J. Math. Phys.* **18**, 235–8.

Baker, G. A., (1961), Application of the Padé approximant method to the investigation of some magnetic properties of the Ising model, *Phys. Rev.* **124**, 768–74.

Broadbent, S. R. and Hammersley, J. M. (1957), Percolation processes, *Proc. Camb. Phil. Soc.* **53**, 629–41.

Buzacott, J. A. (1980), A recursive algorithm for finding reliability measures related to the connection of nodes in a graph, *Networks*, **10**, 311–27.

Cardy, S. (1973), The proof of, and generalisations to, a conjecture of Baker and Essam, *Discrete Mathematics* **4**, 101–22.

Coniglio, A. (1982), Cluster structure near the percolation threshold, *J. Phys. A.* **15**, 3829–44

Crapo, H. H., (1967), A higher invariant for matroids, *J. Comb. Th.* **2**, 406–17.

Den Nijs, M. P. M., (1979), A relation between the temperature exponents of the eight-vertex and q-state Potts model, *Phys. Rev. B* **27**, 1857–68.

Domb, C. (1959), Fluctuation phenomena and stochastic processes, *Nature* **184**, 509–12.

Essam, J. W., (1972), Percolation and cluster size. In *Phase Transitions and Critical Phenomena* Vol. 2, ed. C. Domb and M. S. Green (Academic Press, New York) Chap. 6, 197–270.

Essam, J. W. (1980), Percolation theory, *Rep. Prog. Phys.* **43**, 833–912.

Essam, J. W., De'Bell, K., Adler, J. and Bhatti, F. M. (1986), Analysis of extended series for bond percolation on the directed square lattice, *Phys. Rev. B* **33**, 1982–6.

Fortuin, C. M. and Kasteleyn, P. W. (1972), On the random cluster model I, *Physica* **57**, 536–64.

Grimmett, G. (1987), Long paths and cycles in a random lattice, Ann. Disc. Math. **33**, 69-76.

Harary, F. (1969), *Graph Theory* (Addison-Wesley) Chap. 5, 43–56.

Harris, A. B. (1983), Field theoretic approach to biconnectedness in percolating systems, *Phys. Rev. B* **28**, 2614–29.

Harris, A. B. and Lubensky, T. C. (1983), Field theoretic approaches to biconnectedness in percolating systems, *J. Phys. A.* **16**, L365–73.

Kesten, H. (1982), *Percolation Theory for Mathematicians*, (Birkhäuser, Boston).

Kinzel, W. (1983), Directed percolation, *Annals of the Israel Phys. Soc.* **5**, 426–45.

Martin, J. L. (1974). Computer enumerations. In *Phase Transitions and Critical Phenomena*, eds. C. Domb and M. S. Green (Academic Press, London).

McDiarmid, C. (1980), Clutter percolation and random graphs, *Mathematical programming study* **13**, 17-25, (North-Holland).

McDiarmid, C. (1981), General percolation and random graphs, *Adv. Appl. Prob.* **13**, 40–60.

Russo, L. (1981), On the critical percolation probabilities, *Z. Wahrsch. Verw. Geb.* **56**, 229–37.

Stauffer, D. (1979), Scaling theory of percolation clusters, *Phys. Rep.* **54**, 1–74.

HAMILTONIAN CLOSURE IN RANDOM GRAPHS

John GIMBEL and David KURTZ

Colby College,
Waterville, Maine 04901

Linda LESNIAK

Drew University,
Madison, New Jersey 07940

Edward R. SCHEINERMAN and John C. WIERMAN

The Johns Hopkins University,
Baltimore, Maryland 21218 U.S.A.

The *closure* $C^*(G)$ of a graph G with n vertices is obtained by iteratively adding edges to G between nonadjacent vertices whose degrees sum to at least n. We show that for almost all graphs G with edge probability at least $1/2$, $C^*(G)$ is a complete graph, answering a question posed by Palmer.

Ore (1960) showed that if a graph G with n vertices has nonadjacent vertices u and v with $\deg(u)+\deg(v) \geq n$ then G is Hamiltonian iff $G+uv$ is Hamiltonian. Based on this fact, Bondy and Chvátal (1976) developed an algorithm that can construct a Hamiltonian cycle in G in $O(n)$ steps, given one in $G+uv$.

The Bondy-Chvátal construction is typically applied to the *closure* of G, denoted $C^*(G)$: the graph obtained by repeatedly adding edges between nonadjacent vertices whose degrees sum to at least n. In this process the vertex degrees increase as incident edges are added. The process terminates when the degrees of every pair of nonadjacent vertices sum to less than n. It was shown in Bondy and Chvátal (1976) that for a given G, the closure is unique; that is, the graph $C^*(G)$ is independent of the order in which the edges are added.

To describe the algorithm from an alternative viewpoint, we define the *first closure* of a graph G with n vertices, denoted $C(G)$, to be the spanning supergraph of G obtained by adding all edges uv between nonadjacent vertices u and v for

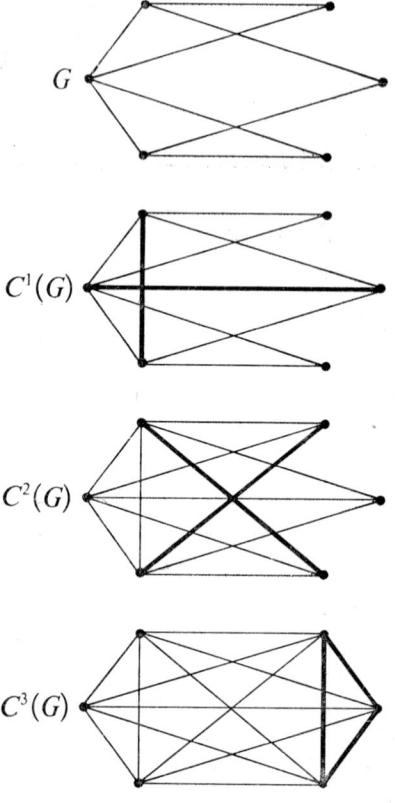

Figure 1.

which $\deg_G(u) + \deg_G(v) \geq n$. Putting $C^1(G) = C(G)$ we define $C^k(G) = C(C^{k-1}(G))$ for $k \geq 2$. We call $C^2(G)$ and $C^3(G)$ the *second* and *third closures* respectively. See Figure 1. Clearly, for k sufficiently large, $C^k(G) = C^*(G)$ (for example, if k exceeds the number of edges in the complement of G).

The operation C is easily implemented on a parallel computer. For each pair of vertices one has a processor. Each processor checks if the sum of the degrees of its vertices is at least n and adds the edge between them as appropriate. How many steps does this parallel algorithm take to compute $C^*(G)$? D. B. West (personal communication) is investigating worst case behavior of this algorithm. We are interested in determining the 'typical' case behavior.

We consider random graphs defined according to a model proposed by Erdös and Rényi (termed "Model A" in Palmer (1985)). The sample space consists of all simple, undirected graphs on n labeled vertices. Each of the $\binom{n}{2}$ possible edges is present with probability p. A statement holds for *almost all* graphs if the probability that it is true tends to 1 as $n \to \infty$. Other graph theory definitions and notation conform to that of Behzad et al. (1986).

When $p < \frac{1}{2}$ it is easy to show that $C^*(G) = C^1(G) = G$ and when $p > \frac{1}{2}$ one has $C^*(G) = C^1(G) = K_n$ for almost all graphs. Palmer (1986) asked: What is the limiting probability that $C^*(G)$ is complete when $p = \frac{1}{2}$? By generating 1000 random graphs with $p = \frac{1}{2}$ on a computer, one observes first that $C^*(G) = K_n$ for most graphs. Table I (see appendix) reports the percentage of graphs on n vertices for which $C^*(G) = K_n$. Moreover, for most graphs $C^2(G) \neq K_n$, but $C^3(G) = K_n$. Table II reports the percentage of graphs for which $C^3(G) = K_n$. These observations led to the following result, which answers Palmer's question in a more precise form.

Theorem. *If $p < \frac{1}{2}$ then for almost all graphs $C^1(G) = G$. If $p = \frac{1}{2}$ then for almost all graphs $C^3(G) = K_n$, but $C^2(G) \neq K_n$. If $p > \frac{1}{2}$ then for almost all graphs $C^1(G) = K_n$.*

As an immediate corollary, we see that the closure of almost all graphs is complete when $p \geq \frac{1}{2}$, but is not complete when $p < \frac{1}{2}$.

Proof. From Erdös and Rényi (1960) we know that for any $\varepsilon > 0$ almost all graphs' smallest and largest degrees satisfy $(1-\varepsilon)pn < \delta(G) \leq \Delta(G) < (1+\varepsilon)pn$. Hence if $p < \frac{1}{2}$ the sum of the degrees of any pair of vertices is less than n and if $p > \frac{1}{2}$ the sum is more than n for almost all graphs. This gives $C^1(G) = G$ for $p < \frac{1}{2}$ and $C^1(G) = K_n$ for $p > \frac{1}{2}$.

For the remainder of this proof we will be concerned only with the case $p = \frac{1}{2}$. We begin by showing that for almost all graphs, $C^3(G) = K_n$.

For a graph G let $B(G)$ denote $\left\{ v \in V(G) : \deg_G(v) \geq \frac{|V(G)|}{2} \right\}$ and $S(G) = V(G) - B(G)$. For $v \in V(G)$ and $X \subset V(G)$ let $d_G(v, X)$ denote the number of edges of G joining v to vertices in X. Finally, choose ε and ε_0 with $0 < \varepsilon_0 < \varepsilon < \frac{1}{8}$. We assert the following claims for almost all graphs G with n vertices:

$$\left| |B(G)| - \frac{n}{2} \right| \leq o(n), \text{ and} \tag{1}$$

For all $v \in V(G)$ we have $\left| d_G(v, B(G)) - \frac{n}{4} \right| \leq \varepsilon_0 n.$ (2)

We also claim that both (1) and (2) hold with $S(G)$ in place of $B(G)$.

Once these claims are established the fact that $C^3(G) = K_n$ for almost all graphs is proved as follows (in Figure 1 $B(G)$ consists of the three vertices on the left): Observe that in $C^1(G)$ every pair of vertices in $B(G)$ are joined by an edge. By

(1) and (2) this implies that their degrees are at least $(\frac{3}{4}-\varepsilon)n$ in $C^1(G)$. Notice now that for $v \in B(G)$ and $w \in S(G)$,

$$\deg_{C^1(G)}(v) + \deg_{C^1(G)}(w) \geq (\tfrac{5}{4} - 2\varepsilon)n$$

using $\deg_{C^1(G)}(w) \geq \deg_G(w) \geq \delta(G) \geq (\frac{1}{2}-\varepsilon)n$ by the minimum degree result quoted earlier. Thus in $C^2(G)$ every vertex in $B(G)$ is connected to every vertex in $S(G)$. The claims imply that the degrees (in $C^2(G)$) of vertices in $S(G)$ are at least $(\frac{3}{4}-2\varepsilon)n$. The only possible pairs of nonadjacent vertices lie in $S(G)$ and since their degrees now sum to at least $(\frac{3}{2}-4\varepsilon)n > n$ we see that $C^3(G) = K_n$.

We now return to prove claims (1) and (2).

Proof of claim (1). First, check that for any vertex v in a random graph G,

$$\left| P\left(\deg(v) \geq \frac{n}{2} \right) - \frac{1}{2} \right| \leq O(n^{-\frac{1}{2}})$$

by noting that $\deg(v)$ has a Binomial distribution $(n, \frac{1}{2})$. Next, let X denote the random variable giving $|B(G)|$. Compute that $E(X) = \frac{n}{2} + O(\sqrt{n})$ and $E(X^2) = \frac{n^2}{4} + O(n^{3/2})$, hence $\text{Var}(X) = O(n^{3/2})$. Chebyshev's inequality gives $P\left[\left| X - \frac{n}{2} \right| \geq n^{4/5} \right] \leq O(n^{-1/10})$ and claim (1) follows. The corresponding statement for $S(G)$ follows from $|B(G)| + |S(G)| = n$.

Proof of claim (2). For each vertex $v \in V(G)$, define

$$B_v = B(G-v) = \left\{ w \in V(G-v): \deg_{G-v}(w) \geq \frac{n-1}{2} \right\}.$$

Our proof proceeds as follows: We begin by showing that $B_v \approx B(G)$. Note that the edges between v and $G-v$ are independent of the edges in $G-v$, which determine the vertices in B_v. We use this independence to establish (2) with B_v in place of $B(G)$, and then apply $B_v \approx B(G)$ to establish our claim.

Observe that $B_v \supset B(G) - \{v\}$. If $w \in B_v - [B(G) - \{v\}]$, then both $\deg_G(w) < \frac{n}{2}$ and $\deg_G(w) \geq \deg_{G-v}(w) \geq \frac{n-1}{2}$ hold, so $\deg_G(w) = \frac{n-1}{2}$. Let A denote the set of vertices in G with degree exactly $\frac{n-1}{2}$. Hence, $B_v \subset B(G) \cup A$. Then, for every $v \in V(G)$, we have

$$|B(G)| - 1 \leq |B_v| \leq |B(G)| + |A|.$$

We assert that $|A|$ is very small relative to $|B(G)|$. Indeed, $|A|=0$ when n is even. If n is odd, the probability that a particular vertex is in A is

$$\binom{n-1}{(n-1)/2} 2^{-(n-1)} \sim kn^{-\frac{1}{2}}$$

for some constant k. Thus, for n odd, $E(|A|) \sim k\sqrt{n}$. Computing that $E(|A|^2) \sim k^2 n$, one obtains $\text{Var}(|A|) = o(n)$. Apply Chebyshev's inequality to show that $|A| = o(n)$ in almost all graphs.

Since $||B(G)| - |B_v|| \leq |A| + 1 = o(n)$, we also have

$$|d_G(v, B(G)) - d_G(v, B_v)| \leq |A| + 1 = o(n)$$

for almost all graphs. Together with $|B_v| = \frac{n}{2} + o(n)$ for all v, these imply

$$\lim_{n \to \infty} P\left[\left| \frac{d_G(v, B(G))}{n/2} - \frac{d_G(v, B_v)}{|B_v|} \right| < \varepsilon_0 \text{ for all } v \in V(G) \right] = 1. \quad (*)$$

We now establish (2) with B_v in place of $B(G)$. We rely on the following large deviation result in probability theory, due to Chernoff (1952) [or see Kotz et al. (1985) pp. 438–439].

Lemma 1. *Let S_n be a Binomial $(n, \frac{1}{2})$ random variable. For each $\varepsilon > 0$ there exists a constant $f = f(\varepsilon) > 0$ such that*

$$\lim_{n \to \infty} \frac{1}{n} \log P\left[\left| \frac{S_n}{n} - \frac{1}{2} \right| \geq \varepsilon \right] = -f$$

or equivalently

$$P\left[\left| \frac{S_n}{n} - \frac{1}{2} \right| \geq \varepsilon \right] = e^{-fn + o(n)}$$

as $n \to \infty$. □

When this result is applied to $|B_v|$ we can establish

$$P\left[\left| |B_v| - \frac{n}{2} \right| \geq \frac{n}{8} \right] \leq e^{-f(1/8) n/2}$$

consequently,

$$P\left[|B_v| \geq \frac{n}{3} \text{ for all } v \in V(G)\right] \geq 1 - ne^{-f(1/8)n/2} \to 1$$

as $n \to \infty$.

We also apply the result conditionally. Note that $d_G(v, B_v)$ has a Binomial $(|B_v|, \frac{1}{2})$ distribution. For fixed ε_0, there exists $f(\varepsilon_0)$ such that

$$P\left[\left|\frac{d_G(v, B_v)}{|B_v|} - \frac{1}{2}\right| \geq \varepsilon_0 \text{ and } |B_v| \geq \frac{n}{3}\right]$$

$$= \sum_{k \geq n/3} P\left[\left|\frac{d_G(v, B_v)}{|B_v|} - \frac{1}{2}\right| \geq \varepsilon_0 \text{ and } |B_v| = k\right]$$

$$= \sum_{k \geq n/3} P\left[\left|\frac{d_G(v, B_v)}{|B_v|} - \frac{1}{2}\right| \geq \varepsilon_0 \,\bigg|\, |B_v| = k\right] P[|B_v| = k]$$

$$= \sum_{k \geq n/3} e^{-f(\varepsilon_0) k/2} P[|B_v| = k]$$

$$\leq \sum_{k \geq n/3} e^{-f(\varepsilon_0) n/6} P[|B_v| = k]$$

$$= e^{-f(\varepsilon_0) n/6} P\left[|B_v| \geq \frac{n}{3}\right] \leq e^{-f(\varepsilon_0) n/6}$$

and so

$$P\left[\left|\frac{d_G(v, B_v)}{|B_v|} - \frac{1}{2}\right| \geq \varepsilon_0\right]$$

$$\leq P\left[\left|\frac{d_G(v, B_v)}{|B_v|} - \frac{1}{2}\right| \geq \varepsilon_0 \text{ and } |B_v| \geq \frac{n}{3}\right] + P\left[|B_v| < \frac{n}{3}\right]$$

$$\leq e^{-f(\varepsilon_0) n/6} P\left[|B_v| \geq \frac{n}{3}\right] + P\left[\left||B_v| - \frac{n}{2}\right| \geq \frac{n}{8}\right]$$

$$\leq e^{-f(\varepsilon_0) n/6} + e^{-f(1/8) n/2}.$$

Therefore,

$$P\left[\left|\frac{d_G(v, B_v)}{|B_v|} - \frac{1}{2}\right| < \varepsilon_0 \text{ for all } v \in V(G)\right]$$

$$\geq 1 - n(e^{-f(\varepsilon_0) n/6} + e^{-f(1/8) n/2}) \to 1 \qquad (**)$$

as $n \to \infty$.

Finally, apply the triangle inequality to the intersection of the events in (*) and (**) to conclude

$$\lim_{n\to\infty} P\left[\left|d_G(v, B(G)) - \frac{n}{4}\right| < \varepsilon_0 n \text{ for all } v \in V(G)\right] = 1.$$

This verifies (2) for $B(G)$. A similar proof works for $S(G)$.

Next we show that $C^2(G) \neq K_n$ for almost all G. Label the vertices $v_1, v_2, ..., v_n$ of G so $d_i = \deg(v_i)$ satisfies $d_1 \geq d_2 \geq ... \geq d_n$. We use the following special cases of results of Bollobás and Matula:

Lemma 2 [Bollobás (1981), Theorem 12]. *Let $m \to \infty$ and $m = o(n)$. Put*

$$k(m) = \left(\frac{n \log(n/m)}{2}\right)^{\frac{1}{2}}$$

and

$$\text{err}(m) = O\left[\log\log\left(\frac{n}{m}\right)\left(\frac{n}{\log(n/m)}\right)^{\frac{1}{2}}\right].$$

Almost all graphs satisfy

$$\left|d_m - \frac{n}{2} - k(m)\right| \leq \text{err}(m). \quad \square$$

Note that $\text{err}(m) = o(\sqrt{n})$. By symmetry, $\left|d_{n+1-m} - \frac{n}{2} + k(m)\right| \leq \text{err}(m)$ for almost all graphs.

Lemma 3 [Matula (1976), see also Palmer (1985), pp. 76–77]. *In almost all graphs the size of the largest complete subgraph is $O(\log n)$.* $\quad \square$

Let v_{-i} denote the vertex v_{n+1-i} and let $d_{-i} = d_{n+1-i}$. Hence, $\Delta(G) = d_1$ and $\delta(G) = d_{-1}$. Let $V^- = \{v_{-1}, v_{-2}, ..., v_{-m}\}$ where $m = n^{1/4}$. Thus V^- contains the "small" degree vertices. Observe,

$$\max_{v \in V^-} \deg(v) = d_{-m} = \frac{n}{2} - k(m) + \text{err}(m)$$

$$= \frac{n}{2} - \left(\frac{3}{8} n \log n\right)^{\frac{1}{2}} + o(\sqrt{n}).$$

Let $V^+ = \{v_1, ..., v_{M-1}\}$ where $M = n^{1/3}$. Then

$$\max_{v \in V - V^+} \deg(v) = d_M = \frac{n}{2} + \left(\frac{1}{3} n \log n\right)^{\frac{1}{2}} + o(\sqrt{n}).$$

If $v \in V^-$ and $w \in V - V^+$ such that v and w are not adjacent in G, then they are not adjacent in $C(G)$ since $\deg_G(v) + \deg_G(w)$ is bounded above by

$$\frac{n}{2} - \left(\frac{3}{8} n \log n\right)^{\frac{1}{2}} + \frac{n}{2} + \left(\frac{1}{3} n \log n\right)^{\frac{1}{2}} + o(\sqrt{n}) < n$$

since $\frac{3}{8} > \frac{1}{3}$.

Thus for any vertex in V^- its degree in $C(G)$ is at most

$$\frac{n}{2} - \left(\frac{3}{8} n \log n\right)^{\frac{1}{2}} + M + o(\sqrt{n}) < \frac{n}{2}$$

for n sufficiently large. Now in almost all graphs V^- is not a clique (by Lemma 3) since $|V^-| = n^{1/4} \gg O(\log n)$. Choose $v, w \in V^-$ with v not adjacent to w in G. Observe that v is not adjacent to w in $C^1(G)$ since they both have degree less than $n/2$. Moreover, they are not adjacent in $C^2(G)$ and again, their degrees in $C^1(G)$ are less than $n/2$. Hence $C^2(G) \neq K_n$ for almost all graphs G. □

ACKNOWLEDGEMENTS

The authors would like to thank M. Alexander for help with computer programming. Edward Scheinerman is supported in part by Office of Naval Research contract No. N00014-85-K0622 and John Wierman is supported in part by National Science Foundation grant No. DMS-8303238.

Appendix

Table I.

Percent of graphs with $C^*(G) = K_n$.

No. Vertices	5	10	15	20	25
Percent	21.2	70.0	96.3	99.7	99.9

Table II.

Percent of graphs with $C^3(G) = K_n$.

No. Vertices	30	40	50	60	70	80	90	100
Percent	74.3	85.5	93.3	96.7	98.5	99.9	99.2	100.0

References

M. Behzad, G. Chartrand and L. Lesniak-Foster (1986), *Graphs and Digraphs* Wadsworth Publishers, California.

B. Bollobás (1981), "Degree sequences of random graphs" *Discrete Math. 33* pp. 1–19.

J. Bondy and V. Chvátal (1976), "A Method in Graph Theory" *Discrete Math. 15* pp. 111–136.

H. Chernoff (1952) "A measure of asymptotic efficiency for tests of a hypothesis based on the sum of observations" *Ann. Math. Stat. 23* pp. 493–507.

P. Erdös and A. Rényi (1960), "On the Evolution of Random Graphs" *Magyar Tud. Akad. Mat. Kutato Int. Kozl. 5* pp. 17–61.

S. Kotz, N. Johnson and C. Read, eds. (1985), *Encyclopedia of Statistical Sciences* Wiley (New York).

D. W. Matula (1976), "The largest clique size in a random graph" Technical Report, Department of Computer Science, Southern Methodist University, Dallas.

O. Ore (1960) "Note on Hamiltonian Cycles" *Amer. Math. Monthly 67* p. 55.

E. Palmer (1985) *Graphical Evolution: An Introduction to the Theory of Random Graphs* Wiley (New York).

E. Palmer (1987), "Unsolved Problems in the Theory of Random Graphs" *Ann. Disc. Math. 33* pp. 227 - 239.

LONG PATHS AND CYCLES IN A RANDOM LATTICE

Geoffrey R. GRIMMETT

School of Mathematics, University of Bristol,
Bristol, United Kingdom

Let B_n be an n by n section of the square lattice, in which each edge has been deleted with probability $1-p$. We study the asymptotic behaviour of the lengths of the longest path and the longest cycle in B_n as $n \to \infty$, showing that these quantities have order $\log n$ if $p < 1/2$ and order n^2 if $p > 1/2$.

1. Introduction

The concept of euclidean space is of little or no importance in most work on random graphs. To one particular area, however, it is fundamental, since the topic in question attempts to describe characteristics of the physical universe. We have in mind the "percolation model", a simple model for randomly positioned impurities in a material. The percolation model is a type of random graph which has as initial graph a crystalline lattice (such as the square lattice). All the difficulties which arise in the study of this model are related directly to the nature of the euclidean space which contains this lattice, and thus the body of useful techniques is rather different from that encountered elsewhere in the theory of random graphs.

Let Z^2 be the square lattice, with vertex set $\{(i,j) : i,j = 0, \pm 1, \pm 2, ...\}$ and edges joining pairs of vertices which are unit euclidean distance apart. The point $(0, 0)$ is the *origin*, denoted by $\mathbf{0}$. The "bond percolation model" with parameter p ($\in [0, 1]$) is obtained by deleting each edge of Z^2 with probability $1-p$, independently of all other edge deletions. Thus each edge remains with probability p, and we denote by ω the (random) subgraph of Z^2 which remains when the deletion process has been completed. The main question is to describe the structure of ω as p varies from 0 to 1. It is known that there are three different regimes, depending on which of the following is valid: $p < \frac{1}{2}$, $p = \frac{1}{2}$, $p > \frac{1}{2}$. If $p < \frac{1}{2}$ then all components of ω are a.s. finite, whereas if $p>\tfrac{1}{2}$ then ω contains a.s. a single infinite component; the case when $p=\tfrac{1}{2}$ has certain features which are characteristic of the transition point.

In this paper we concentrate upon certain properties of the percolation model which are closely related to well-studied features of other random graphs: the existence of long paths and long cycles. To this end we define B_n to the subgraph of $\mathbf{Z}^2$ induced by the vertices (i,j) satisfying $1\leqslant i,j\leqslant n$, and write ω_n for the graph ω restricted to B_n. We shall be concerned with the sizes of the longest path and the longest cycle in ω_n. We shall see that the orders of magnitude of these quantities depend crucially on whether $p<\tfrac{1}{2}$ or $p>\tfrac{1}{2}$: in the former case they have order $\log n$ for large n, whilst in the latter case they have order n^2. In the intermediate case when $p=\tfrac{1}{2}$, it is likely that they have orders n^α for values of α satisfying $0<\alpha<2$. We have several results and conjectures along these lines. In closely related work, Grimmett (1984) has studied the sizes of the largest components of ω_n.

Finally, we discuss the degree of connectivity of ω, and verify a conjecture of Essam (1985) concerning the uniqueness of the critical transition point $p=\tfrac{1}{2}$.

We shall make some use of two correlation inequalities, commonly called the FKG and BK inequalities. These are of great value in the study of random graphs, and we believe that they will prove to have many other important applications.

2. Long paths

Let L_n be the number of edges in the longest path in ω_n. The asymptotic behaviour of L_n for large n is described in the next theorem. We consider mainly convergence in probability here, but it is clear that other modes of convergence are valid also.

Theorem 1. (a) *If $p<\tfrac{1}{2}$, there exists $\lambda(p)$ (>0) such that $L_n/\log n \to \lambda(p)$ in probability as $n\to\infty$.*

(b) *If $p>\tfrac{1}{2}$, there exist $\alpha(p), \beta(p)$ (>0) such that, as $n\to\infty$,*

$$P(\alpha(p)\leqslant n^{-2}L_n\leqslant\beta(p))\to 1.$$

We conjecture that $\lim n^{-2}L_n$ exists for $p>\tfrac{1}{2}$. The situation is rather more delicate when $p=\tfrac{1}{2}$, though existing evidence (Kesten (1982)) indicates that L_n grows asymptotically as n^γ for some γ satisfying $1\leqslant\gamma<2$.

The problem of ascertaining the asymptotic behaviour of L_n is closely related to the study of reliable chains in wafer-scale processor arrays in the presence of defects; see Greene and El Gamal (1984).

Proof. (a) Suppose $p<\frac{1}{2}$. Let π be the length of the longest path in ω which has the origin as an endpoint. Let A_r be the event that $\pi \geqslant r$, and let $B_{m,r}$ be the event that some vertex, whose Manhattan distance from the origin is no greater than r, is the endpoint of a path in ω with length at least m. Then, in the notation of van den Berg and Kesten (1985),

$$A_{m+r} \subseteq A_r \square B_{m,r}$$

giving, by the BK inequality of that paper, that

$$P(A_{m+r}) \leqslant P(A_r) P(B_{m+r}).$$

However

$$P(B_{m,r}) \leqslant 4r^2 P(A_m),$$

so that $g(r) = \log P(A_r)$ satisfies

$$g(m+r) \leqslant g(m) + g(r) + \log(4r^2).$$

Applying Theorem 2 of Hammersley (1962), we deduce that

$$\gamma = -\lim_{r \to \infty} \frac{1}{r} g(r)$$

exists and satisfies $0 \leqslant \gamma \leqslant \infty$. To see that $0 < \gamma < \infty$, note first that $P(A_r) \geqslant p^r$, and secondly that $P(A_r) \leqslant P(W \geqslant r) \leqslant e^{-\delta r}$ for some $\delta > 0$, where W is the size of the component of ω containing the origin (see Kesten (1982)). We conclude that

$$\log P(\pi \geqslant r) \cong -\gamma(p) r,$$

where $\delta \leqslant \gamma \leqslant \log(1/p)$.

Having reached this point, the result of part (a) of the theorem follows exactly as in the proof of Theorem 1 in Grimmett (1984).

(b) Suppose that $p > \frac{1}{2}$. The upper bound for L_n is obvious, since L_n does not exceed the size of the largest component of ω_n (see Grimmett (1984)). To show the lower bound, we proceed as follows. Let η_n (respectively φ_n) be the maximal number of edge-disjoint (respectively vertex-disjoint) paths in ω_n joining the left side of B_n to the right side. It is proved in Grimmett and Kesten (1984) that

$n^{-1}\eta_n \to \eta(p)$ as $n \to \infty$, in probability, for some $\eta(p)$ satisfying $\eta(p)>0$. There is a vertex-cutset of size φ_n separating the left and right hand sides of B_n, and thus there is an edge-cutset of size no larger than $4\varphi_n$, giving that $\eta_n \leq 4\varphi_n$; hence

$$P(n^{-1}\varphi_n \geq \tfrac{1}{5}\eta(p)) \to 1 \text{ as } n \to \infty.$$

Let E_n be the event that there are top-to-bottom crossings of ω_n contained within each of the regions $[0, (\log n)^2] \times [0, n]$ and $[n-(\log n)^2, n] \times [0, n]$, and let F_n be the event that $n^{-1}\varphi_n \geq \tfrac{1}{5}\eta(p)$. By the results of Grimmett (1983), the argument above, and the FKG inequality (see Fortuin, Kasteleyn and Ginibre (1971) or Harris (1960))

$$P(E_n \cap F_n) \geq P(E_n) P(F_n) \to 1 \text{ as } n \to \infty.$$

On $E_n \cap F_n$, it is clear by geometrical considerations that there is a path in ω_n of length at least $(n - 2(\log n)^2)\varphi_n$, giving that

$$P(L_n \geq \tfrac{1}{5}n^2\eta(p)(1+o(1))) \to 1 \text{ as } n \to \infty,$$

as required. □

3. Long cycles

Broadly speaking, we have the same results for the existence of long cycles as we have for long paths. Let C_n be the length of the longest cycle in ω_n.

Theorem 2. (a) *If $p < \tfrac{1}{2}$, there exists $v(p)$ (>0) such that $C_n/\log n \to v(p)$ in probability as $n \to \infty$.*

(b) *If $p > \tfrac{1}{2}$, there exist $\mu(p), \rho(p)$ (>0) such that, as $n \to \infty$,*

$$P(\mu(p) \leq n^{-2}C_n \leq \rho(p)) \to 1.$$

We conjecture that $\lim n^{-2}C_n$ exists for $p > \tfrac{1}{2}$, and that C_n grows in the manner of n^γ for some γ if $p = \tfrac{1}{2}$.

Proof. (a) Suppose $p < \tfrac{1}{2}$. For any cycle C in $\mathbb{Z}^2$, define $b(C)$ (respectively $c(C)$) to be the vertex at the left hand end of the lowest edges in C (respectively the right hand end of the highest edges in C). That is to say, $b(C) = (x, y)$ where

$y=\min\{j:(i,j)\in C$ for some $i\}$ and $x=\min\{i:(i,y)\in C\}$, and $c(C)=(x,y)$ where $y=\max\{j:(i,j)\in C$ for some $i\}$ and $x=\max\{i:(i,y)\in C\}$. Let u_n be the probability that the origin is in a cycle of ω of length n, and let v_n be the probability that the origin is in a cycle C of ω having $b(C)=(0,0)$. We note that $v_n \leqslant u_n$. Furthermore,

$$u_n \leqslant \sum_x P(\text{origin in some cycle } C \text{ with } b(C)=x \text{ and length } n)$$

$$= \sum_x P(\text{origin and } -x \text{ in some cycle } C \text{ with } b(C)=0 \text{ and length } n)$$

$$\leqslant n^2 v_n,$$

where the summations are over all points $x=(x_1, x_2)$ satisfying $x_2 < 0$, or $x_2 = 0$ and $x_1 < 0$.

We prove next that

$$u_{n+m} \geqslant p^2 v_n v_m \text{ for } m, n \geqslant 1.$$

To see this, let E be the event that the origin is in some cycle C of ω with $b(C)=0$ and length m, and let F be the event that $(1,-1)$ is in some cycle C' of ω with $c(C')=(1,-1)$ and length n. The events E and F are independent with probabilities v_m and v_n respectively. If the edges joining $(0,0)$ to $(0,-1)$ and $(0,1)$ to $(1,-1)$ are present in ω, then the origin is in a cycle of length $m+n$, giving that $u_{n+m} \geqslant p^2 v_n v_m$ as required.

Combining the two inequalities above, we obtain

$$v_{n+m} \geqslant v_n v_m (m+n)^{-2} p^2 \text{ for } m, n \geqslant 1,$$

and thus, by Hammersley's theorem again,

$$\gamma = -\lim_{n\to\infty} \frac{1}{n} \log v_n$$

exists and satisfies $0 \leqslant \gamma \leqslant \infty$. The argument in the proof of Theorem 1(a) is easily adapted to complete this proof.

(b) Suppose $p > \frac{1}{2}$. The upper bound for C_n follows in the same way as the corresponding step for paths, and the lower bound follows by a minor amendment to the construction in the proof of Theorem 1(b). □

4. Connectedness

There has been much interest recently in the so-called "backbone" of the infinite component of ω when $p > \frac{1}{2}$, and particularly in the shape of this set as p approaches $\frac{1}{2}$ from above (see Kirkpatrick (1978), Coniglio (1982), Harris (1983), Harris and Lubensky (1983)). With this in mind, Essam (1985) has formulated certain questions relating to the degree of connectedness of subgraphs of ω, and it is the purpose of this section to answer one of his questions.

Let k be an integer satisfying $1 \leqslant k \leqslant 4$, and let N_k be the number of vertices of $\mathbb{Z}^2$ with the property that there exist k edge-disjoint paths from such a vertex to the origin in ω. We define the critical probability

$$\pi(k) = \sup\{p : P(N_k = \infty) = 0\},$$

noting that $\pi(1) = \frac{1}{2}$, the usual critical probability. Essam (1985) has asked whether or not these critical probabilities are all equal to one another. It is clear that $\pi(1) \leqslant \pi(2) \leqslant \pi(3) \leqslant \pi(4)$, and so an affirmative answer is provided by the next result.

Theorem 3. $\pi(4) = \frac{1}{2}$.

Proof. Suppose $p > \frac{1}{2}$, and fix α satisfying $0 < \alpha < \frac{1}{6}$. Let C_1 and C_2 be the two doubly-infinite cones of $\mathbb{Z}^2$ defined as the two disjoint sets of points (x, y) satisfying $|y| \leqslant \alpha |x|$ and $|x| \leqslant \alpha |y|$ respectively. Let E be the event that each of the four singly-infinite subcones of $C_1 \cup C_2$ contains an infinite path in ω starting at the origin; we note that $P(E) > 0$. Let F_i be the event that there is a top-to-bottom crossing in ω of the rectangle $[i, 3i] \times [-\lceil 3\alpha i \rceil, \lceil 3\alpha i \rceil]$, and let $F = \bigcap_{i=1}^{\infty} F_i$. By the result of Grimmett and Kesten (1984), $\log(1 - P(F_i)) \leqslant -\zeta i$ for some $\zeta > 0$, so that

$$P(F) \geqslant \prod_{i=1}^{\infty} P(F_i) \text{ by the FKG inequality } > 0,$$

and furthermore $P(E \cap F) > 0$, again by the FKG inequality.

Let I be the indicator function of the event $E \cap F$, and let τ be a shift of the lattice one unit to the left. Writing $\tau\omega$ for the image of ω under τ, we have that the sequence $\{I(\tau^{4k}\omega) : k = \ldots, -1, 0, 1, \ldots\}$ is stationary, and so

$$\frac{1}{n} \sum_{k=0}^{n} I(\tau^{4k}\omega) \to \xi \text{ a.s.}$$

for some ξ, measurable on the σ-field of events invariant under τ; this σ-field is trivial since edges are present independently of each other, giving that ξ is constant a.s., and so $\xi = E(I) = P(E \cap F) > 0$. Thus $I(\tau^{4k}\omega)$ takes the value 1 for infinitely many values of k a.s. Suppose that $I(\tau^{4K}\omega) = I(\tau^{4L}\omega) = 1$ for some $K < L$. Then we claim that a.s. there are four edge-disjoint paths in ω between $(4K, 0)$ and

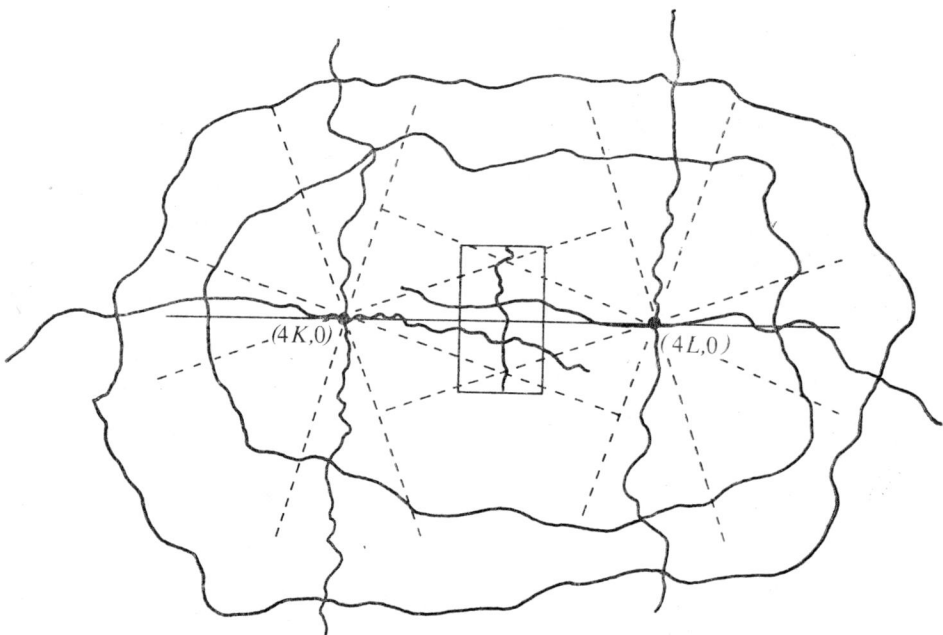

Figure 1. Four edge-disjoint paths joining $(4K, 0)$ to $(4L, 0)$.

$(4L, 0)$. By standard arguments, there exist a.s. infinitely many cycles in ω containing both $(4K, 0)$ and $(4L, 0)$ in their interiors, and the conclusion may be reached easily by inspecting Fig. 1. It follows that there exists a.s. infinitely many vertices on the x-axis such that each pair is joined by four edge-disjoint paths; the density of this set of vertices is positive since $P(E \cap F) > 0$; hence the origin is in this set with positive probability. Thus $P(N_4 = \infty) > 0$ if $p > \frac{1}{2}$, which implies that $\pi(4) = \frac{1}{2}$, as required. □

ACKNOWLEDGEMENT

I am grateful to Professor M. H. Rogers for drawing my attention to the problem of the reliability of wafer-scale arrays of processors, and the paper of Greene and El Gamal (1984).

References

Berg, J. van den and Kesten, H. (1985) Inequalities with applications to percolation and reliability, *Journal of Applied Probability* 22, 556–569.

Coniglio, A. (1982) Cluster structure near the percolation threshold, *Journal of Physics A*: *Mathematical and General* 15, 3829–3844.

Essam, J. W. (1987) Connectedness and connectivity in percolation theory, *Annals of Discrete Mathematics* 33, 41–57.

Fortuin, C., Kasteleyn, P. and Ginibre, J. (1971) Correlation inequalities on some partially ordered sets, *Communications in Mathematical Physics* 22, 89–103.

Greene, J. W. and El Gamal, A. (1984) Configuration of VLSI arrays in the presence of defects, *Journal of the Association for Computing Machinery* 31, 694–717.

Grimmett, G. R. (1983) Bond percolation on subsets of the square lattice, and the threshold between one-dimensional and two-dimensional behaviour, *Journal of Physics A*: *Mathematical and General* 16, 599–604.

Grimmett, G. R. (1984) The largest components in a random lattice, *Studia Scientiarum Mathematicarum Hungarica*, to appear.

Grimmett, G. R. and Kesten, H. (1984) First-passage percolation, network flows and electrical resistances, *Zeitschrift für Wahrscheinlichkeitstheorie und verwandte Gebiete* 66, 335–366.

Hammersley, J. M. (1962) Generalization of the fundamental theorem of subadditive functions, *Proceedings of Cambridge Philosophical Society* 58, 235–238.

Harris, A. B. (1983) Field-theoretic approach to biconnectedness in percolating systems, *Physical Review B* 28, 2614–2629.

Harris, A. B. and Lubensky, T. C. (1983) Field theoretic approaches to biconnectedness in percolating systems, *Journal of Physics A*: *Mathematical and General* 16, L365–L373.

Harris, T. E. (1960) A lower bound for the critical probability in a certain percolation process, *Proceedings of Cambridge Philosophical Society* 56, 13–20.

Kesten, H. (1982) *Percolation theory for mathematicians*, Birkhäuser, Boston.

Kirkpatrick, S. (1978) The geometry of the percolation threshold, in *Electric transport and optical properties of inhomogeneous media*, ed. J. Garland and D. Tanner, American Institute of Physics Conference Proceedings 40, 99–116.

THE DISTRIBUTION OF THE NUMBER OF EMPTY CELLS IN A GENERALIZED RANDOM ALLOCATION SCHEME*

Bernard HARRIS

University of Wisconsin, Madison

Morris MARDEN

University of Wisconsin, Milwaukee

C. J. PARK

San Diego State University

n balls are randomly distributed into N cells, so that no cell may contain more than one ball. This process is repeated m times. In addition, balls may disappear; such disappearances are independent and identically Bernoulli distributed. Conditions are given under which the number of empty cells has an asymptotically ($N \to \infty$) standard normal distribution.

1. Introduction

The distribution of the number of empty cells in the following random allocation process is considered. Let n, N be positive integers with $n \leqslant N$. Assume that n balls are randomly distributed into N cells, so that no cell may contain more than one ball. Then, the probability that each of n specified cells will be occupied is $\binom{N}{n}^{-1}$. This process is repeated m times, so that there are $\binom{N}{n}^m$ random allocations of nm balls among the N cells. In addition, for each ball, let p, $0 \leqslant p \leqslant 1$, be the probability that the ball will not "disappear" from the cell. The "disappearances" are assumed to be stochastically independent for each ball; thus the disappearances constitute a sequence of nm Bernoulli trials.

---

* Partially supported by U.S. Army under contract No. DAAG29-80-C-0041.

Several special cases of this problem have previously been considered. In particular, $p=1$, $n=1$ is the classical occupancy problem, see [2], [3], [11]. The case $p=1$, n arbitrary has been discussed in [4], [5], [8] and [12]. The case $0<p<1$, $n=1$ is treated in C. J. Park [7].

In this paper, we obtain the probability distribution and moments of the number of empty cells. In section 3, we show that the number of empty cells may be represented as a sum of independent Bernoulli random variables. This representation permits us to determine conditions on m, n, p, N such that the number of empty cells is asymptotically normally distributed.

This random allocation process may be viewed as a filing or storage process. Objects are randomly assigned to files or storage bins. From time to time, objects may be missing or have disappeared.

2. The probability distribution and the moments of the number of empty cells

Let m, n, N be positive integers with $n \leqslant N$. m sets, each consisting of n balls, are distributed into N cells at random so that no cell can contain more than one ball from the same set. As each set is distributed, the balls that have been placed during the preceding distributions are left in the cells. Thus, at the end of the process, cells may contain as many as m balls. In addition, each ball may "disappear" with common probability $1-p$, $0 \leqslant p \leqslant 1$. These disappearances are stochastically independent and thus constitute a sequence of mn Bernoulli trials.

Let $P_{m,n,N,p}(j)$ be the probability that exactly j of the N cells are empty. We now establish the following theorem.

Theorem 1.

$$P_{m,n,N,p}(j) = \binom{N}{n}^{-m} \binom{N}{j} \sum_{l=0}^{N-j} (-1)^l \binom{N-j}{l}$$

$$\times \left[\sum_{i=0}^{j+l} (1-p)^i \binom{N-j-l}{n-i} \binom{j+l}{i} \right]^m, \quad 0 \leqslant j \leqslant N. \quad (1)$$

Proof. Let A_v be the event that the vth cell is empty, $v=1, 2, \ldots, N$. Then, for $1 \leqslant v_1 < v_2 < \ldots < v_k \leqslant N$,

$$P(A_{v_1} \cap A_{v_2} \cap \ldots \cap A_{v_k}) = \binom{N}{n}^{-m} \left[\sum_{i=0}^{k} \binom{N-k}{n-i} \binom{k}{i} (1-p)^i \right]^m. \quad (2)$$

Thus, using the inclusion-exclusion method, the probability that exactly j cells are empty is

$$P_{m,n,N,p}(j) = \binom{N}{n}^{-m} \sum_{r=j}^{N} \binom{N}{r}(-1)^{r-j}\binom{r}{j}\left[\sum_{i=0}^{r}\binom{N-r}{n-i}\binom{r}{i}(1-p)^i\right]^m. \quad (3)$$

We can write (3) in the form (1) by letting $r = j + l$.

We now determine the factorial moments of S, the number of empty cells.

Theorem 2. *The vth factorial moment of S,*

$$E(S^{(v)}) = \binom{N}{n}^{-m} N^{(v)}\left[\sum_{j=0}^{v}(1-p)^j\binom{N-v}{n-j}\binom{v}{j}\right]^m. \quad (4)$$

Proof. From J. Riordan [10], p. 53, from (2), it follows immediately that

$$E(S^{(v)}) = \binom{N}{v}v!\binom{N}{n}^{-m}\left[\sum_{i=0}^{v}\binom{N-v}{n-i}\binom{v}{i}(1-p)^i\right]^m. \quad (5)$$

We thus obtain the following.

Corollary.

$$E(S) = N\left(1 - \frac{pn}{N}\right)^m, \quad (6)$$

$$\sigma_S^2 = N(N-1)\left[\frac{(N-n)(N-n-1)}{N(N-1)} + 2(1-p)\frac{n(N-n)}{N(N-1)}\right.$$

$$\left. + (1-p)^2\frac{n(n-1)}{N(N-1)}\right]^m + N\left(1 - \frac{pn}{N}\right)^m\left[1 - N\left(1 - \frac{pn}{N}\right)^m\right]. \quad (7)$$

Proof. From (5)

$$E(S) = N\binom{N}{n}^{-m}\left(\binom{N-1}{n} + \binom{N-1}{n-1}(1-p)\right)^m = N\left(1 - \frac{pn}{N}\right)^m.$$

Since

$$\sigma_S^2 = E(S^{(2)}) + E(S) - (E(S))^2,$$

the conclusion follows readily from (4), after some elementary calculations.

For some purposes, the following equivalent forms of (7) will prove useful.

$$\sigma_S^2 = N(N-1)\left[1 - \frac{np(2(N-1)-p(n-1))}{N(N-1)}\right]^m$$
$$+ N\left(1-\frac{pn}{N}\right)^m\left(1 - N\left(1-\frac{pn}{N}\right)^m\right) \tag{8}$$

and

$$\sigma_S^2 = N^2\left(1-\frac{pn}{N}\right)^{2m}\left\{\left[1 - \frac{np^2(N-n)}{(N-1)(N-pn)^2}\right]^m - 1\right\}$$
$$+ N\left(1-\frac{pn}{N}\right)^m\left\{1 - \left(1-\frac{pn}{N}\right)^m\left[1 - \frac{np^2(N-n)}{(N-1)(N-pn)^2}\right]^m\right\}. \tag{9}$$

From Theorem 2, we readily obtain the following.

Theorem 3. *The factorial moment generating function of S is given by*

$$\varphi_m(t) = E(1+t)^S = \sum_{r=0}^{N} \binom{N}{r} t^r \binom{N}{n}^{-m} \left(\sum_{j=0}^{r}(1-p)^j \binom{N-r}{n-j}\binom{r}{j}\right)^m. \tag{10}$$

Note that $\varphi_m(t)$ is a polynomial in t of degree N. This fact is exploited in the next section, where the asymptotic distribution of S is obtained. In particular,

$$\varphi_0(t) = (1+t)^N \tag{11}$$

and

$$\varphi_1(t) = (1+t)^{N-n}(1+(1-p)t)^n. \tag{12}$$

We now investigate the asymptotic distribution properties of the number of empty cells.

3. The asymptotic distribution of the number of empty cells

In this section, we determine conditions under which the number of empty cells (when suitably normalized) has an asymptotically normal distribution. In order to establish this, a number of preliminary results are required.

Lemma 1. *Let N, n, r be non-negative integers, $r \leqslant n \leqslant N$. Then*

$$\sum_{v=\alpha}^{r} \binom{r}{v}\binom{v}{\alpha}\binom{N-r}{n-v} = \binom{r}{\alpha}\binom{N-\alpha}{n-\alpha}. \tag{13}$$

Proof. Since $\binom{v}{\alpha} = 0$ whenever $v < \alpha$, we can write

$$\sum_{v=\alpha}^{r} \binom{r}{v}\binom{v}{\alpha}\binom{N-r}{n-v} = \sum_{v=0}^{r} \binom{r}{v}\binom{v}{\alpha}\binom{N-r}{n-v}.$$

To obtain the conclusion, note that

$$\sum_{x=0}^{r} \frac{\binom{x}{\alpha}\binom{r}{x}\binom{N-r}{n-x}}{\binom{N}{n}} = E\{X^{(\alpha)}\}/\alpha!,$$

where X has the hypergeometric distribution. From B. Harris [1], p. 105,

$$\sum_{x=0}^{r} \frac{\binom{x}{\alpha}\binom{n}{x}\binom{N-r}{n-x}}{\binom{N}{n}} = \frac{r^{(\alpha)} n^{(\alpha)}}{N^{(\alpha)} \alpha!}.$$

The conclusion follows immediately.

Lemma 2.

$$\sum_{j=0}^{n} \frac{(-1)^j \binom{n}{j}\binom{r}{j} p^j}{\binom{N}{j}} = \sum_{v=0}^{r} \frac{(1-p)^v \binom{N-r}{n-v}\binom{r}{v}}{\binom{N}{n}}. \tag{14}$$

Proof. The right-hand side of (14) may be written

$$\sum_{v=0}^{r} \frac{\binom{N-r}{n-v}\binom{r}{v}}{\binom{N}{n}} \sum_{j=0}^{v} \binom{v}{j}(-1)^j p^j$$

$$= \frac{\sum_{j=0}^{r}(-1)^{j}p^{j}\sum_{v=j}^{r}\binom{v}{j}\binom{N-r}{n-v}\binom{r}{v}}{\binom{N}{n}}.$$

Thus, the coefficient of p^j is

$$(-1)^{j}\sum_{v=j}^{r}\binom{v}{j}\binom{N-r}{n-v}\binom{r}{v}\bigg/\binom{N}{n}.$$

From Lemma 1,

$$(-1)^{j}\sum_{v=j}^{r}\binom{v}{j}\binom{N-r}{n-v}\binom{r}{v}\bigg/\binom{N}{n}=(-1)^{j}\binom{r}{j}\binom{N-j}{n-j}\bigg/\binom{N}{n},$$

from which the conclusion follows immediately. Employing the above lemmas, we can now establish the following theorem.

Theorem 3. *The factorial moment generating function of the number of empty cells $\varphi_m(t)$ (10) satisfies the following differential-difference equation,*

$$\varphi_{m+1}(t)=\left(\sum_{j=0}^{n}\frac{(-1)^{j}\binom{n}{j}(pt)^{j}D^{j}}{N^{(j)}}\right)\varphi_{m}(t), \quad m=0,1,\ldots, \tag{15}$$

where $D^{j}=\dfrac{d^{j}}{dt^{j}}.$

Proof. For $m=0$, $\varphi_0(t)=(1+t)^N$; hence

$$\left(\sum_{j=0}^{n}\frac{(-1)^{j}\binom{n}{j}(pt)^{j}D^{j}}{N^{(j)}}\right)(1+t)^{N}$$

$$=\sum_{j=0}^{n}\frac{(-1)^{j}\binom{n}{j}(pt)^{j}N^{(j)}}{N^{(j)}}[(1+t)^{N-n}(1+t)^{n-j}]$$

$$=(1+t)^{N-n}\sum_{j=0}^{n}(-1)^{j}\binom{n}{j}(pt)^{j}(1+t)^{n-j}$$

$$=(1+t)^{N-n}(1+t-pt)^{n},$$

in agreement with (12).

Assume that (15) holds for $m=1, 2, \ldots, k$. Then, from (10),

$$\left(\sum_{j=0}^{n} \frac{(-1)^j \binom{n}{j}(pt)^j D^j}{N^{(j)}}\right) \sum_{r=0}^{N} \frac{\binom{N}{r} t^r}{\binom{N}{n}^k} \left(\sum_{\alpha=0}^{r} (1-p)^\alpha \binom{N-r}{n-\alpha}\binom{r}{\alpha}\right)^k$$

$$= \sum_{j=0}^{n} \frac{(-1)^j \binom{n}{j}(pt)^j}{N^{(j)}} \sum_{r=0}^{N} \binom{N}{r} r^{(j)} t^{r-j} \left(\frac{\sum_{\alpha=0}^{r} (1-p)^\alpha \binom{N-r}{n-\alpha}\binom{r}{\alpha}}{\binom{N}{n}}\right)^k$$

$$= \sum_{r=0}^{N} t^r \sum_{j=0}^{n} \frac{(-1)^j \binom{n}{j}\binom{r}{j}}{\binom{N}{j}} p^j \left(\frac{\sum_{\alpha=0}^{r} (1-p)^\alpha \binom{N-r}{n-\alpha}\binom{r}{\alpha}}{\binom{N}{n}}\right)^k.$$

The conclusion now follows from Lemma 2.

Let

$$T(f(t)) = \left(\sum_{j=0}^{n} \frac{(-1)^j \binom{n}{j}(pt)^j D^j}{N^{(j)}}\right) f(t), \quad 0 < p < 1. \tag{16}$$

Then, from Theorem 3, we have that

$$\varphi_{m+1}(t) = T(\varphi_m(t)), \quad \varphi_0(t) = (1+t)^N. \tag{17}$$

Lemma 3. *Extend the domain of T to the complex plane, letting $z = x + iy$; x, y real. Let*

$$\psi(z) = \prod_{\alpha=1}^{N} (z - z_\alpha)$$

and

$$\psi_1(z) = T(\psi(z)) = c_1 \prod_{\alpha=1}^{N} (z - z_\alpha^{(1)}). \tag{18}$$

If the zeros of $\psi(z)$ are real and satisfy

$$-b \leq x_\alpha \leq -a, \quad a, b \geq 0,$$

then the zeros of $\psi_1(z)$ are real and satisfy

$$-\frac{b}{(1-p)} \leqslant x_\alpha^{(1)} \leqslant -a. \tag{19}$$

Proof. Let

$$C_\gamma = \{z : |z+(c-i\gamma)| \leqslant [(c-a)^2+\gamma^2]^{\frac{1}{2}}, \ c=\tfrac{1}{2}(a+b)\}. \tag{20}$$

Clearly $-a$ and $-b$ are on the boundary of the circular region C_γ. Consequently all zeros of $\psi(z)$ are in C_γ. Let z^* be a zero of $\psi(z)$. Let

$$\psi_1^*(z_1^{(1)}, z_2^{(1)}, \ldots, z_N^{(1)}) = c_1(z^*-z_1^{(1)})(z^*-z_2^{(1)})\ldots(z^*-z_N^{(1)}).$$

That is, $\psi_1^*(z_1^{(1)}, z_2^{(1)}, \ldots, z_N^{(1)}) = T(\psi(z^*))$ is a linear symmetric function of $(z_1^{(1)}, z_2^{(1)}, \ldots, z_N^{(1)})$. Thus, the conditions of Walsh's theorem (M. Marden [6], p. 62) are satisfied. Thus, if $z_1^{(0)}, z_2^{(0)}, \ldots, z_N^{(0)}$ are points in C_γ, then there is at least one point ζ in C_γ such that

$$T[(z^*-\zeta)^N] = 0,$$

that is, one can set $z_1^{(1)} = \zeta, z_2^{(1)} = \zeta, \ldots, z_N^{(1)} = \zeta$ and preserve the value 0. From (16),

$$T[(z^*-\zeta)^N] = (z^*-\zeta)^{N-n}(z^*-\zeta-pz^*)^n = 0.$$

Thus either $z^* = \zeta$ and therefore z^* is in C_γ or $z^* = \zeta(1-p)^{-1}$ and z^* is in

$$B_{p,\gamma} = \{z : |z+(c-i\gamma)(1-p)^{-1}|\} \leqslant [(c-a)^2+\gamma^2]^{\frac{1}{2}}(1-p)^{-1}. \tag{21}$$

However, γ is real and arbitrary. Hence it is clear that

$$C = \bigcap_{-\infty < \gamma < \infty} C_\gamma = \{z : z \text{ real}, -b \leqslant x \leqslant -a\} \tag{22}$$

and

$$B_p = \bigcap_{-\infty < \gamma < \infty} B_{p,\gamma} = \{z : z \text{ real}, -b(1-p)^{-1} \leqslant x \leqslant -a(1-p)^{-1}\}. \tag{23}$$

Consequently, $C \cup B_p$ is contained in the interval (19), proving the lemma.

We now establish the following theorem.

Theorem 4. *Let*

$$\varphi_m(t) = \sum_{r=0}^{N} \binom{N}{r} t^r \binom{N}{n}^{-m} \left(\sum_{j=0}^{r} (1-p)^j \binom{N-r}{n-j}\binom{r}{j} \right)^m.$$

Let $t_1^{(m)}, t_2^{(m)}, \ldots, t_N^{(m)}$ *be the zeros (not necessarily distinct) of* $\varphi_m(t)$. *Then* $t_j^{(m)}$, $j=1, 2, \ldots, N$ *are all real and*

$$-(1-p)^{-m} \leqslant t_j^{(m)} \leqslant -1, \ j=1,2,\ldots,N; \ m=0,1,\ldots.$$

Proof. From (17),

$$\varphi_{m+1}(t) = T(\varphi_m(t)), \ m=0, 1, \ldots,$$

and from (13),

$$\varphi_0(t) = (1+t)^N.$$

The zeros of $\varphi_0(t)$ are $t_1^{(0)} = t_2^{(0)} = \ldots = t_N^{(0)} = -1$. The zeros of $\varphi_1(t)$ are $t_1^{(1)} = -1, \ldots$, $t_{N-n}^{(1)} = -1$, $t_{N-n+1}^{(1)} = -1/(1-p), \ldots, t_N^{(1)} = -1/(1-p)$. Now apply Lemma 3 with $\psi(z) = \varphi_1(z)$ obtaining $a=1, b=(1-p)^{-1}$. Then, the zeros of $\varphi_2(t)$ are real and satisfy

$$-(1-p)^{-2} \leqslant t_j^{(2)} \leqslant -1, \ j=1, 2, \ldots, N.$$

It then follows readily by induction that the zeros of $\varphi_k(t)$ are real and satisfy

$$-(1-p)^{-k} \leqslant t_j^{(k)} \leqslant -1, \ j=1, 2, \ldots, N, \ k=2, 3, \ldots.$$

Theorem 5. *For* $1 \leqslant n \leqslant N$, $0 \leqslant p \leqslant 1$, $m \geqslant 1$, *S has a representation as the sum of N mutually independent Bernoulli random variables. That is, there exist mutually independent Bernoulli random variables*, $Y_j = Y_j(N, m, p, n), j=1, 2, \ldots, N$, *such that*

$$S = \sum_{i=1}^{N} Y_j \tag{24}$$

and

$$P\{Y_j = 1\} = \gamma_j = 1 - P\{Y_j = 0\}. \tag{25}$$

Proof. Let Y be a Bernoulli random variable with $P\{Y=1\}=\tau$. Then the factorial moment generating function of Y is

$$E_Y\{(1+t)^Y\}=(1+\tau t).$$

If

$$W=\sum_{j=1}^{N} Y_j$$

where $Y_1, Y_2, \ldots, Y_N$ are mutually independent Bernoulli random variables with $P\{Y_j=1\}=\tau_j$, then the factorial moment generating function of W is

$$\xi(t)=E_W\{(1+t)^W\}=\prod_{j=1}^{N} E_{Y_j}\{(1+t)^{Y_j}\}=\prod_{j=1}^{N}(1+\tau_j t), \tag{26}$$

where $0 \leq \tau_j \leq 1, j=1, 2, \ldots, N$. From Theorem 4, the factorial moment generating function of S may be written

$$\varphi_m(t)=(1-p)^{nm}\prod_{j=1}^{N}(t-t_j^{(m)}), \quad m=0, 1, \ldots, \tag{27}$$

where $t_j^{(m)}$ are real and $-(1-p)^{-m} \leq t_j^{(m)} \leq -1, j=1, 2, \ldots, N$.

Since every polynomial of degree N with real roots has a unique representation of the form

$$f(x)=c(x-x_1)(x-x_2)\ldots(x-x_N), \quad x_1 \leq x_2 \leq \ldots \leq x_N,$$

the representation follows by setting $\tau_j=-(t_j^{(m)})^{-1}$ and noting that $\xi(0)=\varphi_m(0)=1$.

Let $\kappa_l=\kappa_l(n, N, m, p)$ be the cumulants of S and let $\kappa_{[\nu]}$ be the factorial cumulants of S. That is,

$$\log \varphi_m(t)=\sum_{\nu=1}^{\infty} \kappa_{[\nu]} t^\nu/\nu!.$$

Then

$$\kappa_l=\sum_{j=1}^{l}\beta_{j,l}\kappa_{[j]}, \quad l \geq 2,$$

where $\beta_{j,l}$ are the Stirling numbers of the second kind.

Then, as $N \to \infty$,

$$V = (S - E(S))/\sigma_S$$

is asymptotically distributed by the standard normal distribution (mean 0, variance unity), whenever

$$\kappa_l/\kappa_2^{l/2} \to 0, \quad l > 2.$$

From (27),

$$\log \varphi_m(t) = nm \log(1-p) + \sum_{i=1}^{N} \log(t - t_j^{(m)}) = \sum_{i=1}^{N} \log(1 + \tau_i t)$$

$$= \sum_{i=1}^{N} \sum_{k=1}^{\infty} \frac{(\tau_i t)^k}{k}(-1)^k.$$

Thus,

$$\frac{\kappa_{[v]}}{v!} = \sum_{i=1}^{N} \frac{(-1)^v}{v} \tau_i^v, \quad 0 < \tau_i \leq 1,$$

and

$$|\kappa_{[v]}|/v! \leq \frac{1}{v} \sum_{i=1}^{N} |\tau_i^v| \leq N/v.$$

Then

$$\left| \sum_{j=1}^{l} \beta_{j,l} \kappa_{[j]} \right| \leq c_l N, \qquad (28)$$

since the $\beta_{j,l}$ do not depend on N, n, m, or p.

We now establish the following theorem.

Theorem 6. $V = (S - E(S))/\sigma_s$ has an asymptotically standard normal distribution as $N \to \infty$, whenever any of the following conditions are satisfied.

1. $\dfrac{mnp}{N} \to 0$, $p \to p^* \neq 1$ and $\dfrac{mnp}{N^{2/3}} \to \infty$;

2. $\dfrac{mnp}{N} \to 0$, $(1-p) \to 0$ so that for some $c > 0$,

$$(1-p) = c\left(\frac{mnp}{N}\right)^\rho + o\left(\left(\frac{mnp}{N}\right)^\rho\right), \quad 0 < \rho < 1, \text{ and}$$

$$mnp N^{-\left(1 - \frac{1}{3(\rho+1)}\right)} \to \infty;$$

3. $\quad \dfrac{mnp}{N} \to 0, \; (1-p) = c\left(\dfrac{mnp}{N}\right)^\rho + o\left(\left(\dfrac{mnp}{N}\right)^\rho\right), \; \rho \geq 1, \text{ and}$

$$\frac{mnp}{N^{5/6}} \to \infty;$$

4. $\quad \dfrac{mnp}{N} \to r > 0;$

5. $\quad \dfrac{mnp}{N} \to \infty \text{ and } \dfrac{3mnp}{N} - \log N \to -\infty.$

Proof. From (9), we can write, for $\alpha \to 0$,

$$\kappa_2 = N(e^{-\alpha})(1 - e^{-\alpha} - \alpha p e^{-\alpha}) + O(np\alpha) + O(p^2\alpha^2)$$

where $\alpha = \dfrac{mnp}{N}$. Then, as $\alpha \to 0$,

$$\kappa_2 = N(1 - \alpha + \alpha^2/2)\left(\alpha - \frac{\alpha^2}{2} - \alpha p + \alpha^2 p\right) + O(N\alpha^3) + O(mn\alpha).$$

Then, if $p \to p^* \neq 1$,

$$\kappa_2 = N\alpha(1-p) + O(N\alpha^2)$$

and

$$\frac{\kappa_2^{3/2}}{N} \to \infty \text{ whenever } \frac{mnp}{N^{2/3}} \to \infty.$$

Similarly, if $(1-p) = c\left(\dfrac{mnp}{N}\right)^\rho + o\left(\left(\dfrac{mnp}{N}\right)^\rho\right), 0 < \rho < 1, c > 0$, then

$$\kappa_2 = N\alpha(1-p) + o(N\alpha(1-p))$$

and

$$\frac{\kappa_2^{3/2}}{N} \to \infty \text{ whenever } mnpN^{-\left(1-\frac{1}{3(\rho+1)}\right)} \to \infty.$$

If

$$(1-p) = c\left(\frac{mnp}{N}\right)^\rho + o\left(\left(\frac{mnp}{N}\right)^\rho\right), \quad \rho \geq 1, \quad c > 0, \text{ then}$$

$$\kappa_2 = N\alpha^2/2 + O(N\alpha^3)$$

and

$$\frac{\kappa_2^{3/2}}{N} \to \infty \text{ whenever } \frac{mnp}{N^{5/6}} \to \infty.$$

The conclusion is obvious whenever $\frac{mnp}{N} \to r > 0$.

If $\alpha \to \infty$ as $N \to \infty$, then

$$\kappa_2 = Ne^{-\alpha} + O(Ne^{-2\alpha})$$

and

$$\frac{\kappa_2^{3/2}}{N} \to \infty \text{ whenever } 3\alpha - \log N \to -\infty.$$

Remark. A referee has pointed out to us that it is possible to establish asymptotic normality using the Berry-Esseen theorem, which would show that $(S-E(S))/\sigma_S$ has a standard normal distribution whenever $\sigma_S^2 \to \infty$ as $N \to \infty$. This method was employed in the work of Vatutin and Mikhailov [12]. The method employed here is somewhat less general than the Berry-Esseen argument but has the merit that it calls attention to the use of the factorial cumulants, which are frequently useful in combinatorial problems and not employed as often as they should be.

References

[1] Harris, B. (1966). *Theory of Probability*, Addison-Wesley Publishing Company, Reading, Mass.

[2] Harris, B. and Park, C. J. (1971). "A note on the asymptotic normality of the distribution of the number of empty cells in occupancy problems", *Ann. Inst. Statist. Math.*, 23, 507–513.

[3] Holst, Lars (1977). "Some asymptotic results for occupancy problems", *Ann. Probability*, 5, 1028-1035.

[4] Holst, Lars (1980). "On matrix occupancy, committee, and capture-recapture problems", *Scand. J. Statist.* 7, 139-146.

[5] Kolchin, V. F., Sevast'yanov, B. A. and Chistyakov, V. P. (1978). *Random Allocations*. V. H. Winstons & Sons, Washington, D.C.

[6] Marden, M. (1966). *Geometry of Polynomials*. Second Edition, Mathematical Surveys, No. 3, American Mathematical Society, Providence, R.I.

[7] Park, C. J. (1972). "A note on the classical occupancy problem", *Ann. Math. Statist.* 43, 1698-1701.

[8] Park, C. J. (1981). "On the distribution of the number of unobserved elements when m-samples of size n are drawn from a finite population", *Comm. Statist., A-Theory Methods*, 10, 371-383.

[9] Rényi, A. (1962). "Three new proofs and a generalization of a theorem of Irving Weiss", *Magyar Tud. Akad. Math. Kutató. Int. Közl. A*, 7, 203-214.

[10] Riordan, J. (1958). *An Introduction to Combinatorial Analysis*, John Wiley and Sons, Inc., New York, N.Y.

[11] Sevast'yanov, B. A. and Chistyakov, V. P. (1964). "Asymptotic normality in the classical ball problem", *Theory of Probability and Its Applications*, 9, 198-211.

[12] Vatutin, V. A. and Mikhailov, V. G. (1982). "Limit theorems for the number of empty cells in an equiprobable scheme for group allocation of particles", *Theory of Probability and Its Applications*, 27, 734-742.

RANDOM SELF-AVOIDING WALKS IN SOME ONE-DIMENSIONAL LATTICES*

Svante JANSON

*Uppsala University,
752 38 Uppsala, Sweden*

Asymptotic results are given for various properties of very long random self-avoiding walks in some "one-dimensional" lattices. The proofs are based on some results from renewal theory which are developed in an appendix.

0. Introduction

The mathematical theory of self-avoiding walks on a lattice is still in its infancy and only a few non-trivial results are known. In sharp contrast to this situation, physicists, chemists and others that do numerical experiments readily make conjectures that go much beyond the results that presently are provable. In high dimensions (≥ 5), Brydges and Spencer [1] have proved important results. In this paper we instead consider some very simple lattices, viz. a class of one-dimensional lattices, and obtain precise asymptotical results for them. (The reader may well wonder whether there is more than one one-dimensional lattice (the linear one) to study. The explanation comes in Section 1.) While the one-dimensional lattices are much simpler than those of higher dimensions, it is hoped that the results of this paper may give an insight in phenomena that appear also in higher dimensions. In particular, the one-dimensional lattices may serve as a test ground for various conjectures. For example, it has been conjectured that the asymptotic distribution of the endpoint of a random self-avoiding walk (as the length $\to \infty$) depends only on the dimension, with the exact form of the lattice determining a scale parameter only. This is true for our lattices (Corollary 2). Similarly, it is encouraging that various limits exist, for example the limit as $n \to \infty$ of the probability that a random self-avoiding walk of length n

---

* This research was partially done during the 26th International Mathematical Olympiad in Joutsa, Finland.

begins in a particular way (Corollary 3), which leads to a reasonable definition of an infinite random self-avoiding walk (Remark 3.2).

The lattices that we will study are defined in Section 1. The self-avoiding walks are analysed in Section 2, and most results are proved in Section 3. Section 4 gives some numerical results for the simplest example as an illustration to the theory.

The proofs are based on some renewal theoretic results. These are developed in Section 5, which can be read independently of the rest of the paper.

1. Dimension

It should be clear that the adequate notion of dimension of a lattice is not a topological but an algebraic or combinatorial concept. For example, the lattice of integer points in the plane with every point connected to the 8 neighbors whose coordinates differ by at most ± 1 should be regarded as a two-dimensional lattice although it is not a planar graph. Similarly, every lattice may be embedded as a graph in R^3, but the distinction between e.g. 3- and 4-dimensional lattices should be upheld. One way to achieve this is by the following definition, see Grimmett [3]. (We consider only simple, undirected edges. Lattices with directed or multiple edges can be similarly treated.)

A d-dimensional lattice is a connected graph with only finitely many edges at each vertex, such that there is a group $\cong Z^d$ of translations (symmetries of the graph without fixpoints) which divide the set of vertices into a finite number of classes (orbits). In other words, the vertices may be represented by $Z^d \times F$, where F is a finite set (the set of classes) and two vertices (m, a) and (n, b) are connected iff $n - m \in V_{ab}$, where for each pair $(a, b) \in F^2$, V_{ab} is a finite subset of Z^d.

Remark 1.1. There may be other symmetries of the lattice than these translations. In particular, all vertices may be equivalent although they are divided into several classes by the definition above.

Remark 1.2. The number of classes and the representation above are not uniquely determined by the lattice. However, the dimension d is determined, e.g. by the argument below.

It is not difficult to show that the above definition implies that the number of vertices that can be reached in at most n steps from a given vertex is $cn^d + O(n^{d-1})$ for some constant c. This provides an alternative definition of the dimension d that also applies to some less regular graphs, cf. Godsil and McKay [2].

Given the definition above, it is easy to construct non-trivial one-dimensional lattices.

Example 1.1. (A strip of finite width W in the simple square lattice.) Let the vertices be all pairs (x, y) with $x \in \mathbf{Z}$, $y \in \{1, 2, \ldots, W\}$, and let every vertex be connected to its immediate neighbors.

Example 1.2. (A cylinder lattice.) Connect the top and bottom rows in the preceding example, i.e. add edges between (x, W) and $(x, 1)$, $x \in \mathbf{Z}$. In this lattice all vertices are equivalent.

Similar constructions in other lattices, e.g. the triangular or cubic lattice, yield similar one-dimensional lattices. For simplicity, we will in the rest of the paper only consider the lattices in Examples 1.1 and 1.2 although our methods apply to many of these other lattices as well.

This type of one-dimensional lattices has earlier been studied by Wall et al. [12], [13], [14], Whittington [15] and Hammersley and Whittington [7].

2. The structure of self-avoiding walks

We consider a lattice of the type defined in Example 1.1 or 1.2 with width $W \geq 2$. A *self-avoiding walk* of length n is a sequence $(v_0, \ldots, v_n)$ of $n+1$ distinct vertices such that v_{i-1} and v_i are connected by an edge of the lattice for $i = 1, \ldots, n$. We say that the walk starts at v_0 and ends at v_n; the edges $(v_0, v_1), \ldots$ etc. are called steps of the walk.

If γ_1 and γ_2 are two self-avoiding walks of lengths n and m respectively (($v_0, \ldots, v_n$) and ($w_0, \ldots, w_m$), say) such that γ_2 starts where γ_1 ends, then $\gamma_1 \oplus \gamma_2$ denotes the composite walk $(v_0, \ldots, v_n = w_0, w_1, w_2, \ldots, w_m)$ of length $n+m$, provided this is a self-avoiding walk. We extend the notion of composition to the case when the endpoint of γ_1 and the starting point of γ_2 have the same second coordinate but different first coordinates by first performing a translation of γ_2. Thus $\gamma_1 \oplus \gamma_2 = \gamma_1 \oplus \gamma_2'$ whenever γ_2' is a translation (in the first coordinate) of γ_2, in the sense that if one side is defined, then so is the other and they are equal.

Choose a starting point S (fixed in the remainder of the paper). We may assume that $S = (0, y)$ for some y. Let $\Gamma(n)$ be the set of all self-avoiding walks of length n that start at S and define further

$$\Gamma = \bigcup_0^\infty \Gamma(n), \quad a(n) = \#\Gamma(n) \quad \text{and}$$

$$A(z) = \sum_0^\infty a(n) z^n. \tag{2.1}$$

It is well-known [6] that $a(n)^{1/n} \to \alpha$ for some positive constant α, known as the connectivity constant of the lattice. In particular, the power series (2.1) has radius of convergence $1/\alpha$ and defines $A(z)$ as an analytic function for $|z| < 1/\alpha$.

We need some more notation before we study the structure of Γ. If $(v_0, \ldots, v_n)$ is a self-avoiding walk, then a *subwalk* is any sequence $(v_k, v_{k+1}, \ldots, v_m)$, with $0 \leq k \leq m \leq n$. We define a *barrier* in a walk γ to be a subwalk of γ of the form $((x, 1), (x, 2), \ldots, (x, W))$ for some $x \in \mathbb{Z}$. The name derives from the obvious fact that a barrier never can be crossed; the part of the walk before (or after) the barrier is confined to one side of it.

We divide the barriers into three types.

A *dexterous barrier* in a self-avoiding walk is a barrier such that the part of the walk before the barrier (if any) lies to the left of the barrier, and the part of the walk after the barrier (if any) to the right of it.

A *sinister barrier* is the opposite of a dexterous barrier. (By our definitions, if $S = (0, 1)$, then $(0, 1), \ldots, (0, W)$ is a both dexterous and sinister barrier in itself. This exception causes no problems when we study longer walks.)

A *reflecting barrier* is a barrier that is neither dexterous nor sinister. Hence the remainder of the walk lies on one side of it.

It is easy to see that a self-avoiding walk never contains more than two reflecting barriers, and that it never contains both dexterous and sinister barriers (except the possible ambidexterous walk of length $W-1$ mentioned above).

We define

$$\Gamma_+ = \{\gamma \in \Gamma : \gamma \text{ contains at least one dexterous barrier}\},$$
$$\Gamma_- = \{\gamma \in \Gamma : \gamma \text{ contains at least one sinister barrier}\}, \text{ and}$$
$$\Gamma_0 = \Gamma \setminus (\Gamma_+ \cup \Gamma_-).$$

We further define

$$\Gamma_+(n) = \Gamma_+ \cap \Gamma(n), \quad a_+(n) = \#\Gamma_+(n), \quad A_+(z) = \sum_0^\infty a_+(n) z^n, \text{ etc.}$$

By the obvious left-right symmetry, $A_+(z) = A_-(z)$.

Let us study Γ_+ more closely. We decompose an arbitrary self-avoiding walk $\gamma \in \Gamma_+$ as follows. First comes a *head* consisting of the part of γ up to and including the first dexterous barrier; then comes zero, one or several *segments*, each containing exactly one dexterous barrier and ending with it; finally comes a (possibly empty) *tail* devoid of dexterous barriers. We define

$$\Gamma_h = \{\gamma \in \Gamma : \gamma \text{ contains exactly one dexterous barrier}$$
$$\text{and ends with it}\}$$

and, if Γ' is the set of all self-avoiding walks that start at $(0, W)$ and, except for the starting point, are included in $\{(x, y) : x > 0\}$

$$\Gamma_s = \{\gamma \in \Gamma' : \gamma \text{ contains exactly one dexterous barrier}$$
$$\text{and ends with it}\},$$

$$\Gamma_t = \{\gamma \in \Gamma : \gamma \text{ contains no dexterous barrier}\}.$$

A moment's consideration shows that the head, segments and tail of a walk in Γ_+ after suitable translations belong to Γ_h, Γ_s and Γ_t, respectively, and conversely, that an element of Γ_h, a number of elements of Γ_s and an element of Γ_t always may be composed to a walk in Γ_+ of which they are head, segments and tail. Thus

$$\Gamma_+ = \{\gamma_h \oplus \gamma_1 \oplus \ldots \oplus \gamma_j \oplus \gamma_t : j \geq 0, \gamma_h \in \Gamma_h, \gamma_t \in \Gamma_t, \text{ and}$$
$$\gamma_1, \ldots, \gamma_j \in \Gamma_s\}. \tag{2.2}$$

Consequently, with obvious notations,

$$A_+(z) = \sum_{j=0}^{\infty} A_h(z) A_s(z)^j A_t(z) = \frac{A_h(z) A_t(z)}{1 - A_s(z)} \tag{2.3}$$

and thus

$$A(z) = A_+(z) + A_-(z) + A_0(z) = 2\frac{A_h(z) A_t(z)}{1 - A_s(z)} + A_0(z). \tag{2.4}$$

(For simplicity we ignore the correction term $-z^{W-1}$ that is necessary when $S = (0, 1)$ and thus $\Gamma_+ \cap \Gamma_- \neq \emptyset$.)

We now invoke a theorem by Kesten [9] that (in our situation) says that walks without barriers are rare. (Kesten formulated his theorem for the simple cubic lattice in $d \geq 2$ dimensions and for subwalks that fills a cube of a given size, but the proof holds in our case too without essential changes.)

Lemma.

$$\limsup_{n\to\infty} \#\{\gamma \in \Gamma(n) : \gamma \text{ contains no barrier}\}^{1/n} < \alpha.$$

Proof. See Kesten [9, Section 5]. □

Let $x_0 = 1/\alpha$. Recall that a walk in Γ_0 contains at most two barriers. The lemma and the easily proven fact $\limsup \#\{\gamma \in \Gamma(n) : \gamma \text{ contains at most two barriers}\}^{1/n} = \limsup \{\#\gamma \in \Gamma(n) : \gamma \text{ contains no barrier}\}^{1/n}$ (see also [9, Lemma 4, p. 967]) show that

$$\limsup_{n\to\infty} a_0(n)^{1/n} < \alpha \tag{2.5}$$

i.e. $A_0(z)$ has radius of convergence $>x_0$. Similarly, $A_h(z)$, $A_s(z)$ and $A_t(z)$ have radii of convergence $>x_0$. Consequently, the right hand side of (2.4) defines a meromorphic function in some disc $D(0,r)$ with $r>x_0$. This function extends $A(z)$ and we denote the extension too by $A(z)$. Since (2.1) has radius of convergence exactly x_0, $A(z)$ has a pole on the circle $|z|=x_0$, but no poles inside it. Every pole is a root of $A_s(z)=1$. This equation has a root on the positive real axis that has strictly smaller absolute value than any other root (because $A_s(0)=0$ and the coefficients $a_s(n)$ are non-negative and $a_s(n)>0$ for $n \geqslant W$). Since A_h and A_t are nonzero on the positive real axis, this positive root is a pole of $A(z)$; hence it has to be x_0. We have already proved that $A(z)$ is meromorphic in $D(0,r)$. Decreasing r if necessary, we may assume that there are no other poles besides x_0 in $D(0,r)$ and that $A(z)$ is continuous at the boundary ($|z|=r$). Since $A_s'(x_0)>0$, (2.3) shows that $A(z)$ has the singular part $2A_h(x_0)A_t(x_0)(A_s'(x_0))^{-1}(x_0-z)^{-1}$ at x_0. Hence $\sum_0^\infty a(n)z^n - \sum_0^\infty 2A_h(x_0)A_t(x_0)(A_s'(x_0))^{-1}x_0^{-n-1}z^n$ is a bounded analytic function in $D(0,r)$, and Cauchy's estimates complete the proof of the following theorem.

Theorem 1. *Let x_0 be the (unique) positive root of $A_s(x_0)=1$. Then $\alpha=1/x_0$. Furthermore, there exists $r>x_0$ such that $A(z)$ can be extended to a meromorphic function in $D(0,r)$ with a simple pole at x_0 and no other poles, and, with*

$$a = 2\frac{A_h(x_0)A_t(x_0)}{x_0 A_s'(x_0)} > 0, \tag{2.6}$$

$$a(n) = a\alpha^n + O(\alpha_1^n) \tag{2.7}$$

where $\alpha_1 = 1/r < \alpha$. □

In particular, $a(n)/\alpha^n$ converges as $n \to \infty$. (In higher dimensions it is conjectured that $a(n) \sim$ const. $n^\beta \alpha^n$ for some β that depends on the dimension. For our lattices this thus holds and $\beta = 0$.) Another consequence of (2.7) is that $a(n+1)/a(n) \to \alpha$ as $n \to \infty$. Again, this is not known in higher dimensions, but Kesten [9] proved something very close.

3. Random self-avoiding walks

By definition, a random self-avoiding walk of length n is a random, uniformly distributed, element of $\Gamma(n)$. The probability that such a walk belongs to Γ_0 is $a_0(n)/a(n)$, which by (2.5) and (2.7) vanishes exponentially fast. Hence it suffices for most purposes to consider random elements of $\Gamma_+(n) \cup \Gamma_-(h)$, and thus, utilizing the symmetry, uniformly distributed random elements of $\Gamma_+(n)$. We will construct such random elements of $\Gamma_+(n)$ by a method that links our problems to renewal theory. We keep the notation of section 2; in particular $x_0 = 1/\alpha$. Let $l(\gamma)$ denote the length of $\gamma \in \Gamma$.

Let $X(n)$ denote a uniformly distributed random element of $\Gamma(n)$, and let $X_+(n)$ denote a uniformly distributed random element of $\Gamma_+(n)$ (when n is large enough so that $\Gamma_+(n) \neq \emptyset$). Furthermore, let Y_h be a random element of Γ_h with the distribution

$$P(Y_h = \gamma) = x_0^{l(\gamma)}/A_h(x_0), \quad \gamma \in \Gamma_h. \tag{3.1}$$

(These probabilities add up to one by the definition of A_h.) Similarly, let Y_t and Y_s be random elements of Γ_t and Γ_s with

$$P(Y_t = \gamma) = x_0^{l(\gamma)}/A_t(x_0), \quad \gamma \in \Gamma_t, \tag{3.2}$$

$$P(Y_s = \gamma) = x_0^{l(\gamma)}, \quad \gamma \in \Gamma_s. \tag{3.3}$$

(Recall that $A_s(x_0) = 1$.) Let $Y_1, Y_2, \ldots$ be independent copies of Y_s and assume that Y_h and Y_t are independent of each other and of $\{Y_i\}_1^\infty$. Define $\tau(n)$ $(n \geq 1)$ by

$$\tau(n) = \min\left\{m \geq 0 : l(Y_h) + l(Y_t) + \sum_1^m l(Y_i) \geq n\right\}. \tag{3.4}$$

The random self-avoiding walk $Y_h \oplus Y_1 \oplus \ldots \oplus Y_{\tau(n)} \oplus Y_t$ takes values in a subset of Γ_+ that contains $\Gamma_+(n)$, and any possible value γ is taken with probability $x_0^{l(\gamma)}(A_h(x_0)A_t(x_0))^{-1}$. In particular, all elements of $\Gamma_+(n)$ are taken with the same probability, i.e.

Theorem 2. *The conditional distribution of* $Y_h \oplus Y_1 \oplus \ldots \oplus Y_{\tau(n)} \oplus Y_t$ *given*

$$l(Y_h) + l(Y_1) + \ldots + l(Y_{\tau(n)}) + l(Y_t) = n$$

equals the distribution of $X_+(n)$, *i.e. the uniform distribution on* $\Gamma_+(n)$. □

Consider now a real-valued function φ on Γ_+ that is *additive* in the sense

$$\varphi(\gamma_h \oplus \gamma_1 \oplus \ldots \oplus \gamma_m \oplus \gamma_t) = \varphi(\gamma_h) + \varphi(\gamma_1) + \ldots + \varphi(\gamma_m) + \varphi(\gamma_t)$$

whenever

$$m \geq 0, \; \gamma_h \in \Gamma_h, \; \gamma_t \in \Gamma_t \quad \text{and} \quad \gamma_1, \ldots, \gamma_m \in \Gamma_s. \tag{3.5}$$

Define

$$\xi_0 = l(Y_h) + l(Y_t), \quad \eta_0 = \varphi(Y_h) + \varphi(Y_t)$$

and for $i = 1, 2, \ldots$,

$$\xi_i = l(Y_i), \quad \eta_i = \varphi(Y_i).$$

Define further

$$U_m = \sum_0^m \xi_i \quad \text{and} \quad V_m = \sum_0^m \eta_i.$$

Thus

$$\tau(n) = \min\{m \geq 0 : U_m \geq n\}.$$

With these notations, Theorem 2 yields immediately the following.

Corollary 1. *Let φ be an additive function on Γ_+. Then the distribution of $\varphi(X_+(n))$ equals the conditional distribution of $V_{\tau(n)}$ given $U_{\tau(n)} = n$.* □

We are now ready to apply renewal theory. For example, if $E\eta_1$ exists, $\frac{1}{n} V_{t(n)} \xrightarrow{P} E\eta_1/E\xi_1$ [5, Theorem IV.2.1] and, by (5.7) below, the same holds for the conditioned random variables $\frac{1}{n} V_{\tau(n)}$ given $U_{\tau(n)} = n$, and thus $\varphi(X_+(n))/n \xrightarrow{P}$

$\xrightarrow{P} E\eta_1/E\xi_1$. A more refined result, proved in Section 5 below, yields our main theorem.

Theorem 3. *Let φ be an additive function on Γ_+ such that*

$$|\varphi(\gamma)| \leq Cl(\gamma)^k \tag{3.6}$$

for some $C, k < \infty$. Then, with $b = E\eta_1/E\xi_1 = E\varphi(Y_s)/El(Y_s)$ and σ^2 given by (5.5),

$$(\varphi(X_+(n)) - bn)/\sqrt{n} \xrightarrow{d} N(0, \sigma^2) \quad as \quad n \to \infty. \tag{3.7}$$

In particular,

$$\frac{1}{n}\varphi(X_+(n)) \xrightarrow{P} b \quad as \quad n \to \infty. \tag{3.8}$$

Furthermore, all moments of the left hand sides of (3.7) *and* (3.8) *converge to the corresponding moments of the right hand sides.*

Proof. It was shown in Section 2 that A_h has radius of convergence strictly greater than x_0. Hence

$$E(e^{tl(Y_h)}) = \sum a_h(n) e^{tn} x_0^n / A_h(x_0)$$
$$= A_h(e^t x_0)/A_h(x_0) < \infty$$

for some $t > 0$; in particular, all moments of $l(Y_h)$ are finite. By (3.6), all moments of $\varphi(Y_h)$ are finite. The same argument applies to Y_t and Y_s; hence all moments of $\xi_0, \xi_1, \eta_0, \eta_1$ are finite. Theorem 5 and Corollary 1 yield (3.7), from which (3.8) follows. The moments converge by Theorem 6. □

Example 3.1. The first application of Theorem 3 concerns the distribution of the endpoint of a random self-avoiding walk. Let $e_1(\gamma)$ denote the first coordinate of the endpoint of γ. By Theorem 3, $\frac{1}{n}e_1(X_+(n)) \xrightarrow{P} b$ and $(e_1(X_+(n)) - bn)/\sigma\sqrt{n} \xrightarrow{d} N(0, 1)$ (with convergence of moments) as $n \to \infty$. By symmetry and (2.5) this implies

Corollary 2. *There exists $b > 0$ such that*

$$\frac{1}{n}e_1(X(n)) \xrightarrow{d} B \quad as \quad n \to \infty,$$

where B has the two-point distribution $P(B=\pm b)=1/2$. Furthermore, the conditional distribution of $(e_1(X(n))-bn)/\sqrt{n}$ given $e_1(X(n))>0$ converges to $N(0,\sigma^2)$ for some $\sigma^2>0$. In both cases all moments converge correspondingly. □

In particular, the mean-square end-to-end distance is $\sim b^2 n^2$, a sharpening of a result by Wall et al. [12], [13]. We obtain further

$$(|e_1(X(n))|-bn)/\sqrt{n} \xrightarrow{d} N(0,\sigma^2).$$

Example 3.2. We similarly see that there exist numbers $p_1,\ldots,p_W$ such that the proportion of the vertices that have second coordinate j in a random self-avoiding walk $X(n)$ (or $X_+(n)$) converges in probability to p_j as $n\to\infty$. The fluctuations of this proportion about p_j are asymptotically normal, and in fact by the Cramér-Wold device, see Remark 5.2, these fluctuations multiplied by $\sqrt{n}$ converge in distribution to a joint (degenerate) d-dimensional normal distribution.

Example 3.3. The proportion of the steps in $X_+(n)$ that point in a given direction (left, right, up or down) converge in probability to some constant as $n\to\infty$. For $X(n)$ this holds for the directions up and down, while the proportion of steps that point e.g. left converges to a two-point distribution.

Example 3.4. Generalizing Examples 3.2 and 3.3, let ω be a fixed self-avoiding walk and define $\varphi(\gamma)$ as the number of translates of ω that appear as subwalks in γ. Clearly, φ is an additive function when $l(\omega)\leqslant 1$, but in general φ is not additive since a subwalk of γ may consist e.g. of pieces from two adjacent segments. When $l(\omega)\leqslant W$, this is no serious obstacle because all segments end the same way and we can modify $\varphi(\gamma)$ for $\gamma\in\Gamma_s\cup\Gamma_t$ by counting also the subwalks that appear at the junction of γ and some head or segment preceding it. Hence we obtain the convergence (in distribution and with all moments) of $\frac{1}{n}\varphi(X_+(n))$ to some constant b, and of $\frac{1}{n}\varphi(X(n))$ to the same constant or to a two-point distribution, as well as convergence of $\sqrt{n}(\varphi(X_+(n))/n-b)$ to a normal distribution. In fact, these conclusions hold for any ω. In the general case they are proved using the extension of Theorem 5 described in Remark 5.3. (For a related, but weaker, result in higher dimensions, see Kesten [9, Theorem 1].)

These results show that the typical behaviour of a long random self-avoiding walk is described by Y_s. On the other hand, the ends of it are described by Y_h and Y_t.

Theorem 4. *Let $X_h(n)$ and $X_t(n)$ denote the head and tail of $X_+(n)$. Then*

$$(X_h(n), X_t(n)) \xrightarrow{d} (Y_h, Y_t) \quad \text{as} \quad n \to \infty.$$

Proof. By Theorem 2, for any $\gamma_h \in \Gamma_h$ and $\gamma_t \in \Gamma_t$,

$$P(X_h(n) = \gamma_h \text{ and } X_t(n) = \gamma_t)$$
$$= P(Y_h = \gamma_h \text{ and } Y_t = \gamma_t | U_{\tau(n)} = n)$$
$$= \frac{P(Y_h = \gamma_h \text{ and } Y_t = \gamma_t)}{P(U_{\tau(n)} = n)} P(U_{\tau(n)} = n | Y_h = \gamma_h \text{ and } Y_t = \gamma_t). \tag{3.9}$$

By (5.7) below, $P(U_{\tau(n)} = n) \to \pi > 0$ ($\pi = 1/El(Y_s)$), and by the same argument,

$$P(U_{\tau(n)} = n | Y_h = \gamma_h \text{ and } Y_t = \gamma_t)$$
$$= P(\sum_1^m \xi_i = n - l(\gamma_h) - l(\gamma_t) \text{ for some } m) \to \pi.$$

Hence the right hand side of (3.9) converges to $P(Y_h = \gamma_h \text{ and } Y_t = \gamma_t)$ as $n \to \infty$. □

Remark 3.1. If $f(Y_h, Y_t)$ is a function with $Ef(Y_h, Y_t) < \infty$, then it is an easy consequence of (3.9) and (5.7) that $\{f(X_h(n), X_t(n))\}$ is uniformly integrable. Hence $Ef(X_h(n), X_t(n)) \to Ef(Y_h, Y_t)$. Convergence of higher moments follows similarly.

Example 3.5. The second coordinate of the endpoint of $X_+(n)$ converges in distribution to that of Y_t. By symmetry, the second coordinate of the endpoint of a random self-avoiding walk $X(n)$ converges in distribution to the same limit. Furthermore, it follows from the arguments above that the first coordinate of the endpoint (suitably normalized) and the second coordinate are asymptotically independent.

Example 3.6. The proportion of the self-avoiding walks in $\Gamma_+(n)$ that start in a given direction converges to the probability that Y_+ starts in that direction. More generally, let ω be a self-avoiding walk with the same starting point S as the walks we study. If e.g. ω contains no barrier, then $P(X_+(n) \text{ starts with } \omega) \to P(Y_h \text{ starts with } \omega)$ as $n \to \infty$. If ω contains m barriers, a generalization of Theorem 4 yields $P(X_+(n) \text{ starts with } \omega) \to P(Y_h \oplus Y_1 \oplus \ldots \oplus Y_m \text{ starts with } \omega)$. Hence

Corollary 3. *The limit* $\lim_{n\to\infty} P(X(n)$ *starts with* $\omega)$ *exists for any self-avoiding walk* ω. □

Remark 3.2. Another formulation of the result in Example 3.6 is that $X_+(n)$ converges in distribution in the pointwise topology to the random infinite self-avoiding walk $Y_h \oplus Y_1 \oplus \ldots$. By reflecting this infinite walk leftwards with probability 1/2 (independent of everything else), we obtain the limit of $X(n)$ and a good definition of an infinite random self-avoiding walk.

Results similar to Examples 3–6 hold for the end of a random self-avoiding walk. Recall that, by definition, Y_h and Y_t are independent. Hence the beginning and end of $X_+(n)$ are asymptotically independent. However, the beginning and end of $X(n)$ are (even in the limit) dependent, because both are influenced by the main direction of the self-avoiding walk, i.e. whether it belongs to Γ_+ or Γ_-.

The results of this section may informally be summarized by saying that a long random self-avoiding walk may be approximated by the composition of a head Y_h, a large number of segments Y_i and a tail Y_t (the parts being independent), with the result reflected in the opposite direction with probability 1/2.

4. An example

In order to illustrate the arguments above, let us study the simplest non-trivial case: a strip of width $W=2$, see also Wall et al. [14]. We choose the starting point S as $(0, 2)$. A walk in Γ_s begins with some (≥ 1) steps right, then comes one step down, some further steps right and finally one step up, or in self-explanatory notation

$$\Gamma_s = \{\to^i \downarrow \to^j \uparrow : i \geq 1, j \geq 1\}.$$

Thus

$$A_s(z) = \sum_{i,j \geq 1} z^{i+1+j+1} = \frac{z^4}{(1-z)^2}. \tag{4.1}$$

Hence x_0 is given by $x_0^4/(1-x_0)^2 = 1$, which gives $x_0 = \frac{\sqrt{5}-1}{2}$ and

$$\alpha = 1/x_0 = \frac{\sqrt{5}+1}{2} \approx 1.618. \tag{4.2}$$

Furthermore, in the above notation,

$$\Gamma_h = \Gamma_s \cup \{\downarrow \to^i \uparrow : i \geqslant 1\} \cup \{\leftarrow^i \downarrow \to^{i+j} \uparrow : i \geqslant 1, j \geqslant 1\}$$

and thus

$$A_h(z) = \frac{z^4}{(1-z)^2} + \frac{z^3}{1-z} + \frac{z^5}{(1-z)(1-z^2)}. \tag{4.3}$$

Similarly,

$$A_t(z) = \left(1 + \frac{z^2}{1-z}\right)\left(1 + \frac{z}{1-z} + \frac{z^4}{(1-z)(1-z^2)}\right)$$

$$= \frac{(1-z^2+z^4)(1-z+z^2)}{(1-z)^2(1-z^2)}. \tag{4.4}$$

Thus $A_h(x_0) = 2$, $A_t(x_0) = 4/x_0 = 2\sqrt{5} + 2$, $El(Y_s) = x_0 A_s'(x_0) = 5 + \sqrt{5}$ and by (2.6), $a(n)\left(\frac{1+\sqrt{5}}{2}\right)^{-n} \to \frac{8}{\sqrt{5}}$ as $n \to \infty$.

Theorem 4 and the expressions above yield

$$P(X_+(n) \text{ starts with } \to) \to \frac{x_0^4}{(1-x_0)^2}/A_h(x_0) = 1/2,$$

$$P(X_+(n) \text{ starts with } \downarrow) \to x_0/2 \approx 0.309,$$

$$P(X_+(n) \text{ starts with } \leftarrow) \to (1-x_0)/2 \approx 0.191.$$

By symmetry, the same holds for the last step. Furthermore, P(The first and last steps in $X(n)$ are $\to$ or $\leftarrow$ and they are equal) $\to 1/4 + (1-x_0)^2/4 = \frac{3}{4}(1-x_0)$, while P(The first and last steps in $X(n)$ are $\to$ or $\leftarrow$, and they are unequal) $\to \frac{1}{2}(1-x_0)$, which illustrates the dependence between the two ends of a 1-dimensional random self-avoiding walk.

The asymptotic distribution of the endpoint is given by Example 3.1 (Corollary 2) with $b = 1 - 2/El(Y_s) = (5+\sqrt{5})/10 \approx 0.724$ and $\sigma^2 = 1/10\sqrt{5}$. The proportion of steps $\leftarrow$ in $X_+(n)$ converges (in probability and in L^p for all p) to 0, because Y_s contains no such steps. The proportion of steps $\to$ in $X_+(n)$ converges to $b = (5+\sqrt{5})/10$. The probability that a randomly chosen step in a long random self-avoiding walk is $\to$ or $\leftarrow$ is thus ≈ 0.724, while the probability that the first step is $\to$ or $\leftarrow$ is ≈ 0.691.

Explicit results can be obtained by this method also for $W = 3$, cf. [14], and for higher values too, but the computations become much more complex.

5. Appendix: some renewal theory

In this section we prove the renewal theoretical results that were used above. Since these results are of independent interest, we give a general, self-contained treatment.

Let $(\xi_1, \eta_1), (\xi_2, \eta_2), \ldots$ be a sequence of i.i.d. two-dimensional random variables, and let (ξ_0, η_0) be an additional two-dimensional variable, independent of $\{(\xi_i, \eta_i)\}_{i=1}^{\infty}$ and possibly with a different distribution. We assume that ξ_1 and η_1 have finite variances and that $E\xi_1 > 0$. Define

$$U_m = \sum_{i=0}^{m} \xi_i \tag{5.1}$$

$$V_m = \sum_{i=0}^{m} \eta_i \tag{5.2}$$

and, for every real number t,

$$\tau(t) = \inf\{m \geq 0 : U_m \geq t\}. \tag{5.3}$$

We are interested in the random variables $V_{\tau(t)}$, i.e. the sum of the $\eta:s$ stopped when the sum of the $\xi:s$ reaches a given level. These variables are studied e.g. in [11], [4], [5, Chapter IV], where a central limit result is given. If

$$W_t = \left(V_{\tau(t)} - \frac{E\eta_1}{E\xi_1} t\right)/\sqrt{t} \tag{5.4}$$

and

$$\sigma^2 = \mathrm{Var}((E\xi_1)\eta_1 - (E\eta_1)\xi_1)/(E\xi_1)^3 \tag{5.5}$$
$$= ((E\xi_1)^2 \mathrm{Var}\,\eta_1 + (E\eta_1)^2 \mathrm{Var}\,\xi_1 - 2E\xi_1 E\eta_1 \mathrm{Cov}(\xi_1, \eta_1))/(E\xi_1)^3,$$

then

$$W_t \xrightarrow{d} N(0, \sigma^2) \quad \text{as} \quad t \to \infty, \tag{5.6}$$

see [5, Theorem IV.2.3]. (Actually, in these references $\xi_0 = \eta_0 = 0$, but the general case follows by simple modifications, or as in the proof below.)

We will prove a conditional version of this result.

Theorem 5. *Assume that ξ_0 and ξ_1 take only integer values. Then the conditional distribution of W_n given $\{U_{\tau(n)}=n\}$ converges to $N(0,\sigma^2)$ as $n\to\infty$. (We ignore any n for which $P(U_{\tau(n)}=n)=0$.)*

Proof. We first prove the assertion for the case $\xi_0=\eta_0=0$. The result then follows from Lalley [10, Theorem 3], but as a preparation for the generalization in Remark 5.3 below, we give a complete proof. We may assume that the distribution of ξ_1 is not concentrated on any set $\{nd\}_{n\in\mathbb{Z}}$ with $d>1$. (Otherwise we may replace ξ_i by ξ_i/d and n by n/d.) Then there exists $\pi>0$ such that

$$P(U_{\tau(n)}=n)\to\pi \quad \text{as} \quad n\to\infty. \tag{5.7}$$

(See [17, Theorem 2.3] and [5, section III.10]. Alternatively, if $\xi_1\geqslant 1$ a.s., as in the applications in Section 3, Blackwell's renewal theorem [5, Theorem II.4.2] implies (5.7) with $\pi=1/E\xi_1$; the general case follows by using ladder variables.)

Define $n_1=n-2[n^{1/3}]$ and $n_2=n-[n^{1/3}]$. $(U_{\tau(n)},V_{\tau(n)})$ is obtained by summing (ξ_i,η_i) until U_m reaches (at least) n. We may do that in two steps by first summing until U_m reaches n_1, and then adding additional terms until n is reached. The first step gives $(U_{\tau(n_1)},V_{\tau(n_1)})$. The terms used in the second step are independent of the first step (because $\tau(n_1)$ is a stopping time) and given $U_{\tau(n_1)}=k$, the number of terms used in the second steps is distributed as $\tau(n-k)$. Hence we obtain, for any real x,

$$P(W_{n_1}\leqslant x \text{ and } U_{\tau(n)}=n)$$
$$=\sum_{k=n_1}^{\infty}P(W_{n_1}\leqslant x \text{ and } U_{\tau(n_1)}=k \text{ and } U_{\tau(n)}=n)$$
$$=\sum_{k=n_1}^{\infty}P(W_{n_1}\leqslant x \text{ and } U_{\tau(n_1)}=k)P(U_{\tau(n-k)}=n-k) \tag{5.8}$$

and, for $k\geqslant n_1$,

$$E(|\sqrt{n}W_n-\sqrt{n_1}W_{n_1}||U_{\tau(n_1)}=k)$$
$$=E\left(\left|V_{\tau(n)}-V_{\tau(n_1)}-(n-n_1)\frac{E\eta_1}{E\xi_1}\right||U_{\tau(n_1)}=k\right)$$
$$=E|V_{\tau(n-k)}-(n-n_1)E\eta_1/E\xi_1|$$
$$\leqslant E\tau(n-k)E|\eta_1|+(n-n_1)|E\eta_1|/E\xi_1$$
$$\leqslant C_1 E\tau(n-n_1)+C_2(n-n_1), \tag{5.9}$$

where C_1 and C_2 are some constants. Since $E\tau(n)/n$ converges as $n\to\infty$, and thus is bounded, see e.g. [5, Theorem III.6.1], (5.9) yields

$$E|\sqrt{n}\,W_n - \sqrt{n_1}\,W_{n_1}| \leq C_1 E\tau(n-n_1) + C_2(n-n_1) \leq C_3(n-n_1), \quad (5.10)$$

and thus, using (5.7), for n large enough

$$E\left(\left|W_n - \sqrt{\frac{n_1}{n}}\,W_{n_1}\right| \Big| U_{\tau(n)} = n\right)$$

$$\leq P(U_{\tau(n)} = n)^{-1} E\left|W_n - \sqrt{\frac{n_1}{n}}\,W_{n_1}\right|$$

$$\leq C_4 n^{-\frac{1}{2}}(n-n_1) \to 0 \quad \text{as} \quad n\to\infty. \quad (5.11)$$

Hence, Cramér's theorem (applied to the conditioned variables) shows that the sought assertion $\mathscr{L}(W_n|U_{\tau(n)}=n) \to N(0, \sigma^2)$ is equivalent to $\mathscr{L}\left(\sqrt{\frac{n_1}{n}}\,W_{n_1}\Big|U_{\tau(n)}=n\right) \to N(0, \sigma^2)$ and, since $n_1/n \to 1$, to $\mathscr{L}(W_{n_1}|U_{\tau(n)}=n) \to N(0, \sigma^2)$.

We prove the latter assertion as follows. By (5.8),

$$|P(W_{n_1} \leq x \text{ and } U_{\tau(n)} = n) - P(W_{n_1} \leq x)\pi|$$

$$= \left|\sum_{n_1}^{\infty} P(W_{n_1} \leq x \text{ and } U_{\tau(n_1)} = k)(P(U_{\tau(n-k)} = n-k) - \pi)\right|$$

$$\leq \sum_{n_1}^{\infty} P(U_{\tau(n_1)} = k) |P(U_{\tau(n-k)} = n-k) - \pi|$$

$$\leq \sum_{n_1}^{n_2} P(U_{\tau(n_1)} = k) \sup_{k \leq n_2} |P(U_{\tau(n-k)} = n-k) - \pi|$$

$$+ \sum_{n_2+1}^{\infty} P(U_{\tau(n_1)} = k)$$

$$\leq \sup_{j \leq n-n_2} |P(U_{\tau(j)} = j) - \pi| + P(U_{\tau(n_1)} - n_1 > n_2 - n_1). \quad (5.12)$$

The first term on the right hand side converges to 0 by (5.7) as $n\to\infty$ and thus $n-n_2 \to \infty$. The second term also converges to 0 as $n\to\infty$, because $E(U_{\tau(n_1)} - n_1)$ is bounded [5, Theorem III.10.5] and $n_2 - n_1 \to \infty$.

Since $P(W_{n_1} \leq x) \to \Phi(x/\sigma)$ by (5.6), (5.12) yields

$$P(W_{n_1} \leq x \text{ and } U_{\tau(n)} = n) \to \pi \Phi(x/\sigma) \quad \text{as} \quad n\to\infty,$$

which together with (5.7) yields

$$P(W_{n_1} \leq x | U_{\tau(n)} = n) \to \Phi(x/\sigma), \quad \text{i.e.}$$

$$\mathscr{L}(W_{n_1} | U_{\tau(n)} = n) \to N(0, \sigma^2)$$

as desired.

This completes the proof when $\xi_0 = \eta_0 = 0$. In general we have, if primed variables denote the result when ξ_0 and η_0 are replaced by 0, $\tau(n) = \tau'(n - \xi_0)$, $U_{\tau(n)} = \xi_0 + U'_{\tau'(n-\xi_0)}$ and $V_{\tau(n)} = \eta_0 + V'_{\tau'(n-\xi_0)}$.

Since $\mathscr{L}(W'_{n-k} | U'_{\tau'(n-k)} = n - k) \to N(0, \sigma^2)$ as $n \to \infty$ for every fixed integer k by the case already proved, and $\{W'_n\}$ is independent of ξ_0,

$$\mathscr{L}(W'_{n-\xi_0} | U_{\tau(n)} = n)$$
$$= \mathscr{L}(W'_{n-\xi_0} | U'_{\tau'(n-\xi_0)} = n - \xi_0) \to N(0, \sigma^2). \tag{5.13}$$

The proof is completed by Cramér's theorem and the easily proved fact

$$\mathscr{L}\left(W_n - \sqrt{\frac{n-\xi_0}{n}} W'_{n-\xi_0} \bigg| U_{\tau(n)} = n\right) \to 0.$$

We omit the details. □

Remark 5.1. The same argument yields $\mathscr{L}(W_n | U_{\tau(n)} - n = k) \to N(0, \sigma^2)$ as $n \to \infty$ for every integer $k \geq 0$. Hence W_n and $U_{\tau(n)} - n$ are asymptotically independent. In this formulation the theorem extends (with essentially the same proof) to the "continuous" case (ξ_1 is non-arithmetical). Siegmund [11] has given a result of this type, closely related to ours. The special case $\eta_1 = 1$ a.s. yields the result that $\tau(t)$ (properly normalized) and $U_{\tau(t)} - t$ are asymptotically independent [16].

Remark 5.2. Theorem 5 generalizes immediately to vector-valued η_i (in a finite dimensional space) by the Cramér-Wold device. Cf. [5, Section IV.2.].

It is easy to show moment convergence.

Theorem 6. *Assume further that $\xi_0, \xi_1, \eta_0, \eta_1 \in L^r$ for some $r \geq 2$. Then, for every p, $0 < p \leq r$,*

$$E(|W_n|^p | U_{\tau(n)} = n) \to \sigma^p E|Z|^p \quad \text{as} \quad n \to \infty, \quad \text{where } Z \sim N(0, 1), \tag{5.14}$$

and, if $p\leqslant r$ is an odd integer,

$$E(W_n^p|U_{\tau(n)}=n)\to 0 \quad \text{as} \quad n\to\infty. \tag{5.15}$$

Proof. The family $\{|W_n|^r\}$ is uniformly integrable by [5, Theorem IV.2.3 (ii)] (which easily is extended to allow for ξ_0 and $\eta_0\neq 0$). By (5.7), the family of conditioned random variables $\{|W_n|^r|U_{\tau(n)}=n\}$ then is uniformly integrable, which together with Theorem 5 proves the assertions. □

Remark 5.3. Let $\zeta_0, \zeta_1, \ldots$ be a sequence of independent random variables in some space Ω such that $\zeta_1, \zeta_2, \ldots$ are identically distributed. Let $d\geqslant 0$ be an integer and let $f: \Omega\to Z$ and $g_i: \Omega^{i+1}\to R$, $i=0,\ldots,d$, be measurable functions. Define $\xi_i=f(\zeta_i)$, $i\geqslant 1$ and $\eta_i=g_i(\zeta_0,\ldots,\zeta_i)$ for $0\leqslant i\leqslant d$, $\eta_i=g_d(\zeta_{i-d},\ldots,\zeta_i)$ for $i>d$. If $d=0$ we are in the situation studied above. If $d>0$, $\{\xi_i\}$ is a sequence of independent random variables, and $\{\eta_i\}$ is a sequence of d-dependent variables. The theorems above extend to this situation too, with essentially the same proofs (the estimates in [8, section 1] are useful); (5.5) is replaced by

$$\sigma^2 = \sum_{i=j-d}^{j+d} \text{Cov}\big((E\xi_i)\eta_i-(E\eta_i)\xi_i,\ (E\xi_j)\eta_j-(E\eta_j)\xi_j\big)/(E\xi_1)^3 \tag{5.16}$$

for any $j>d$.

References

[1] D. Brydges and T. Spencer, Self-avoiding walk in 5 or more dimensions. Commun. Math. Phys. 97 (1985), 125–148.

[2] C. D. Godsil and B. D. McKay, The dimension of a graph. Quart. J. Math. Oxford (2) 31 (1980), 423–427.

[3] G. R. Grimmett, Multidimensional lattices and their partition functions. Quart. J. Math. Oxford (2) 29 (1978), 141–157.

[4] A. Gut and S. Janson, The limiting behaviour of certain stopped sums and some applications. Scand. J. Statist. 10 (1983), 281–292.

[5] A. Gut, Stopped random walks. Limit theorems and applications. Springer 1987. To appear.

[6] J. M. Hammersley, Percolation processes II. The connective constant. Proc. Camb. Phil. Soc. 53 (1957), 642–645.

[7] J. M. Hammersley and S. G. Whittington, Self-avoiding walks in wedges. J. Phys. A 18 (1985), 101–111.

[8] S. Janson, Renewal theory for m-dependent variables. Ann. Probab. 11 (1983), 558–568.

[9] H. Kesten, On the number of self-avoiding walks, J. Math. Phys. 4 (1963), 960–969.

[10] S. P. Lalley, Limit theorems for first-passage times in linear and nonlinear renewal theory. Adv. Appl. Prob. 16 (1984), 766–803.

[11] D. Siegmund, The time until ruin in collective risk theory. Mitt. Verein. Schweiz. Versicherungsmath. 75 (1975), 157–165.

[12] F. T. Wall, J. C. Chin and F. Mandel, Configurations of macromolecular chains confined to strips or tubes. J. Chem. Phys. 66 (1977), 3066–3069.

[13] F. T. Wall and D. J. Klein, Self-avoiding random walks on lattice strips. Proc. Nat. Acad. Sci. USA 76 (1979), 1529–1531.

[14] F. T. Wall, W. A. Seitz, J. C. Chin and F. Mandel, Self-avoiding walks subject to boundary constraints. J. Chem. Phys. 67 (1977), 434–438.

[15] S. G. Whittington, Self-avoiding walks with geometrical constraints. J. Statist. Phys. 30 (1983), 449–457.

[16] M. Woodroofe, A renewal theorem for curved boundaries and moments of first passage times. Ann. Probab. 4 (1976), 67–80.

[17] M. Woodroofe, Nonlinear renewal theory in sequential analysis. CBMS-NSF Reg. Conf. Ser. in Appl. Math. 39, SIAM, Philadelphia, 1982.

ON A RANDOM DIGRAPH

Jerzy JAWORSKI*

*Adam Mickiewicz University,
Poznań, Poland*

Ipe H. SMIT

*Free University,
Amsterdam, Holland*

We consider a random digraph on a vertex set $V=\{1, 2, ..., n\}$ such that each vertex chooses independently its set of images as follows. First each vertex $i \in V$ chooses its out-degree $d^+(i)$ according to a prescribed probability distribution $\{P_j\}_{j=0}^{n-1}$ and then chooses at random its set of images from the family of all $d^+(i)$-element subsets of $V-\{i\}$. If the prescribed distribution is degenerate in d then we obtain the case of random out-regular digraph $D(n, d)$. One result of our study is concerned with relations between the general model and $D(n, d_i)$, $i=0, 1$ when

$$\sum_{j=d_0}^{d_1} P_j = 1 + o(1/n) \quad (n \to \infty).$$

Another result treats the asymptotic behavior of the strength of connectedness of our general model if the above condition holds where d_0 and d_1 are constants. We prove that under some additional conditions its vertex connectivity, edge connectivity and minimum degree have asymptotically the same distribution.

1. Introduction

Let $\tilde{P}=(P_0, P_1, ..., P_{n-1})$ be an n-tuple of non-negative real numbers such that $\sum_m P_m = 1$ (so $\tilde{P}$ constitutes a probability distribution). Let $D(n, \tilde{P})$ be a random digraph with a vertex set $V=\{1, 2, ..., n\}$ and such that for every $i \in V$

$$\Pr\{d^+(i)=m\}=P_m, \quad m=0, 1, ..., n-1$$

* The research was partially carried out during a visit in April 1985 to the Free University in Amsterdam.

and where for each $S \subseteq V - \{i\}$ with $|S| = m$

$$p_m = P_m \bigg/ \binom{n-1}{m} = \Pr\{S \text{ is the set of images of } i\}$$

(j is an image of i iff there is an arc from i to j; $d^+(i)$ denotes the out-degree of vertex i). Moreover, in this model we assume that each vertex chooses its images independently of other vertices.

Denote by $\lambda_v(G)$ *vertex connectivity* of a graph G, i.e., the minimum number of vertices the deletion of which disconnects G. The *edge connectivity* $\lambda_e(G)$ is defined similarly in terms of edges. Finally, $\delta(G)$ denotes the *minimum vertex degree* of G. If D is digraph, by $\lambda_v(D)$, $\lambda_e(D)$ and $\delta(D)$ we simply mean $\lambda_v(G)$, $\lambda_e(G)$ and $\delta(G)$ respectively, where G is the underlying simple graph of D. The *underlying simple graph* G is obtained from D by omission of the orientation of arcs and replacement of eventual double edges by a single one. The asymptotic behavior of λ_v, λ_e and δ has been studied in some special cases of the model defined above.

First consider $D(n, \tilde{P})$ such that $P_d = 1$ for some d (we denote $D(n, d)$). Clearly, $D(n, d)$ can be defined as an element chosen at random from the family of all $\binom{n-1}{d}^n$ digraphs with n labelled vertices of out-degree d. Fenner and Frieze [2] and Palmer [5] showed that for $d \geq 2$, $\lambda_v(D(n, d))$, $\lambda_e(D(n, d))$ as well as $\delta(D(n, d))$ are almost surely equal to d.

Another special case that has been studied in this sense is related with directed multigraphs. Consider a random element of the family of all n^{nd} directed multigraphs on n labelled vertices such that exactly d distinguishable arcs emanate from each vertex. By ignoring "colours" of arcs, deleting loops and replacing each group of arcs with the same head and tail by a single arc, we obtain a random digraph that also obeys the axioms of our model $D(n, \tilde{P})$ with

$$P_m = n^{-d} \binom{n-1}{m} \sum_{k=0}^{d-m} \binom{d}{k} \sum_{j=0}^{m} \binom{m}{j}(-1)^j(m-j)^{d-k}, \quad m = 0, 1, \ldots, d.$$

This case has been investigated by Jaworski and Karoński [4]. They proved that for $d \geq 2$ these three characteristics have asymptotically the same two-point distribution over $\{d-1, d\}$.

In section 3 we prove in Theorem 2 that if there are constants w_1 and w_2, $w_1 \leq w_2$, and

$$\sum_{m=w_1}^{w_2} P_m = 1 + o\left(\frac{1}{n}\right),$$

then, under mild additional conditions, λ_v, λ_e, and δ of $D(n, \tilde{P})$ have asymptotically the same distribution as well. This distribution is also made explicit. It is easy to check that our result is a generalization of the results in [2], [4], and [5] cited above.

The nature and scope of section 2 is a little different. Consider $D(n, \tilde{P})$ with $\sum_{m=d_0}^{d_1} P_m = 1 + o\left(\frac{1}{n}\right)$. It seems intuitively clear that $D(n, \tilde{P})$ is in some sense related to $D(n, d_0)$ and $D(n, d_1)$. Theorem 1 and Corollary record that relation in terms of so-called monotone properties. Since there are a lot of papers concerning $D(n, d)$ (see also Fenner and Frieze [3], and Shamir and Upfal [6]) it seems worthwhile to consider such a relation. Some results on $D(n, d)$ appear to be directly translatable into results on $D(n, \tilde{P})$ by this theorem; for example, it will be used in the studies of strength of connectedness in section 3.

Finally it is worthwhile to mention that if we assume that $\tilde{P}$ is the binomial distribution with parameters $n-1$ and p^*, then our general model is equivalent to a random digraph on n labelled vertices in which each of the $n(n-1)$ possible arcs appears independently with probability p^*. An underlying simple graph of such a digraph is a random graph $K(n, p)$ obtained by independent deletion of edges of a complete graph on n vertices so that each edge has the same probability $p = 2p^* - (p^*)^2$ of being present. Therefore the well known random graph $K(n, p)$ can be treated as a special case of $D(n, \tilde{P})$.

2. Relations with $D(n, d)$

In this section we consider the relations between our general model and its special case $D(n, d)$. Let A be a graph or digraph property. This property is *increasing* (*decreasing*) if from the fact that a graph G has property A (denote $G \in A$) it follows that every spanning supergraph (subgraph) has this property also. In general such properties are called monotone.

Theorem 1. *Let $0 \leq d_0 \leq d_1 \leq n-1$ and let A be an increasing property. Then*

$$\Pr\{D(n, d_0) \in A\} \left(\sum_{k=d_0}^{n-1} P_k\right)^n \leq \Pr\{D(n, \tilde{P}) \in A\}$$

$$\leq \Pr\{D(n, d_1) \in A\} + 1 - \left(\sum_{k=0}^{d_1} P_k\right)^n.$$

Proof. Denote by Δ^+ and δ^+ the maximum and minimum out-degree of $D(n, \tilde{P})$ respectively. Obviously

$$\Pr\{D(n,\tilde{P})\in A\} = \Pr\{D(n,\tilde{P})\in A \text{ and } \varDelta^+\leqslant d_1\}$$
$$+\Pr\{D(n,\tilde{P})\in A \text{ and } \varDelta^+ > d_1\}$$
$$\leqslant \Pr\{D(n,\tilde{P})\in A \text{ and } \varDelta^+\leqslant d_1\}+\Pr\{\varDelta^+ > d_1\}$$
$$= \Pr\{D(n,\tilde{P})\in A \text{ and } \varDelta^+\leqslant d_1\}+1-(\sum_{k=0}^{d_1} P_k)^n$$

and

$$\Pr\{D(n,\tilde{P})\in A\} = \Pr\{D(n,\tilde{P})\in A \text{ and } \delta^+\geqslant d_0\}$$
$$+\Pr\{D(n,\tilde{P})\in A \text{ and } \delta^+ < d_0\}$$
$$\geqslant \Pr\{D(n,\tilde{P})\in A \text{ and } \delta^+\geqslant d_0\}.$$

Therefore to prove our theorem it is enough to show that

(I) $\quad \Pr\{D(n,\tilde{P})\in A \text{ and } \varDelta^+\leqslant d_1\}\leqslant \Pr\{D(n,d_1)\in A\}$

and

(II) $\quad \Pr\{D(n,\tilde{P})\in A \text{ and } \delta^+\geqslant d_0\}\geqslant \Pr\{D(n,d_0)\in A\}(\sum_{k=d_0}^{n-1} P_k)^n.$

Denote by $A(n_0, n_1, \ldots, n_{n-1})$ the number of digraphs (without loops and multiple arcs) having an increasing property A, with n_k vertices of out-degree k, $k=0, 1, \ldots, n-1$. Clearly we have

$$\Pr\{D(n,d_0)\in A\} = A(0,0,\ldots,0,n_{d_0},0,\ldots,0)\binom{n-1}{d_0}^{-n}$$

and

$$\Pr\{D(n,d_1)\in A\} = A(0,0,\ldots,0,n_{d_1},0,\ldots,0)\binom{n-1}{d_1}^{-n}$$

where $n_{d_0} = n_{d_1} = n$.

Let $\sum_{k=0}^{d_1} n_k = n$, $A(n_0, n_1,\ldots,n_{d_1}) = A(n_0, n_1, \ldots, n_{d_1}, 0, \ldots, 0)$. One can check that if A is an increasing property then for $1\leqslant k\leqslant d_1$,

$$(n-k)n_{k-1} A(n_0, n_1, \ldots, n_{d_1})$$
$$\leqslant (n_k+1)k A(n_0, \ldots, n_{k-1}-1, n_k+1, \ldots, n_{d_1})$$

which leads to

$$A(n_0, n_1, \ldots, n_{d_1}) \leq \binom{n}{n_0, n_1, \ldots, n_{d_1}} \prod_{k=0}^{d_1} \binom{n-1}{k}^{n_k} \Pr\{D(n, d_1) \in A\}.$$

Moreover

$$\Pr\{D(n, \tilde{P}) \in A \text{ and } \Delta^+ \leq d_1\}$$

$$= \sum_{(n_0, n_1, \ldots, n_{d_1})} A(n_0, n_1, \ldots, n_{d_1}) \prod_{k=0}^{d_1} p_k^{n_k}$$

where the sum is over all ordered (d_1+1)-tuples $(n_0, n_1, \ldots, n_{d_1})$ such that $\sum_{k=0}^{d_1} n_k = n$, $n_k \geq 0$.

Therefore

$$\Pr\{D(n, \tilde{P}) \in A \text{ and } \Delta^+ \leq d_1\}$$

$$\leq \Pr\{D(n, d_1) \in A\} \left[\sum_{k=0}^{d_1} \binom{n-1}{k} p_k \right]^n$$

$$= \Pr\{D(n, d_1) \in A\} \left[\sum_{k=0}^{d_1} P_k \right]^n \leq \Pr\{D(n, d_1) \in A\},$$

and (I) is proved.

Similarly, let $\sum_{k=d_0}^{n-1} n_k = n$ and

$$A(n_{d_0}, n_{d_0+1}, \ldots, n_{n-1}) = A(0, 0, \ldots, 0, n_{d_0}, n_{d_0+1}, \ldots, n_{n-1}).$$

Then for an increasing property A and $d_0 + 1 \leq k \leq n-1$ we have

$$k n_k A(n_{d_0}, n_{d_0+1}, \ldots, n_{n-1})$$

$$\geq (n_{k-1} + 1)(n - k) A(n_{d_0}, \ldots, n_{k-1} + 1, n_k - 1, \ldots, n_{n-1})$$

which leads to the inequality

$$A(n_{d_0}, \ldots, n_{n-1})$$

$$\geq \binom{n}{n_{d_0}, \ldots, n_{n-1}} \prod_{k=d_0}^{n-1} \binom{n-1}{k}^{n_k} \Pr\{D(n, d_0) \in A\}.$$

Moreover

$$\Pr\{D(n,\tilde{P})\in A \text{ and } \delta^+\geq d_0\}$$

$$= \sum_{(n_{d_0},n_{d_0+1},\ldots,n_{n-1})} A(n_{d_0},\ldots,n_{n-1}) \prod_{k=d_0}^{n-1} p_k^{n_k}$$

where the sum is over all ordered $(n-d_0)$-tuples $(n_{d_0}, n_{d_0+1}, \ldots, n_{n-1})$ such that $\sum_{k=d_0}^{n-1} n_k = n$, $n_k \geq 0$.

Therefore

$$\Pr\{D(n,\tilde{P})\in A \text{ and } \delta^+\geq d_0\}$$

$$\geq \Pr\{D(n,d_0)\in A\} \left[\sum_{k=d_0}^{n-1} \binom{n-1}{k} p_k \right]^n,$$

and (II) is proved as well. □

Corollary. *Let $0 \leq d_0 \leq d_1 \leq n-1$ (d_0 and d_1 may depend on n) and let A be an increasing property. If*

$$\sum_{k=d_0}^{d_1} P_k = 1 - w_n, \quad w_n = o(1/n),$$

then

$$\Pr\{D(n,d_0)\in A\} - nw_n \leq \Pr\{D(n,\tilde{P})\in A\} \leq \Pr\{D(n,d_1)\in A\} + nw_n.$$

Furthermore, if A is decreasing, then

$$\Pr\{D(n,d_0)\in A\} + nw_n \geq \Pr\{D(n,\tilde{P})\in A\}$$
$$\geq \Pr\{D(n,d_1)\in A\} - nw_n.$$

Proof. Notice that if a property A is decreasing, then "not A" is increasing. Hence it is sufficient to prove our corollary for increasing properties only.

It is easy to check that

$$(1-w_n)^n > 1 - nw_n.$$

Therefore for an increasing property A our corollary follows from Theorem 1 immediately. □

3. The strength of connectedness of $D(n, \tilde{P})$

Let $\lambda_v = \lambda_v(D(n, \tilde{P}))$, $\lambda_e = \lambda_e(D(n, \tilde{P}))$ and $\delta = \delta(D(n, \tilde{P}))$. We have the following asymptotic result.

Theorem 2. Let w_1 and w_2 be constants (with respect to n), $0 \leq w_1 \leq w_2 \leq n-1$ and let $D(n, \tilde{P})$ be defined such that

(i) $\quad \sum_{i=w_1}^{w_2} P_i = 1 + o(1/n) \quad (n \to \infty).$

(ii) $\quad \lim_{n \to \infty} nP_i = c_i \quad \text{or} \quad nP_i \to \infty \quad \text{as} \quad n \to \infty \quad \text{for each } i \text{ such that}$
$$w_1 \leq i \leq w_2.$$

(iii) $\quad \lim_{n \to \infty} P_i$ exists for each i such that $0 \leq i \leq n-1$.

Let

$$d = \min\{i; nP_i \to \infty\},$$

$$k = \min\{i; nP_i \to \infty \text{ or } nP_i \to c_i > 0\}, \quad \text{and}$$

$$E = \lim_{n \to \infty} \sum_{i=1}^{n-1} iP_i.$$

Then $k \leq d$ of course, and

if $k = d$, then

(1) $\quad \lim_{n \to \infty} \Pr\{\delta = d\} = 1,$

if $k < d$, then

(2) $\quad \lim_{n \to \infty} \Pr\{\delta = k\} = 1 - \exp(-c_k e^{-E}),$

(3) $\quad \lim_{n \to \infty} \Pr\{\delta = d\} = \exp\left\{ -\left(\sum_{l=0}^{d-1-k} \frac{E^l}{l!} \sum_{i=k}^{d-1-l} c_i \right) e^{-E} \right\},$

and if $k<j<d$

(4) $$\lim_{n\to\infty} \Pr\{\delta=j\} = \exp\left\{-\left(\sum_{l=0}^{j-1-k} \frac{E^l}{l!} \sum_{i=k}^{j-1-l} c_i\right)e^{-E}\right\}$$

$$\times \left(1-\exp\left\{-\left(\sum_{i=k}^{j} c_i \frac{E^{j-i}}{(j-i)!}\right)e^{-E}\right\}\right).$$

Moreover, if $k \geqslant 2$, then

(5) $$\lim_{n\to\infty} \Pr\{\lambda_v = \lambda_e = \delta\} = 1,$$

i.e., the asymptotic distributions of λ_v and λ_e are identical with that of δ.

Proof. We proceed with a brief sketch of the proof. First we show that

(6) $$\lim_{n\to\infty} \Pr\{k \leqslant \delta \leqslant d\} = 1.$$

Obviously (6) implies (1). Let X_k be the number of vertices of out-degree k and in-degree 0 in $D(n, \tilde{P})$. We prove that for $k<d$ X_k is asymptotically (as $n\to\infty$) Poisson distributed with parameter $c_k e^{-E}$:

(7) $$X_k \rightsquigarrow \text{Po}(c_k e^{-E}).$$

We show next that X_k and Y_k — the number of vertices of degree k in the underlying simple graph of $D(n, \tilde{P})$, have asymptotically the same distribution, i.e.,

(8) $$Y_k \rightsquigarrow \text{Po}(c_k e^{-E}).$$

Therefore we obtain (2):

$$\lim_{n\to\infty} \Pr\{\delta=k\} = 1 - \lim_{n\to\infty} \Pr\{Y_k=0\}$$

$$= 1 - \exp\{-c_k e^{-E}\}.$$

Let $k<j<d$ and U_j be the number of vertices of degree $\leq j$ in the underlying simple graph of $D(n, \tilde{P})$. We prove that

(9) $$U_j \rightsquigarrow \mathrm{Po}\left(\sum_{i=k}^{j} c_i \sum_{l=0}^{j-i} \frac{E^l}{l!} e^{-E}\right).$$

Notice that for j, $k<j<d$

$$\Pr\{\delta=j\} = \Pr\{\delta \leq j\} - \Pr\{\delta \leq j-1\}$$
$$= \Pr\{U_j > 0\} - \Pr\{U_{j-1} > 0\}$$
$$= \Pr\{U_{j-1} = 0\} - \Pr\{U_j = 0\}$$

and by (9)

$$\lim_{n \to \infty} \Pr\{U_j=0\} = \exp\left\{-\left(\sum_{i=k}^{j} c_i \sum_{l=0}^{j-i} \frac{E^l}{l!}\right) e^{-E}\right\}$$
$$= \lim_{n \to \infty} \Pr\{U_{j-1}=0\} \exp\left\{-\left(\sum_{l=k}^{j} c_l \frac{E^{j-l}}{(j-l)!}\right) e^{-E}\right\},$$

where $U_k = X_k$.

Moreover, by (6)

$$\lim_{n \to \infty} \Pr\{\delta=d\} = \lim_{n \to \infty} \Pr\{U_{d-1}=0\}.$$

Therefore we arrive at the assertions about asymptotic distribution of δ stated in (3) and (4).

Finally, following a classical approach introduced by Erdös and Rényi in [1] (see also [2], [4], [5]) we prove (5).

Hence, to complete the proof of our theorem we have to show that (6), (7), (8), (9), and (5) hold. We start with three useful formulas. Let t be a constant. Then, by (i)

$$\sum_{i=0}^{n-1-t} \binom{n-1-t}{i} p_i = \sum_{i=0}^{n-1-t} \frac{(n-1-t)_i}{(n-1)_i} P_i$$
$$\leq \sum_{i=0}^{n-1-t} \left(1 - \frac{t}{n-1}\right)^i P_i$$

$$= \sum_{i=0}^{n-1-t} P_i \sum_{j=0}^{i} \binom{i}{j}(-1)^j \left(\frac{t}{n-1}\right)^j$$

$$= 1 - \frac{t}{n-1} \sum_{i=0}^{n-1-t} i P_i + o(1/n)$$

$$= 1 - \frac{t}{n-1} E + o(1/n),$$

and, similarly,

$$\sum_{i=0}^{n-1-t} \binom{n-1-t}{i} P_i = \sum_{i=0}^{w_2} \frac{(n-1-t)_i}{(n-1)_i} P_i + o(1/n)$$

$$\geqslant \sum_{i=0}^{w_2} \left(1 - \frac{t}{n-w_2}\right)^i P_i + o(1/n) = 1 - \frac{t}{n-1} E + o(1/n).$$

It follows that for a constant t

(10) $$\sum_{i=0}^{n-1-t} \binom{n-1-t}{i} P_i = 1 - \frac{t}{n-1} E + o(1/n).$$

With the aid of (i), after simple manipulations, we get

(11) $$\sum_{i=0}^{n-2} \binom{n-2}{i} P_{i+1} = \frac{E}{n-1} + o(1/n)$$

and

(12) $$\sum_{i=0}^{n-3} \binom{n-3}{i} P_{i+2} = o(1/n).$$

Proof of (6). Let X_d be a random variable (r.v.) — the number of vertices of out-degree d and in-degree 0. Then, by (10) with $t=1$

$$E(X_d) = n P_d \left\{ \sum_{i=0}^{n-1} \binom{n-2}{i} P_i \right\}^{n-1} = n P_d e^{-E}(1 + o(1))$$

and by (10) with $t=2$

$$E_2(X_d) = E(X_d(X_d-1))$$
$$= n(n-1)\left[\binom{n-2}{d}p_d\right]^2\left[\sum_{i=0}^{n-3}\binom{n-3}{i}p_i\right]^{n-2}$$
$$= n^2 P_d^2 e^{-2E}(1+o(1)).$$

Therefore, by Chebyshev's inequality we obtain

$$\Pr\{X_d=0\} \leqslant \frac{\operatorname{Var}(X_d)}{(E(X_d))^2} = o(1)$$

since by definition $nP_d \to \infty$ as $n \to \infty$.

Hence

$$\lim_{n\to\infty} \Pr\{\delta \leqslant d\} = \lim_{n\to\infty} \Pr\{X_d > 0\} = 1.$$

Furthermore, by (i), the obvious facts that: "$\delta \geqslant k$" is an increasing property, and $\Pr\{\delta(D(n,k)) \geqslant k\} = 1$, using the Corollary we have

$$\lim_{n\to\infty} \Pr\{\delta \geqslant k\} = 1 \text{ for } k \leqslant d.$$

So, finally we obtain (6).

Proof of (7). Let $k < d$. Then by (10) with a fixed t

$$E_t(X_k) = (n)_t \left[\binom{n-t}{k}p_k\right]^t \left[\sum_{i=0}^{n-1-t}\binom{n-1-t}{i}p_i\right]^{n-t}$$
$$= (nP_k)^t\left[1 - \frac{t}{n-1}E + o(1/n)\right]^{n-t}(1+o(1)),$$

and consequently

$$\lim_{n\to\infty} E_t(X_k) = [c_k e^{-E}]^t.$$

Therefore, by a standard probabilistic argument we obtain (7).

Proof of (8). Let $k<d$ and let Z_k be a r.v. — the number of vertices of out-degree k and in-degree $\geqslant 1$ chosen by their images only. It follows by the Corollary that $\delta^+ \geqslant k$ almost surely (a.s.). Therefore Y_k and $X_k + Z_k$ have asymptotically the same distribution. However, by (10) and (11)

$$\Pr\{Z_k>0\} \leqslant E(Z_k) \leqslant nP_k \left[\sum_{i=0}^{n-2}\binom{n-2}{i}p_i\right]^{n-k}$$

$$\times k \sum_{i=0}^{n-2}\binom{n-2}{i}p_{i+1}$$

$$= c_k e^{-E} k \frac{E}{n-1}(1+o(1)) = O(1/n).$$

Therefore Y_k and X_k have asymptotically the same distribution and (7) implies (8).

Proof of (9). Let $k<j<d$, $v \in V$. Then

$$\Pr\{d(v)\leqslant j\} = \sum_{i=0}^{j}\Pr\{d(v)=i\}$$

$$= \sum_{i=0}^{j}\sum_{t=0}^{i}\Pr\{d(v)=i \mid d^+(v)=t\}\Pr\{d^+(v)=t\}$$

$$= \sum_{i=0}^{j}\sum_{t=0}^{i}\binom{n-1-t}{i-t}P_t\left(\sum_{l=0}^{n-2}\binom{n-2}{l}p_{l+1}\right)^{i-t}\left(\sum_{l=0}^{n-2}\binom{n-2}{l}p_l\right)^{n-1-i}$$

$$= \sum_{t=0}^{j}P_t \sum_{i=t}^{j}\binom{n-1-t}{i-t}\left(\frac{E^*}{n-1}\right)^{i-t}\left(1-\frac{E^*}{n-1}\right)^{n-1-i}$$

$$= \sum_{t=0}^{j}P_t \sum_{l=0}^{j-t}\binom{n-1-t}{l}\left(\frac{E^*}{n-1}\right)^{l}\left(1-\frac{E^*}{n-1}\right)^{n-1-t-l}$$

where $E^* = \sum_{i=1}^{n-1} iP_i$.

Therefore

$$\lim_{n\to\infty} E(U_j) = \sum_{i=k}^{j} c_i \sum_{l=0}^{j-i} \frac{E^l}{l!} e^{-E}.$$

To prove (9) we show that for a fixed t, $t \geqslant 2$

$$\lim_{n\to\infty} E_t(U_j) = \left(\sum_{i=k}^{j} c_i \sum_{l=0}^{j-i} \frac{E^l}{l!} e^{-E}\right)^t.$$

Let $\{v_1, v_2, \ldots, v_t\}$ be a fixed t-element subset V_t of V and let us introduce the following events:

A_1: "there are no arcs between any two vertices of V_t"

A_2: "the sets of images $I(v_1), I(v_2), \ldots, I(v_t)$ of $v_1, v_2, \ldots, v_t$ are pairwise disjoint"

A_3: "every vertex of $V - V_t$ has at most one image in V_t"

A_4: "no vertex from $I(v_1) \cup I(v_2) \cup \ldots \cup I(v_t)$ has an image in V_t"

$A = A_1 \cap A_2 \cap A_3 \cap A_4.$

First we prove that

(13) $\quad E_t(U_j) = (n)_t \Pr\{\text{"}d(v_1) \leq j, d(v_2) \leq j, \ldots, d(v_t) \leq j\text{"} \cap A\} + o(1).$

— Let $F_{v_1 v_2}$ be the event that there is an arc from v_1 to v_2, then

$$\Pr\{\text{"}d(v_1) \leq j, \ldots, d(v_t) \leq j\text{"} \cap \bar{A}_1\}$$

$$\leq \Pr\{\text{"}d^+(v_1) \leq j, \ldots, d^+(v_t) \leq j\text{"} \cap \bar{A}_1\}$$

$$= O(n^{-t+1}) \Pr\{\text{"}d^+(v_1) \leq j\text{"} \cap F_{v_1 v_2}\}$$

$$= O(n^{-t+1}) \sum_{i=0}^{j-1} \binom{n-2}{i} p_{i+1} = O(n^{-t-1})$$

since "$d^+(v_1) \leq j$" $\cap F_{v_1 v_2}$, "$d^+(v_2) \leq j$", …, "$d^+(v_t) \leq j$" are independent, and $\Pr\{d^+(v_l) \leq j\} = O(n^{-1})$ for $l = 1, 2, \ldots t$, and $P_i = O(1/n)$ for $i \leq j$.

— Let $D_{v_1 v_2}$ be the event that v_1 and v_2 have a common image, then similarly

$$\Pr\{\text{"}d(v_1) \leq j, \ldots, d(v_t) \leq j\text{"} \cap \bar{A}_2\}$$

$$= O(n^{-t+2}) \Pr\{\text{"}d^+(v_1) \leq j, d^+(v_2) \leq j\text{"} \cap D_{v_1 v_2}\}$$

$$= O(n^{-t+3}) \left(\sum_{i=0}^{j-1} \binom{n-2}{i} p_{i+1}\right)^2 = O(n^{-t-1}).$$

— Let B_v denote the event that the vertex $v \in V - V_t$ has more than one image in V_t. Then by (12)

$$\Pr\{``d(v_1)\leqslant j, \ldots, d(v_t)\leqslant j"\cap \bar{A}_3\}$$
$$\leqslant n\Pr\{``d^+(v_1)\leqslant j, \ldots, d^+(v_t)\leqslant j"\cap B_v\}$$
$$= O(n^{-t+1})O\left(\sum_{i=0}^{n-3}\binom{n-3}{i}p_{i+2}\right) = o(n^{-t}).$$

— Finally by (11)
$$\Pr\{``d(v_1)\leqslant j, \ldots, d(v_t)\leqslant j"\cap \bar{A}_4\}$$
$$= O(n^{-t})O\left(\sum_{i=0}^{n-2}\binom{n-2}{i}p_{i+1}\right) = O(n^{-t-1}).$$

Hence (13) holds. Moreover

$$(n)_t\Pr\{``d(v_1)\leqslant j, \ldots, d(v_t)\leqslant j"\cap A\}$$
$$= (n)_t \sum_{i_1=0}^{j}\sum_{i_2=0}^{j}\cdots\sum_{i_t=0}^{j}\Pr\{``d(v_1)=i_1, \ldots, d(v_t)=i_t"\cap A\}$$
$$= (n)_t \sum_{i_1=0}^{j}\cdots\sum_{i_t=0}^{j}\sum_{k_1=0}^{i_1}\cdots\sum_{k_t=0}^{i_t}\Pr\{``d(v_1)=i_1, d^+(v_1)=k_1, \ldots,$$
$$d(v_t)=i_t, d^+(v_t)=k_t"\cap A\}$$
$$= (n)_t \sum_{i_1=0}^{j}\cdots\sum_{i_t=0}^{j}\sum_{k_1=0}^{i_1}\cdots\sum_{k_t=0}^{i_t}\binom{n-t}{k_1}p_{k_1}\binom{n-t-k_1}{k_2}p_{k_2}\cdots$$
$$\binom{n-t-k_1-\ldots-k_{t-1}}{k_t}p_{k_t}$$
$$\times\binom{n-t-\sum k_i}{i_1-k_1}\binom{n-t-\sum k_i-i_1+k_1}{i_2-k_2}\cdots\binom{n-t-\sum_{l=1}^{t-1}i_l-k_t}{i_t-k_t}$$
$$\times\left(\sum_{i=0}^{n-t-1}\binom{n-t-1}{i}p_{i+1}\right)^{\sum_{l=1}^{t}(i_l-k_l)}\left(\sum_{i=0}^{n-t-1}\binom{n-t-1}{i}p_i\right)^{n-t-\Sigma(i_l-k_l)}.$$

Therefore by (10) and (11)

$$(n)_t\Pr\{``d(v_1)\leqslant j, \ldots, d(v_t)\leqslant j"\cap A\}$$
$$= (n)_t \sum_{i_1}\cdots\sum_{i_t}\sum_{k_1}\cdots\sum_{k_t}\frac{(n-t)!}{k_1!\ldots k_t!(i_1-k_1)!\ldots(i_t-k_t)!(n-t-\sum(i_l-k_l))!}$$
$$\times p_{k_1}\ldots p_{k_t}\left(\frac{1}{n}E\right)^{\Sigma(i_l-k_l)}e^{-tE}+o(1)$$

$$= e^{-tE} \sum_{k_1} \cdots \sum_{k_t} c_{k_1} \cdots c_{k_t} \sum_{l_1=0}^{j-k_1} \cdots \sum_{l_t=0}^{j-k_t} \frac{E^{l_1}}{l_1!} \cdots \frac{E^{l_t}}{l_t!} + o(1)$$

$$= e^{-tE} \left(\sum_{i=k}^{j} c_i \sum_{l=0}^{j-i} \frac{E^l}{l!} \right)^t + o(1).$$

Hence

$$\lim_{n \to \infty} E_t(U_j) = \left(e^{-E} \sum_{i=k}^{j} c_i \sum_{l=0}^{j-i} \frac{E^l}{l!} \right)^t$$

and (9) is proved.

Proof of (5). Consider now r.v. λ_v — the vertex connectivity and λ_e — the edge connectivity. It is well known that in general the following relation holds

$$\lambda_v \leqslant \lambda_e \leqslant \delta.$$

So obviously

$$\lim_{n \to \infty} \Pr\{\lambda_v \leqslant d\} = \lim_{n \to \infty} \Pr\{\lambda_e \leqslant d\} = 1.$$

Denote by S_t the event that there is a set $K \subseteq V$, $|K| = r$, which after removal from $D(n, \tilde{P})$ separates a set $T \subset V - K$, $|T| = t$, from the remainder of vertices $V - (K \cup T)$. Without loss of generality we may assume that $1 \leqslant t \leqslant \frac{n-r}{2}$.

Then

$$\Pr\{S_t \cap ``\delta^+ \geqslant k"\}$$

$$\leqslant \binom{n}{r}\binom{n-r}{t}\left(\sum_{i=k}^{t+r-1}\binom{t+r-1}{i}p_i\right)^t\left(\sum_{i=k}^{n-t-1}\binom{n-t-1}{i}p_i\right)^{n-r-t}$$

$$= \binom{n}{r}\binom{n-r}{t}\left(\frac{t+r}{n}\right)^{kt}\left(\frac{n-t}{n}\right)^{(n-r-t)k}\left(\sum_{i=k}^{t+r-1}\frac{n^k}{(t+r)^k}\frac{(t+r-1)_i}{(n-1)_i}p_i\right)^t$$

$$\times \left(\sum_{i=k}^{n-t-1}\frac{n^k(n-t-1)_i}{(n-t)^k(n-1)_i}p_i\right)^{n-r-t}.$$

But

$$\frac{n^k}{(t+r)^k}\frac{(t+r-1)_i}{(n-1)_i} \leqslant \frac{n^k(t+r)^i}{n^i(t+r)^k} = \left(\frac{t+r}{n}\right)^{i-k} \leqslant 1 \quad \text{for} \quad i \geqslant k$$

and similarly

$$\left(\frac{n}{n-t}\right)^k \frac{(n-t-1)_i}{(n-1)_i} \leqslant 1 \quad \text{for} \quad i \geqslant k.$$

Therefore for $k \geqslant 2$ one can check (see Palmer [5] p. 297) that

$$\Pr\{\bigcup_{t>r/(k-1)} S_t\} = \Pr\{\bigcup_{t>r/(k-1)} \{S_t \cap \text{``}\delta^+ \geqslant k\text{''}\}\}$$
$$+ \Pr\{\bigcup_{t>r/(k-1)} S_t \cap \text{``}\delta^+ < k\text{''}\}$$
$$\leqslant \sum_{t>r/(k-1)} \Pr\{S_t \cap \text{``}\delta^+ \geqslant k\text{''}\} + \Pr\{\delta^+ < k\} = o(1).$$

It follows that after removal of $r = k-1$ vertices from $D(n, \tilde{P})$ at most a single vertex can be separated from the remaining vertices a.s.. But

$$\lim_{n\to\infty} \Pr\{\delta \geqslant k\} = 1, \quad \text{so we have}$$

$$\lim_{n\to\infty} \Pr\{\lambda_v \geqslant k\} = \lim_{n\to\infty} \Pr\{\lambda_e \geqslant k\} = 1,$$

and so, the ranges of λ_v, λ_e and δ are asymptotically equal. Furthermore, in case $k = d$ there is nothing left to prove. Let $k \leqslant r < j \leqslant d$, $1 < t \leqslant \frac{r}{k-1}$ and assume that $\delta = j$ and $\lambda_v = r$. Then

$$\Pr\{S_t \cap \text{``}\delta = j\text{''} \cap \text{``}\delta^+ \geqslant k\text{''}\}$$
$$\leqslant n^{r+t}\left[\sum_{i=k}^{r+t-1}\binom{r+t-1}{i}p_i\binom{r+t-1-i}{j-i}\left(\sum_{i=k}^{n-2}\binom{n-2}{i}p_i\right)^{j-i}\right]^t$$
$$\leqslant C n^{r+t}\left[n^{-j}\sum_{i=k}^{r+t-1}\frac{n^i}{i!}p_i\right]^t \leqslant C_1 n^{r+t-jt}$$
$$\leqslant C_1 n^{-(j-1)(t-1)} = o(1) \quad \text{since} \quad r \leqslant j-1,\ j \geqslant 2,\ t \geqslant 2.$$

Therefore $\Pr\{\delta \leqslant \lambda_v\} = 1 - o(1)$ and consequently, since $\lambda_v \leqslant \lambda_e \leqslant \delta$, we obtain (5), and the theorem is proved. □

Remark. For $k<2$, in general (5) doesn't hold and it is an interesting open problem to find asymptotic distributions of λ_v and λ_e for such k.

ACKNOWLEDGEMENT

We wish to express our thanks to A. Ruciński for many suggestions concerning the exposition.

References

[1] P. Erdös, A. Rényi, On the strength of connectedness of a random graph, *Acta Math. Sci. Hung.*, **12**, (1961), 261–267.

[2] T. I. Fenner, A. M. Frieze, On the connectivity of random m-orientable graphs and digraphs, *Combinatorica*, **2**, (1982), 347–360.

[3] T. I. Fenner, A. M. Frieze, On the existence of hamiltonian cycles in a class of random graphs, *Discrete Mathematics*, **45**, (1983).

[4] J. Jaworski, M. Karoński, On the connectivity of graphs generated by a sum of random mappings, to appear.

[5] D. V. Palmer, On the connectivity of regular random digraphs, *Utilitas Mathematica*, **20**, (1981), 293–299.

[6] E. Shamir, E. Upfal, One-factor in random graphs based on vertex choice, *Discrete Mathematics*, **41**, (1982), 281–286.

ADDITIVE WEIGHTS OF NON-REGULARLY DISTRIBUTED TREES

Rainer KEMP

Johann Wolfgang Goethe-Universität,
D-6000 Frankfurt a.M., West Germany

In searching algorithms, trees of the same size are not equally likely to occur. Given a particular probability distribution over trees with the same number of nodes, the paper presents a general approach to the computation of the expected value of a general additive weight defined on these trees; special cases of this weight are various types of path lengths and free search cost measures. Generally, the presented method yields a system of formal equations satisfied by the generating function of the average weights whose actual interpretation represents a system of algebraic and functional-differential equations. The method is explicitly demonstrated for the class of random t-ary trees, binary search trees, t-ary digital search trees and Patricia trees.

1. Introduction

Let $\mathscr{F}$ be a family of unlabelled rooted planar trees and let $T=(N(T), r(T)) \in \mathscr{F}$ be a tree with the set of nodes $N(T)$ and the root $r(T)$. The set of all subtrees of T is denoted by SUB (T). The *degree* $D(x)$ of a node $x \in N(T)$ is the number of subtrees of x. The set DEG $(\mathscr{F}) := \{d \in \mathsf{N}_0 | (\exists T \in \mathscr{F})(\exists x \in N(T)) (D(x)=d)\}$ consists of all allowed node-degrees appearing in the trees $T \in \mathscr{F}$. Note that $0 \in \text{DEG}(\mathscr{F})$ is always valid. The family $\mathscr{F}$ is said to be *bounded* if $|\text{DEG}(\mathscr{F})|$ is finite.

A tree $T=(N(T), r(T)) \in \mathscr{F}$ is called λ-*tree* if $D(r(T))=\lambda$. The set of all λ-trees in $\mathscr{F}$ with n nodes is denoted by $\mathscr{F}_\lambda(n)$. A λ-tree $T=(N(T), r(T))$ with the subtrees T_i, $1 \leq i \leq \lambda$, in left-to-right order is said to be of *type* $\langle (n, \lambda); (m_1, d_1), \ldots, (m_\lambda, d_\lambda) \rangle$ if $|N(T)|=n$, $|N(T_i)|=m_i$ and $D(r(T_i))=d_i$, $1 \leq i \leq \lambda$. The set of all trees $T \in \mathscr{F}$ of type $\langle (n, \lambda); (m_1, d_1), \ldots, (m_\lambda, d_\lambda) \rangle$ is denoted by $\mathscr{F}_{n, \lambda}(\vec{m}, \vec{d})$, where $\vec{m}=(m_1, \ldots, m_\lambda)$ and $\vec{d}=(d_1, \ldots, d_\lambda)$. Note that always $\sum_{1 \leq i \leq \lambda} m_i = n-1$.

Now, let $f: \mathsf{N}_0^3 \to \mathsf{R}$ be a given mapping, the so-called *weight function*. The weight $w_f(T)$ of a tree $T \in \mathscr{F}_{n, \lambda}(\vec{m}, \vec{d})$ is recursively defined by $w_f(T) := \omega_f(r(T))$, where

$$\omega_f(r(T)) := \text{IF } \lambda = 0 \quad \text{THEN } 0$$
$$\text{ELSE } \sum_{1 \leq i \leq \lambda} [\omega_f(r(T_i)) + f(\lambda, m_i, d_i)];.$$

Choosing a particular weight function f, the weight $w_f(T)$ of a tree $T \in \mathscr{F}$ corresponds to a well-known parameter which is directly related to the analysis of sorting and searching algorithms ([8], [9]). Typical examples are ([4])

- the total path length $p(T)$ (choose $f(\lambda, m, d) = m$)
- the total degree path length $pd(T)$ (choose $f(\lambda, m, d) = \lambda m$)
- the number $N_s(T)$ of nodes of degree s (choose $f(\lambda, m, d) = \lambda^{-1} \delta_{\lambda s}$).

In all these cases, the weight function f is a polynomial of degree $N \leq 1$ in the variable m with coefficients depending on λ.

In [3] the so-called *free search cost measure* FCOST(T) of a tree $T \in \mathscr{F}$ has been introduced. It is defined as follows: Let the *distance* dist(u, v) from node u to node v be the length of the shortest path from u to v in the tree $T \in \mathscr{F}$. Then

$$\text{FCOST}(T) = \sum_{(u,v) \in N(T) \times N(T)} \text{dist}(u, v).$$

This quantity plays an important part in the analysis of algorithms using the representation of trees in major-minor loop configurations of bubble memories. It is not hard to see ([5]) that FCOST(T), $T \in \mathscr{F}_{n,\lambda}(\vec{m}, \vec{d})$, has the inductive definition FCOST(T) = $|N(T)| p(T) - F(T)$, where

$$F(r(T)) = \sum_{1 \leq i \leq \lambda} [F(r(T_i)) + m_i^2].$$

Note that $F(T)$ also satisfies the above definition of a weight; we have to choose $f(\lambda, m, d) = m^2$. In this case, the weight function f is a polynomial of degree $N = 2$ in the variable m.

Assuming that $\mathscr{F}$ is a simply generated family of tree ([10]) and that all trees $T \in \mathscr{F}$ with n nodes are equally likely, a general approach to the computation of the average weight $\underline{w}_f(n)$ of such a tree has been presented in [5]. Furthermore, the asymptotic behaviour of $\underline{w}_f(n)$ for a particular class of weight functions has been derived in [5], [6]. Explicit and asymptotic expressions for typical weights (e.g. various types of path lengths or of free search cost measures) and for particular families of trees (e.g. t-ary trees, s, $s+1$-trees, ordered trees, etc.) has been computed in [4], [5], [6].

In this paper we shall present a simple approach to the computation of the average weight $w_f(n)$ of a tree $T \in \mathscr{F}$ with n nodes under the assumption that these trees are distributed according to a particular probability distribution which depends on the node-degrees and on the number of nodes of the subtrees appearing in a tree $T \in \mathscr{F}$ with n nodes. Choosing a particular probability distribution and a particular weight function f, the presented mathematical model enables us to derive expressions for the average weight $w_f(n)$ of various families of trees which play an important part in the analysis of searching algorithms. In particular, we shall demonstrate the method for the class of random t-ary trees, binary search trees, t-ary digital search trees and Patricia trees. In all cases, the derivation seems somewhat more straightforward and direct than the known methods: the mathematical model yields an algebraic or a functional-differential equation satisfied by the generating function of the average weights $w_f(n)$ which can be solved by an application of the Laplace Transformation. Many of the known results can be generalized by the approach described in the subsequent sections.

2. A general approach to the computation of the average weight

Let $\mathscr{F}$ be a bounded family of trees, $\mathscr{F}_\lambda(n) \subset \mathscr{F}(n)$ be the set of all λ-trees with n nodes and let $\mathscr{F}_{n,\lambda}(\vec{m}, \vec{d})$ be the class of all trees $T \in \mathscr{F}$ of type $\langle(n, \lambda); (m_1, d_1), \ldots, (m_\lambda, d_\lambda)\rangle$. Henceforth, we shall assume that the probability $p_\lambda(n, \vec{m}, \vec{d})$ that a λ-tree $T \in \mathscr{F}_\lambda(n)$ is of type $\langle(n, \lambda); (m_1, d_1), \ldots, (m_\lambda, d_\lambda)\rangle$ has a representation of the form

$$p_\lambda(n, \vec{m}, \vec{d}) = |q(n, \lambda)|^{-1} \prod_{1 \leq i \leq \lambda} q_i^{(\lambda)}(m_i, d_i), \tag{1}$$

where $q_i^{(\lambda)}(m, 0) = \delta_{m1}$, $m \in \mathbb{N}$. We will constantly use the obvious convention that the one-node tree has the probability 1, that is $p_0(1, \vec{m}, \vec{d}) = 1$. Note that for all $\lambda \in \mathrm{DEG}(\mathscr{F})$

$$\sum_{d_1, \ldots, d_\lambda \in \mathrm{DEG}(\mathscr{F})} \sum_{\substack{m_1, \ldots, m_\lambda : m_i \geq d_i + 1, 1 \leq i \leq \lambda \\ m_1 + m_2 + \ldots + m_\lambda = n-1}} p_\lambda(n, \vec{m}, \vec{d}) = 1. \tag{2}$$

Introducing the generating functions ($\lambda, l \in \mathrm{DEG}(\mathscr{F})$)

$$Q_\lambda(z) = \sum_{n \geq \lambda+1} q(n, \lambda) z^n \tag{3a}$$

Table 1.

Characteristics of a family of trees $\mathscr{F}$ with DEG $(\mathscr{F}) = \{0, 1, 2\}$. Here, $A(z) = z(1-z)^{-2} - z$ and $B(z) = A(z) - 2z^2$; internal nodes are marked by "●", leaves by "■" and arbitrary subtrees of $\mathscr{F}$ by "△".

λ	j	l	$q_j^{(\lambda)}(m, l)$	$Q_{j,l}^{(\lambda)}(z)$	Structure of the tree characterized by λ, j, l
1	1	0	$\delta_{m,1}$ $(m \geq 1)$	z	
1	1	1	$6m$ $(m \geq 2)$	$6A(z)$	
1	1	2	$5m$ $(m \geq 3)$	$5B(z)$	
2	1	0	$\delta_{m,1}$ $(m \geq 1)$	z	
2	1	1	m $(m \geq 2)$	$A(z)$	
2	1	2	$4m$ $(m \geq 3)$	$4B(z)$	
2	2	0	$\delta_{m,1}$ $(m \geq 1)$	z	
2	2	1	$2m$ $(m \geq 2)$	$2A(z)$	
2	2	2	$3m$ $(m \geq 3)$	$3B(z)$	

and

$$Q_{j,l}^{(\lambda)}(z) = \sum_{m \geq l+1} q_j^{(\lambda)}(m, l) z^m, \quad j \in [1:\lambda], \tag{3b}$$

the equations (1) and (2) immediately imply

$$z \prod_{1 \leq j \leq \lambda} \sum_{v \in \mathrm{DEG}(\mathcal{F})} Q_{j,v}^{(\lambda)}(z) = Q_\lambda(z), \quad \lambda \in \mathrm{DEG}(\mathcal{F}). \tag{4}$$

Thus, $p_\lambda(n, \vec{m}, \vec{d})$ is uniquely determined by the numbers $q_j^{(\lambda)}(m, l)$ and therefore by the functions $Q_{j,l}^{(\lambda)}(z)$, $l, \lambda \in \mathrm{DEG}(\mathcal{F})$, $1 \leq j \leq \lambda$. Note that always $Q_{j,0}^{(\lambda)}(z) = z$. Henceforth, the functions $Q_{j,l}^{(\lambda)}(z)$, $l, \lambda \in \mathrm{DEG}(\mathcal{F})$, $1 \leq j \leq \lambda$, are called *characteristics of $\mathcal{F}$*.

Since the sets $\mathcal{F}_\lambda(n)$, $\lambda \in \mathrm{DEG}(\mathcal{F})$, are a partition of the set $\mathcal{F}(n)$ of all trees $T \in \mathcal{F}$ with n nodes, each tree $T \in \mathcal{F}(n)$ belongs to exactly one class $C(T) := \mathcal{F}_{n,\lambda}(\vec{m}, \vec{d})$ with $\lambda \in \mathrm{DEG}(\mathcal{F})$, $\vec{d} \in \mathrm{DEG}(\mathcal{F})^\lambda$ and $\sum_{1 \leq i \leq \lambda} m_i = n-1$; if $C(T) \equiv \mathcal{F}_{n,\lambda}(\vec{m}, \vec{d})$, then $p_\lambda(n, \vec{m}, \vec{d})$ is denoted by $p(C(T))$. Given the probabilities $p_\lambda(n, \vec{m}, \vec{d})$ with $\lambda \in \mathrm{DEG}(\mathcal{F})$, $\vec{d} \in \mathrm{DEG}(\mathcal{F})^\lambda$ and $\sum_{1 \leq i \leq \lambda} m_i = n-1$, it is not hard to see (the probabilities of the subtrees are independent) that the probability $p_\lambda(T)$ of a tree $T \in \mathcal{F}_\lambda(n)$ is given by

$$p_\lambda(T) = \prod_{T' \in \mathrm{SUB}(T)} p(C(T')). \tag{5}$$

For example, choosing $p_\lambda(n, \vec{m}, \vec{d}) := |\mathcal{F}_{n,\lambda}(\vec{m}, \vec{d})|/|\mathcal{F}_\lambda(n)|$, it is easily verified by induction, that $p_\lambda(T) = 1/|\mathcal{F}_\lambda(n)|$, $T \in \mathcal{F}_\lambda(n)$, in other words, all λ-trees with n nodes are equally likely.

Only for purposes of illustration, consider the family of trees $\mathcal{F}$ with $\mathrm{DEG}(\mathcal{F}) = \{0, 1, 2\}$ and the characteristics given in Table 1. In order to obtain the probabilities $p_\lambda(n, \vec{m}, \vec{d})$, we have to compute the functions $Q_\lambda(z)$ defined by (4). We find

$$Q_0(z) = z$$
$$Q_1(z) = z \sum_{v \in \{0,1,2\}} Q_{1,v}^{(1)}(z) = z[z + 6A(z) + 5B(z)]$$
$$= z^2 + 12z^3 + \sum_{n \geq 4} 11(n-1) z^n \tag{5a}$$

and

$$Q_2(z) = z \Big[\sum_{v \in \{0,1,2\}} Q_{1,v}^{(2)}(z) \Big] \Big[\sum_{v \in \{0,1,2\}} Q_{2,v}^{(2)}(z) \Big]$$

$$= z[z+A(z)+4B(z)][z+2A(z)+3B(z)]$$

$$= z^3 + 6z^4 + 38z^5 + \sum_{n \geq 6} \left[25\binom{n}{3} - 110n + 290 \right] z^n. \tag{5b}$$

Thus by (3a)

$$q(n,1) = \begin{cases} 1 & \text{if } n=2 \\ 12 & \text{if } n=3 \\ 11(n-1) & \text{if } n \geq 4 \end{cases}$$

$$q(n,2) = \begin{cases} 1 & \text{if } n=3 \\ 6 & \text{if } n=4 \\ 38 & \text{if } n=5 \\ 25\binom{n}{3} - 110n + 290 & \text{if } n \geq 6. \end{cases} \tag{6}$$

Now, the probability $p_\lambda(n, \vec{m}, \vec{d})$ that a λ-tree $T \in \mathscr{F}(n)$ is of type $\langle (n,\lambda); (m_1, d_1), \ldots, (m_\lambda, d_\lambda) \rangle$ is given by (1). An elementary computation yields

$$p_1(n, m_1, d_1) = \begin{cases} 5/11 & \text{if } (n, m_1, d_1) \in \{(n, n-1, 2) | n \geq 4\} \\ 6/11 & \text{if } (n, m_1, d_1) \; \{(n, n-1, 1) | n \geq 4\} \\ 1 & \text{if } (n, m_1, d_1) \in \{(2,1,0), (3,2,1)\} \\ 0 & \text{otherwise} \end{cases}$$

and

$$p_2(3, (m_1, m_2), (d_1, d_2)) = \begin{cases} 1 & \text{if } ((m_1, m_2), (d_1, d_2)) = ((1,1), (0,0)) \\ 0 & \text{otherwise} \end{cases}$$

$$p_2(4, (m_1, m_2), (d_1, d_2)) = \begin{cases} 1/3 & \text{if } ((m_1, m_2), (d_1, d_2)) = ((2,1), (1,0)) \\ 2/3 & \text{if } ((m_1, m_2), (d_1, d_2)) = ((1,2), (0,1)) \\ 0 & \text{otherwise} \end{cases}$$

$$p_2(5, (m_1, m_2), (d_1, d_2)) = \begin{cases} 3/38 & \text{if } ((m_1, m_2), (d_1, d_2)) = ((3,1), (1,0)) \\ 3/19 & \text{if } ((m_1, m_2), (d_1, d_2)) = ((1,3), (0,1)) \\ 4/19 & \text{if } ((m_1, m_2), (d_1, d_2)) = ((2,2), (1,1)) \\ 9/38 & \text{if } ((m_1, m_2), (d_1, d_2)) = ((1,3), (0,2)) \\ 6/19 & \text{if } ((m_1, m_2), (d_1, d_2)) = ((3,1), (2,0)) \\ 0 & \text{otherwise} \end{cases}$$

and for $n \geq 6$

$$p_2(n,(m_1,m_2),(d_1,d_2)) = \begin{cases} \dfrac{m_1}{a_n} & \text{if } ((m_1,m_2),(d_1,d_2)) = ((n-2,1),(1,0)) \\[2mm] \dfrac{4m_1}{a_n} & \text{if } ((m_1,m_2),(d_1,d_2)) = ((n-2,1),(2,0)) \\[2mm] \dfrac{2m_1 m_2}{a_n} & \text{if } ((m_1,m_2),(d_1,d_2)) \in \{((1,n-2),(0,1))\} \\ & \quad \cup \{((i,n-1-i),(1,1)) \mid i \in [2:n-3]\} \\[2mm] \dfrac{3m_1 m_2}{a_n} & \text{if } ((m_1,m_2),(d_1,d_2)) \in \{((1,n-2),(0,2))\} \\ & \quad \cup \{((i,n-1-i),(1,2)) \mid i \in [2:n-3]\} \\[2mm] \dfrac{8m_1 m_2}{a_n} & \text{if } ((m_1,m_2),(d_1,d_2)) = ((i,n-1-i),(2,1)) \\ & \quad i \in [2:n-3] \\[2mm] \dfrac{12m_1 m_2}{a_n} & \text{if } ((m_1,m_2),(d_1,d_2)) = ((i,n-1-i),(2,2)) \\ & \quad i \in [2:n-3] \\[2mm] 0 & \text{otherwise} \end{cases}$$

where $a_n = 25\binom{n}{3} - 110n + 290$. All trees $T \in \mathcal{F}(n)$, $n \leq 6$, together with the probabilities $p_\lambda(T)$ and $p_\lambda(n, \vec{m}, \vec{d})$ are listed in Table 2.

The following theorem describes a general approach to the computation of the average weight of a tree $T \in \mathcal{F}_\lambda(n)$.

Theorem 1. *Let $\mathcal{F}$ be a bounded family of trees with the characteristics $Q_{j,l}^{(\lambda)}(z)$, $l, \lambda \in \mathrm{DEG}(\mathcal{F})$, $1 \leq j \leq \lambda$, and let f be a given weight function. If*

$$E^{(l)}(z) = \sum_{n \geq l+1} \mathsf{E}[w_f^{(l)}(n)] z^n, \quad l \in \mathrm{DEG}(\mathcal{F}),$$

denotes the generating function of the average weight $\mathsf{E}[w_f^{(l)}(n)]$ of an l-tree $T \in \mathcal{F}$ with n nodes, then we have for all $\lambda \in \mathrm{DEG}(\mathcal{F})$

$$E^{(\lambda)}(z) \odot Q_\lambda(z) = \sum_{1 \leq j \leq \lambda} [Q_\lambda(z) / \sum_{v \in \mathrm{DEG}(\mathcal{F})} Q_{j,v}^{(\lambda)}(z)]$$
$$\times [\sum_{u \in \mathrm{DEG}(\mathcal{F})} \{E^{(u)}(z) + F_\lambda^{(u)}(z)\} \odot Q_{j,u}^{(\lambda)}(z)],$$

where $F_\lambda^{(l)}(z)$ is the generating function of the values of the weight function f

Table 2.

All trees $T \in \mathcal{F}(n)$, $n \leq 6$, of our running example. The number attached to a tree denotes its probability $p_\lambda(T)$. Trees of the same type are characterized by "⎵".

n	$\mathcal{F}_0(n)$	$\mathcal{F}_1(n)$
1	$p_0(1,0,0)=1$	
2		$p_1(2,1,0)=1$
3		$p_1(3,2,1)=1$
4		$p_1(4,3,1)=\frac{6}{11}$, $p_1(4,3,2)=\frac{5}{11}$
5		$p_1(5,4,1)=\frac{6}{11}$ [$\frac{36}{121}$, $\frac{30}{121}$] ; $p_1(5,4,2)=\frac{5}{11}$ [$\frac{5}{33}$, $\frac{10}{33}$]
6		$p_1(6,5,1)=\frac{6}{11}$ [$\frac{216}{1331}$, $\frac{180}{1331}$, $\frac{10}{121}$, $\frac{20}{121}$] ; $p_1(6,5,2)=\frac{5}{11}$ [$\frac{15}{418}$, $\frac{20}{209}$, $\frac{15}{209}$, $\frac{30}{209}$, $\frac{45}{418}$]

Weights of non-regularly distributed trees

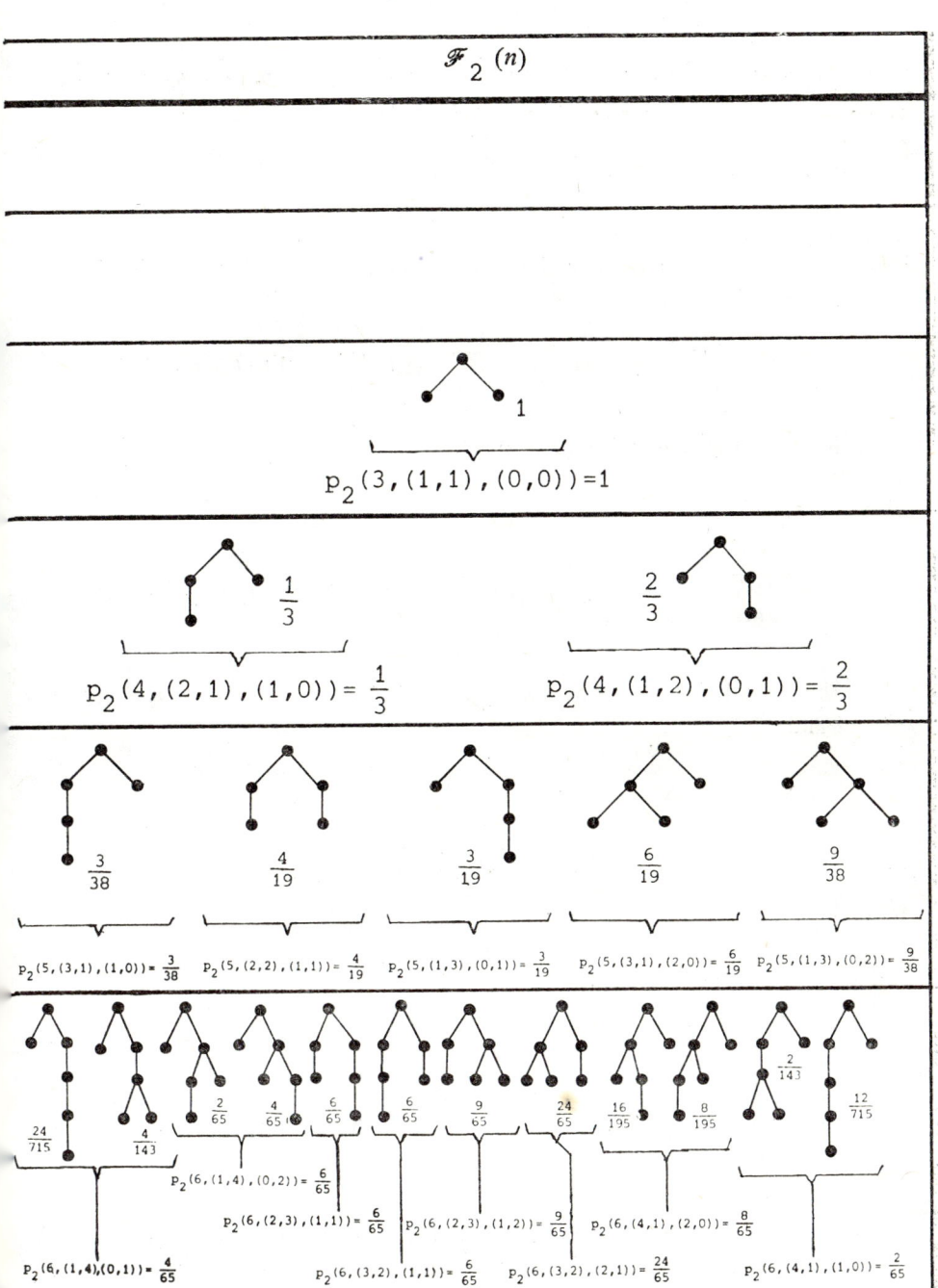

$\mathscr{F}_2(n)$

$p_2(3,(1,1),(0,0))=1$

$p_2(4,(2,1),(1,0))=\frac{1}{3}$ $p_2(4,(1,2),(0,1))=\frac{2}{3}$

$p_2(5,(3,1),(1,0))=\frac{3}{38}$ $p_2(5,(2,2),(1,1))=\frac{4}{19}$ $p_2(5,(1,3),(0,1))=\frac{3}{19}$ $p_2(5,(3,1),(2,0))=\frac{6}{19}$ $p_2(5,(1,3),(0,2))=\frac{9}{38}$

$p_2(6,(1,4),(0,1))=\frac{4}{65}$ $p_2(6,(1,4),(0,2))=\frac{6}{65}$ $p_2(6,(2,3),(1,1))=\frac{6}{65}$ $p_2(6,(2,3),(1,2))=\frac{9}{65}$ $p_2(6,(4,1),(2,0))=\frac{8}{65}$ $p_2(6,(3,2),(1,1))=\frac{6}{65}$ $p_2(6,(3,2),(2,1))=\frac{24}{65}$ $p_2(6,(4,1),(1,0))=\frac{2}{65}$

given by

$$F_\lambda^{(l)}(z) = \sum_{n \geq l+1} f(\lambda, n, l) z^n, \quad l, \lambda \in \text{DEG}(\mathcal{F}),$$

and $Q_\lambda(z)$ is the function defined in (4). Here, $h_1(z) \odot h_2(z)$ denotes the Hadamard product of the two power series $h_i(z) = \sum_{n \geq 0} h_{in} z^n$, $i \in \{1, 2\}$, defined by

$$h_1(z) \odot h_2(z) = \sum_{n \geq 0} h_{1n} h_{2n} z^n.$$

Proof. Let $T \in \mathcal{F}_{n,\lambda}(\vec{m}, \vec{d})$ with the subtrees $T_1, \ldots, T_\lambda$. We have by the definition of $w_f(T)$

$$\mathsf{E}[w_f^{(\lambda)}(n) | T \in \mathcal{F}_{n,\lambda}(\vec{m}, \vec{d})] = \sum_{1 \leq j \leq \lambda} \{\mathsf{E}[w_f^{(d_j)}(m_j)] + f(\lambda, m_j, d_j)\}.$$

Using (1), the unconditional expectation is given by $(\vec{m} = (m_1, \ldots, m_\lambda)$, $\vec{d} = (d_1, \ldots, d_\lambda))$

$$\mathsf{E}[w_f^{(\lambda)}(n)]$$
$$= \sum_{\substack{d \in \text{DEG}(\mathcal{F}) \\ 1 \leq i \leq \lambda}} \sum_{m_1, \ldots, m_\lambda} \sum_{1 \leq j \leq \lambda} \{\mathsf{E}[w_f^{(d_j)}(m_j)] + f(\lambda, m_j, d_j)\} p_\lambda(n, \vec{m}, \vec{d})$$

where the second sum ranges over all $m_i \geq d_i + 1$, $1 \leq i \leq \lambda$, with $m_1 + m_2 + \ldots + m_\lambda = n - 1$ and

$$p_\lambda(n, \vec{m}, \vec{d}) = \{q(n, \lambda)\}^{-1} \prod_{1 \leq \mu \leq \lambda} q_\mu^{(\lambda)}(m_\mu, d_\mu).$$

Multiplying this equation by $q(n, \lambda) z^n$ and taking the sum over all $n \geq \lambda + 1$, we further find

$$\sum_{n \geq \lambda+1} \mathsf{E}[w_f^{(\lambda)}(n)] q(n, \lambda) z^n = E^{(\lambda)}(z) \odot Q_\lambda(z)$$
$$= \sum_{\substack{d_i \in \text{DEG}(\mathcal{F}) \\ 1 \leq i \leq \lambda}} \sum_{1 \leq j \leq \lambda} H_{\lambda, j, \vec{d}}(z),$$

where

$$H_{\lambda, j, \vec{d}}(z)$$
$$= \sum_{n > \lambda} z^n \sum_{m_1, \ldots, m_\lambda} \{\mathsf{E}[w_f^{(d_j)}(m_j)] + f(\lambda, m_j, d_j)\} \prod_{1 \leq \mu \leq \lambda} q_\mu^{(\lambda)}(m_\mu, d_\mu)$$

and the second sum ranges over all $m_i \geq d_i+1$, $1 \leq i \leq \lambda$, with $m_1+m_2+\ldots+m_\lambda = n-1$. Thus

$$H_{\lambda,j,\vec{d}}(z) = z \Big[\sum_{m_j \geq d_j+1} \{\mathsf{E}[w_f^{(d_j)}(m_j)] + f(\lambda, m_j, d_j)\} q_j^{(\lambda)}(m_j, d_j) z^{m_j} \Big]$$
$$\times \prod_{\substack{1 \leq \mu \leq \lambda \\ \mu \neq j}} \sum_{m_\mu \geq d_\mu+1} q_\mu^{(\lambda)}(m_\mu, d_\mu) z^{m_\mu}$$
$$= z [\{E^{(d_j)}(z) + F_\lambda^{(d_j)}(z)\} \odot Q_{j,d_j}^{(\lambda)}(z)] \prod_{\substack{1 \leq \mu \leq \lambda \\ \mu \neq j}} Q_{\mu,d_\mu}^{(\lambda)}(z).$$

Hence

$$E^{(\lambda)}(z) \odot Q_\lambda(z)$$
$$= z \sum_{\substack{d_i \in \mathrm{DEG}(\mathscr{F}) \\ 1 \leq i \leq \lambda}} \sum_{1 \leq j \leq \lambda} [\{E^{(d_j)}(z) + F_\lambda^{(d_j)}(z)\} \odot Q_{j,d_j}^{(\lambda)}(z)] \prod_{\substack{1 \leq \mu \leq \lambda \\ \mu \neq j}} Q_{\mu,d_\mu}^{(\lambda)}(z)$$
$$= z \sum_{1 \leq j \leq \lambda} \Big[\sum_{u \in \mathrm{DEG}(\mathscr{F})} \{E^{(u)}(z) + F_\lambda^{(u)}(z)\} \odot Q_{j,u}^{(\lambda)}(z) \Big]$$
$$\times \prod_{\substack{1 \leq \mu \leq \lambda \\ \mu \neq j}} \sum_{v \in \mathrm{DEG}(\mathscr{F})} Q_{\mu,v}^{(\lambda)}(z)$$
$$= \sum_{1 \leq j \leq \lambda} \Big[\sum_{u \in \mathrm{DEG}(\mathscr{F})} \{E^{(u)}(z) + F_\lambda^{(u)}(z)\} \odot Q_{j,u}^{(\lambda)}(z) \Big]$$
$$\times \Big[Q_\lambda(z) \Big/ \sum_{v \in \mathrm{DEG}(\mathscr{F})} Q_{j,v}^{(\lambda)}(z) \Big].$$

Here, we have used the relation (4) in the last step. This completes the proof of our theorem. □

Considering again our running example with $f(\lambda, m, d) := m$ — this choice of f corresponds to the total path length — we find by (6)

$$E^{(1)}(z) \odot Q_1(z) = \sum_{n \geq 2} \mathsf{E}[w_f^{(1)}(n)] q(n,1) z^n$$
$$= \mathsf{E}[w_f^{(1)}(2)] z^2 + 12 \mathsf{E}[w_f^{(1)}(3)] z^3$$
$$+ 11 \sum_{n \geq 4} (n-1) \mathsf{E}[w_f^{(1)}(n)] z^n \qquad (7a)$$

and

$$E^{(2)}(z) \odot Q_2(z) = \sum_{n \geq 3} \mathsf{E}[w_f^{(2)}(n)] q(n,2) z^n$$
$$= \mathsf{E}[w_f^{(2)}(3)] z^3 + 6 \mathsf{E}[w_f^{(2)}(4)] z^4 + 38 \mathsf{E}[w_f^{(2)}(5)] z^5$$

$$+ \sum_{n \geq 6} \left\{ 25\binom{n}{3} - 110n + 290 \right\} E[w_f^{(2)}(n)] z^n. \tag{7b}$$

Using the results summarized in Table 2, we obtain the initial values

$$E[w_f^{(0)}(1)] = 0,$$
$$E[w_f^{(1)}(1)] = 0, \ E[w_f^{(1)}(2)] = 1, \ E[w_f^{(1)}(3)] = 3,$$
$$E[w_f^{(2)}(1)] = E[w_f^{(2)}(2)] = 0, \ E[w_f^{(2)}(3)] = 2, \ E[w_f^{(2)}(4)] = 4,$$
$$E[w_f^{(2)}(5)] = 237/38.$$

Introducing the functions

$$G_i(z) = \sum_{n \geq 1} E[w_f^{(i)}(n)] z^n, \quad i \in \{1, 2\},$$

the relations (7a) and (7b) are equivalent to

$$E^{(1)}(z) \odot Q_1(z) = 11z G_1'(z) - 11 G_1(z) - 10z^2 - 30z^3 \tag{8a}$$

and

$$E^{(2)}(z) \odot Q_2(z) = \tfrac{25}{6} z^3 G_2'''(z) - 110z G_2'(z) + 290 G_2(z) + 32 z^3$$
$$+ 224 z^4 + \tfrac{5688}{19} z^5. \tag{8b}$$

On the other hand, we find by the results of Table 1, (5a) and (5b)

$$[Q_1(z)/\sum_{v \in \{0,1,2\}} Q_{1,v}^{(1)}(z)][\sum_{u \in \{0,1,2\}} \{E^{(u)}(z) + F_1^{(u)}(z)\} \odot Q_{1,u}^{(1)}(z)]$$
$$= z[\sum_{n \geq 1} \{E[w_f^{(0)}(n)] + n\} \delta_{n1} z^n + 6 \sum_{n \geq 2} \{E[w_f^{(1)}(n)] + n\} n z^n$$
$$+ 5 \sum_{n \geq 3} \{E[w_f^{(2)}(n)] + n\} n z^n]$$
$$= 6z^2 G_1'(z) + 5z^2 G_2'(z) + 11z^2(1+z)/(1-z)^3 - 10z^2 - 20z^3 \tag{9a}$$

and in an analogous way

$$\sum_{1 \leq j \leq 2} [Q_2(z)/\sum_{v \in \{0,1,2\}} Q_{j,v}^{(2)}(z)][\sum_{u \in \{0,1,2\}} \{E^{(u)}(z) + F_2^{(u)}(z)\} \odot Q_{j,u}^{(2)}(z)]$$
$$= z^3 G_1'(z) \left[\frac{15}{(1-z)^2} - 12 - 22z \right] + z^3 G_2'(z) \left[\frac{35}{(1-z)^2} - 28 - 48z \right]$$

$$+50z^3+\frac{1+z}{(1-z)^5}-10z^3\frac{(1+z)(4+7z)}{(1-z)^3}-20z^3\frac{(2+7z)}{(1-z)^2}$$

$$+192z^5+168z^4+32z^3. \tag{9b}$$

Applying Theorem 1, the left sides of the equations (8a) and (9a) and those of (8b) and (9b) are equal. Thus, the generating functions $G_1(z)$ and $G_2(z)$ of the average total path lengths $E[w_f^{(1)}(n)]$ and $E[w_f^{(2)}(n)]$ of the trees $T \in \mathscr{F}_1(n)$ and $T \in \mathscr{F}_2(n)$ are given by the system of differential equations

$$z(11-6z)G_1'(z)-11G_1(z)-5z^2G_2'(z)=10z^3+11z^2\frac{1+z}{(1-z)^3}$$

$$z^3\left[\frac{15}{(1-z)^2}-12-22z\right]G_1'(z)-\frac{25}{6}z^3G_2'''(z)-290G_2(z)$$

$$+z\left[\frac{35z^2}{(1-z)^2}-28z^2-48z^3+110\right]G_2'(z)$$

$$=-50z^3\frac{1+z}{(1-z)^5}+10z^3\frac{(1+z)(4+7z)}{(1-z)^3}+20z^3\frac{2+7z}{(1-z)^2}$$

$$+56z^4+\frac{2040}{19}z^5,$$

where the initial conditions are $G_1(0)=G_2(0)=G_2'(0)=G_2''(0)=0$.

Specializing Theorem 1 to the case of t-ary trees, $t \geq 2$, we obtain the following basic result.

Corollary 1. Let $t \in \mathbb{N}$, $t \geq 2$, $\mathscr{F}$ be the family of t-ary trees with the characteristics $Q_{j,l}^{(\lambda)}(z)$, $l, \lambda \in \{0, t\}$, $1 \leq j \leq \lambda$, and f be a given weight function. If

$$E^{(t)}(z) = \sum_{n \geq t+1} E[w_f^{(t)}(n)]z^n$$

denotes the generating function of the average weight $E[w_f^{(t)}(n)]$ of a tree $T \in \mathscr{F}$ with n nodes, then

$$E^{(t)}(z) \odot Q_t(z) = \sum_{1 \leq j \leq t} [Q_t(z)/\{z+Q_{j,t}^{(t)}(z)\}]$$

$$\times [f(t,1,0)z+\{E^{(t)}(z)+F_t^{(t)}(z)\} \odot Q_{j,t}^{(t)}(z)],$$

where

$$F_t^{(t)}(z) = \sum_{n \geq t+1} f(t, n, t) z^n \text{ and}$$

$$Q_t(z) = z \prod_{1 \leq j \leq t} \{z + Q_{j,t}^{(t)}(z)\}. \quad \square$$

Various algorithms for searching are based on t-ary tree structures, where the trees with n nodes are distributed according to a particular probability distribution. In the subsequent section, we shall apply the above corollary to important classes of t-ary trees in order to compute the expected value of some weights playing a central part in the analysis of searching algorithms.

3. Applications

In this section we shall consider the following classes of t-ary trees:

- t-ary digital search trees,
- Binary search trees,
- Regularly distributed t-ary trees,
- Patricia trees.

For each of these families of trees $\mathscr{F}$, we shall present an explicit expression for the expected value of the weight $w_f(T)$, $T \in \mathscr{F}(n)$, where the weight function f is a polynomial of degree N in the numbers of nodes. Only in the case of t-ary digital search trees, we shall derive the corresponding result in all its details; for the details of the remaining cases, the reader is referred to [7]. For technical reasons, we assume that f is given by

$$f(\lambda, m, d) = \sum_{0 \leq l \leq N} a_l f_l(m), \quad \lambda, d \in \{0, t\}, \ m \in \mathbb{N}, \ N \in \mathbb{N}_0, \ a_N \neq 0, \tag{10a}$$

where

$$f_l(m) = \frac{1}{l! t^l} \prod_{1 \leq i \leq l} (m - 1 - t(i-1)). \tag{10b}$$

Obviously, each polynomial of degree N has a unique representation of the form (10a). As we have seen in Section 1, the weight $w_f(T)$ of a tree $T \in \mathscr{F}(n)$ is equal to the total path length $p(T)$, if $f(\lambda, m, d) = m$ (i.e. $N=1$, $a_0=1$, $a_1=t$), and to the weight $F(T) = np(T) - \text{FCOST}(T)$, if $f(\lambda, m, d) = m^2$ (i.e. $N=2$, $a_0=1$, $a_1 = t^2 + 2t$, $a_2 = 2t^2$).

Now, let $T=(I, L, r)$ be an arbitrary ordered tree with the set of internal nodes I, the set of leaves L and the root r. The above weights $p(T)$ and FCOST(T) can be refined by introducing the (see [5], [8])

- internal path length $p_I(T)$ of T defined by $p_I(T) = \sum_{v \in I} \text{dist}(r, v)$;

- external path length $p_L(T)$ of T defined by $p_L(T) = \sum_{v \in L} \text{dist}(r, v)$;

- internal free search cost measure FCOST$_I(T)$ of T defined by FCOST$_I(T)$ $= \frac{1}{2} \sum_{(u,v) \in I \times I} \text{dist}(u, v)$;

- external free search cost measure FCOST$_L(T)$ of T defined by FCOST$_L(T)$ $= \frac{1}{2} \sum_{(u,v) \in L \times L} \text{dist}(u, v)$;

- internal-external free search cost measure FCOST$_{IL}(T)$ of T defined by FCOST$_{IL}(T) = \sum_{(u,v) \in I \times L} \text{dist}(u, v)$.

If the root r of the tree T has the λ subtrees $T_i = (I_i, L_i, r_i)$, $1 \leq i \leq \lambda$, then each of these weights has the following inductive definition (see [5], [8], [9]):

$$p_I(T) = \sum_{1 \leq i \leq \lambda} \{p_I(T_i) + |I_i|\}; \tag{11a}$$

$$p_L(T) = \sum_{1 \leq i \leq \lambda} \{p_L(T_i) + |L_i|\}; \tag{11b}$$

$$\text{FCOST}_I(T) = |I| p_I(T) - I(T), \text{ where } I(T) = \sum_{1 \leq i \leq \lambda} \{I(T_i) + |I_i|^2\}; \tag{11c}$$

$$\text{FCOST}_L(T) = |L| p_L(T) - L(T), \text{ where } L(T) = \sum_{1 \leq i \leq \lambda} \{L(T_i) + |L_i|^2\}; \tag{11d}$$

$$\text{FCOST}_{IL}(T) = |I| p_L(T) + |L| p_I(T) - IL(T), \text{ where}$$

$$IL(T) = \sum_{1 \leq i \leq \lambda} \{IL(T_i) + 2|L_i||I_i|\}. \tag{11e}$$

Since a t-ary tree T has always $(tr+1)$, $r \in \mathbf{N}_0$, nodes (r internal nodes, $r(t-1)+1$ leaves), the inductive definition of $p_I(T)$ and $p_L(T)$ immediately yields the well-known relation ([8]) $p_L(T) = (t-1)p_I(T) + rt$; since $p(T) = p_I(T) + p_L(T)$, we further obtain for a t-ary tree T with $(tr+1)$ nodes, $r \in \mathbf{N}_0$

$$p_I(T) = t^{-1} p(T) - r \tag{12a}$$

and

$$p_L(T) = (1 - t^{-1}) p(T) + r. \tag{12b}$$

In a similar way, the inductive definition of $p_I(T)$, $p_L(T)$, $I(T)$, $L(T)$ and $IL(T)$ implies $L(T)=(t-1)^2 I(T)+2(t-1)p_I(T)+rt$ and $IL(T)=2(t-1)I(T)+2p_I(T)$; since $F(T)=I(T)+L(T)+IL(T)$, where $F(T)=(tr+1)p(T)-\text{FCOST}(T)$ is the weight introduced in Section 1, we further find by (12a) and (12b)

$$I(T)=t^{-2}F(T)-2t^{-2}p(T)+rt^{-1}, \tag{13a}$$

$$L(T)=(t-1)^2 t^{-2}F(T)+2(t-1)t^{-2}p(T)+rt^{-1}, \tag{13b}$$

$$IL(T)=2(t-1)t^{-2}F(T)-2(t-2)t^{-2}p(T)-2rt^{-1}. \tag{13c}$$

Therefore, we obtain for a t-ary tree T with $(tr+1)$, $r \in \mathbf{N}_0$, nodes by an application of (11c–e), (12a–b) and (13a–c)

$$\text{FCOST}_I(T)=t^{-2}\text{FCOST}(T)+t^{-2}p(T)-r(tr+1)t^{-1}, \tag{14a}$$

$$\text{FCOST}_L(T)=(t-1)^2 t^{-2}\text{FCOST}(T)-(t-1)t^{-2}p(T)+r(t-1)(tr+1)t^{-1}, \tag{14b}$$

$$\text{FCOST}_{IL}(T)=2(t-1)t^{-2}\text{FCOST}(T)+(t-2)t^{-2}p(T)-r(t-2)(tr+1)t^{-1}. \tag{14c}$$

Thus, if we are able to derive an expression for the expected value of $p(T)$ and $\text{FCOST}(T)$ of a t-ary tree $T \in \mathscr{F}_t(tr+1)$, $r \in \mathbf{N}_0$, then we also have expressions for the expected values of the refined weights defined in (11a–c). In the following subsections, we shall compute explicit expressions for the average weight $p(T)$ and $\text{FCOST}(T)$, where T is a tree belonging to a particular class of t-ary trees.

3.1. t-ary digital search trees

Consider the class $\mathscr{F}_t$ of all t-ary trees with the characteristics $Q_{j,t}^{(t)}(z)=z\exp(z^t)-z$, $1 \leq j \leq t$. Formula (3b) and Corollary 1 show that $q_j^{(t)}(tm_j+1,t)=\langle z^{tm_j+1}; Q_{j,t}^{(t)}(z)\rangle^{1)}=1/m_j!$, $1 \leq j \leq t$, and $Q_t(z)=z^{t+1}\exp(tz^t)$. Therefore, $q(tr+1,t)=\langle z^{tr+1}; Q_t(z)\rangle=t^{r-1}/(r-1)!$, $r \in \mathbf{N}$. Using (1), we find the probability $p_t(tr+1, t\vec{m}+1, \vec{d})$ that a tree $T \in \mathscr{F}_t(tr+1)$, $r \in \mathbf{N}$, is of type $\langle (tr+1, t); (tm_1+1, d_1), \ldots, (tm_t+1, d_t)\rangle$, $\vec{d} \in \{0,t\}^t$. We obtain

$$p_t(tr+1, t\vec{m}+1, \vec{d})=\frac{(r-1)!}{t^{r-1}}\prod_{1 \leq j \leq t} m_j!^{-1}.$$

Using this relation together with (5), a simple induction shows that the prob-

---

[1] Throughout this paper, $\langle z^n; h(z)\rangle$ denotes the coefficient of z^n in the expansion of the function $h(z)$.

ability $p_t(T)$ of a tree $T \in \mathscr{F}_t(tr+1)$, $r \in \mathbb{N}$, is given by

$$p_t(T) = |T|! \prod_{T' \in \mathrm{SUB}(T)} (|T'|/t^{|T'|-1})^{-1},$$

where $|T|$ denotes the number of internal nodes of a tree $T \in \mathscr{F}_t$. This is the probability of a t-ary digital search tree ([3]). Thus, the class $\mathscr{F}_t$ of all t-ary trees together with the characteristics $Q_{j,t}^{(t)}(z) = z \exp(z^t) - z$, $1 \leq j \leq t$, is equal to the class of t-ary digital search trees. Let us now turn to the average weight $\mathsf{E}[w_f^{(t)}(tr+1)]$ for these trees.

Theorem 2. *Let $\mathscr{F}_t$ be the set of all t-ary digital search trees and let f be the weight function defined in (10a). The average weight $\mathsf{E}[w_f^{(t)}(tr+1)]$ is given by*

$$\mathsf{E}[w_f^{(t)}(tr+1)] = \sum_{0 \leq l \leq N} a_l t^{-l+1} \sum_{\mu \geq 0} \binom{r}{\mu+l+1} \prod_{1 \leq i \leq \mu} \left(\frac{1}{t^{l+i-1}} - 1\right).$$

Proof. Since

$$E^{(t)}(z) \odot Q_t(z) = \sum_{r \geq 1} \mathsf{E}[w_f^{(t)}(tr+1)] \frac{t^{r-1}}{(r-1)!} z^{tr+1}$$

and

$$\sum_{1 \leq j \leq t} [Q_t(z)/\{z + Q_{j,t}^{(t)}(z)\}][f(t,1,0)z + \{E^{(t)}(z) + F_t^{(t)}(z)\} \odot Q_{j,t}^{(t)}(z)]$$

$$= z^t e^{(t-1)zt} \left[a_0 z + \sum_{r \geq 1} \mathsf{E}[w_f^{(t)}(tr+1)] \frac{1}{r!} z^{tr+1} \right.$$

$$\left. + \sum_{0 \leq l \leq N} a_l \sum_{r \geq 1} \binom{r}{l} \frac{1}{r!} z^{tr+1} \right],$$

an application of Corollary 1 leads to

$$H_t'(zt) = te^{(t-1)z} \left[H_t(z) + e^z \sum_{0 \leq l \leq N} \frac{a_l}{l!} z^l \right], \tag{15}$$

where

$$H_t(z) = \sum_{r \geq 1} \mathsf{E}[w_f^{(t)}(tr+1)] \frac{z^r}{r!}.$$

Thus, $E[w_f^{(t)}(tr+1)] = r!\langle z^r; H_t(z)\rangle$, where $H_t(z)$ satisfies the functional-differential equation given by (15). In order to compute the solution of this equation, we take the Laplace Transforms on both sides of (15). If

$$h_t(s) = \int_0^\infty e^{-sz} H_t(z)\,dz$$

denotes the Laplace Transform of $H_t(z)$, then formula (15) can be translated by well-known operations (e.g. [1]) into the recursion formula $(\mathrm{Re}(s)>t)$

$$st^{-2}h_t(s/t) - t^{-1}H_t(0) = th_t(s-t+1) + t\sum_{0\leq l\leq N} a_l(s-t)^{-l-1} \qquad (16)$$

where

$$h_t(s) = \sum_{r\geq 1} E[w_f^{(t)}(tr+1)]s^{-r-1}. \qquad (17)$$

Since $H_t(0) = 0$, formula (17) is equivalent to

$$h_t(s) = t^2 s^{-1} h_t(t(s-1)+1) + t^2 s^{-1} \sum_{0\leq l\leq N} a_l(st-t)^{-l-1},$$

$\mathrm{Re}(s) > 1$.

Iterating this equation, we immediately obtain for $p\in\mathbb{N}$ and $\mathrm{Re}(s)>1$

$$h_t(s) = s^{-1} t^{2p} h_t(t^p(s-1)+1) \prod_{1\leq k<p} \frac{1}{t^k(s-1)+1}$$

$$+ \sum_{0\leq l\leq N} \frac{a_l}{(s-1)^{l+1}} \sum_{1\leq k\leq p} \frac{1}{st^{(l-1)k}} \prod_{1\leq i<k} \frac{1}{t^i(s-1)+1}. \qquad (18)$$

Now, it is easily verified that the first term in (18) tends to zero for $p\to\infty$. Therefore,

$$h_t(s) = \sum_{0\leq l\leq N} \frac{a_l}{(s-1)^{l+1}} \sum_{k\geq 0} \frac{1}{t^{(l-1)(k+1)}} \prod_{0\leq i\leq k} \frac{1}{t^i(s-1)+1}. \qquad (19)$$

By (17), we have $E[w_f^{(t)}(tr+1)] = \langle s^{-r-1}; h_t(s)\rangle$. Thus, we must evaluate the function $h_t(s)$. This can be done by means of the q-binomial coefficients defined by

$$\binom{n}{p}_q := \prod_{1\leq i\leq p} \frac{1-q^{n-i+1}}{1-q^i} \qquad 0<q<1.$$

Using the general identity ([2; p. 118])

$$\prod_{0 \leq r \leq M} \frac{1}{1-xq^r} = \sum_{j \geq 0} (-1)^j q^{j(M+1)+\binom{j}{2}} \binom{-M-1}{j}_q x^j$$

with $x := -(s-1)^{-1}$, $q := 1/t$ and $M := k$, the product appearing in (19) can be replaced. We obtain

$$h_t(s) = \sum_{0 \leq l \leq N} \frac{a_l}{(s-1)^{l+1}} \sum_{k \geq 0} \sum_{j \geq 0} \frac{1}{(s-1)^{k+1+j}} t^{-(l-1)(k+1)-\binom{k+j+1}{2}} \binom{-k-1}{j}_{1/t}$$

$$= \sum_{0 \leq l \leq N} \frac{a_l}{(s-1)^{l+1}} \sum_{k \geq 0} \frac{1}{(s-1)^{k+1}} t^{-\binom{k+1}{2}} \sum_{0 \leq j \leq k} t^{-(l-1)(k-j+1)} \binom{j-k-1}{j}_{1/t}.$$

Using the definition of the q-binomial coefficients, it is easily verified that

$$q^{\binom{k+1}{2}} \binom{j-k-1}{j}_q = (-1)^j q^{\binom{k-j+1}{2}} \binom{k}{k-j}_q.$$

Thus, $h_t(s)$ can further be transformed into

$$h_t(s) = \sum_{0 \leq l \leq N} \frac{a_l}{(s-1)^{l+1}} \sum_{k \geq 0} \frac{1}{(s-1)^{k+1}}$$

$$\times \sum_{0 \leq j \leq k} (-1)^j t^{-(l-1)(k-j+1)-\binom{k-j+1}{2}} \binom{k}{k-j}_{1/t}.$$

Reversing the summation over j and using the general identity ([2; p. 118])

$$\prod_{0 \leq r < M} (1+xq^r) = \sum_{0 \leq j \leq M} q^{\binom{j}{2}} \binom{M}{j}_q x^j$$

with $M := k$, $q := t^{-1}$ and $x := -t^{-l}$, we finally obtain

$$h_t(s) = \sum_{0 \leq l \leq N} \frac{a_l}{(s-1)^{l+1}} t^{-(l-1)} \sum_{k \geq 0} \frac{1}{(s-1)^{k+1}} \prod_{1 \leq i \leq k} \left(\frac{1}{t^{l+i-1}} - 1 \right).$$

Since generally for $\mathrm{Re}(s) > 1$

$$\frac{1}{(s-1)^{m+1}} = \sum_{j \geq 0} \binom{m+j}{m} \frac{1}{s^{m+j+1}},$$

we can use this relation with $m := l+k+2$ and obtain the desired evaluation

$$h_t(s) = \sum_{k \geq 1} \frac{1}{s^{k+1}} \sum_{0 \leq l \leq N} a_l t^{-l+1} \sum_{\mu \geq 0} \binom{k}{\mu+l+1} \prod_{1 \leq i \leq \mu} \left(\frac{1}{t^{l+i-1}} - 1 \right).$$

This completes the proof of our theorem. □

Choosing $(N, a_0, a_1) = (1, 1, t)$ and $(N, a_0, a_1, a_2) = (2, 1, t^2+2t, 2t^2)$, Theorem 2 implies the following corollary.

Corollary 2. *The average total path length $\underline{p}(tr+1)$ and the average free search cost measure $\underline{\text{FCOST}}(tr+1)$ of a t-ary digital search tree $T \in \mathscr{F}_t(tr+1)$ is given by*

(i) $\quad \underline{p}(tr+1) = tr + t \sum_{\mu \geq 0} \binom{r}{\mu+2} \prod_{1 \leq r \leq \mu} \left(\frac{1}{t^r} - 1 \right)$

(ii) $\quad \underline{\text{FCOST}}(tr+1) = \frac{rt^2(tr-2r+1)}{t-1} + \frac{t\{(t-1)(tr-t+1)+2\}}{t-1}$

$$\times \sum_{\mu \geq 0} \binom{r}{\mu+2} \prod_{1 \leq i \leq \mu} \left(\frac{1}{t^i} - 1 \right). \quad \Box$$

Using (12a) and part (i) of this corollary, the corresponding result concerning the average internal path length $\underline{p}_I(tr+1)$ of a t-ary digital search tree $T \in \mathscr{F}_t(tr+1)$ has been derived in ([9; pp. 497, 686]) by an alternative method. The average number of bit inspections in a random successful search is just equal to $\underline{p}_I(tr+1)$, divided by r. The asymptotic behaviour of $\underline{p}(tr+1)$ and $\underline{\text{FCOST}}(tr+1)$ can easily be derived by the methods developed in ([9; pp. 499, 684, 686]). There is proved that

$$\sum_{k \geq 0} \binom{N}{k+2} \prod_{1 \leq i \leq k} \left(\frac{1}{t^i} - 1 \right)$$

$$= (N+1)\log_t(N) + N \left[\frac{\gamma-1}{\ln(t)} + \frac{1}{2} - \alpha_t' + f_{-1}(t, N) \right] + O(1),$$

where $\gamma = 0.577215\ldots$ is Euler's constant, α_t is the constant

$$\alpha_t = \sum_{j \geq 0} (-1)^j \frac{t^{j+1}}{t^{j+1}-1} \prod_{1 \leq i \leq j+1} \frac{1}{t^i - 1}$$

and $f_s(t, N)$ is the bounded oscillating function with $f_s(t, N) = f_s(t, tN)$ defined by

$$f_s(t, N) = \frac{2}{\ln(t)} \sum_{k \geq 1} \operatorname{Re}\left(\Gamma\left(s - \frac{2\pi i k}{\ln(t)}\right)\right) \exp(2\pi i k \log_t(N)). \tag{20}$$

Here, Γ denotes the complete gamma-function. Using this approximation, we immediately obtain

$$\underline{p}(tr+1) = t(r+1)\log_t(r) + r\left[\frac{\gamma-1}{\ln(t)} + \frac{1}{2} + t - \alpha_t + f_{-1}(t, r)\right] + O(1)$$

and

$$\underline{\operatorname{FCOST}}(tr+1) = \frac{t(r+1)\{(t-1)(tr-t+1)+2\}}{t-1}\log_t(r)$$

$$+ r^2 t^2 \left[\frac{3t-5}{2(t-1)} + \frac{\gamma-1}{\ln(t)} - \alpha_t + f_{-1}(t, r)\right] + O(r).$$

Note that $\underline{\operatorname{FCOST}}(tr+1) \sim tr\underline{p}(tr+1)$.

3.2. Regularly distributed t-ary trees

Consider the class $\mathscr{F}_t$ of all t-ary trees with the characteristics $Q^{(t)}_{j,t}(z) = T(z) - z$, $1 \leq j \leq t$, where $T(z)$ is the enumerator of $\mathscr{F}_t$ given by

$$T(z) = \sum_{r \geq 0} |\mathscr{F}_t(tr+1)| z^{tr+1}.$$

Thus, formula (3b) and Corollary 1 yield $q^{(t)}_j(tm_j+1, t) = \langle z^{tm_j+1} ; Q^{(t)}_{j,t}(z)\rangle = |\mathscr{F}_t(tm_j+1)|$, $1 \leq j \leq t$, $r \in \mathbb{N}$, and $Q_t(t) = zT^t(z)$. Since $\mathscr{F}_t$ is a simply generated family of trees ([10]), the enumerator $T(z)$ satisfies the equation $T(z) = z + zT^t(z)$. Therefore, $Q_t(z) = T(z) - z$ which implies $q(tr+1, t) = |\mathscr{F}_t(tr+1)|$, $r \in \mathbb{N}$. Inserting phese expressions for $q^{(t)}_j(tm_j+1, t)$ and $q(tr+1, t)$ into (1), we find the probability $t_t(tr+1, t\vec{m}+1, \vec{d})$ that a tree $T \in \mathscr{F}_t(tr+1)$, $r \in \mathbb{N}$, is of type $\langle (tr+1, t); (tm_1+1, d_1), \ldots, (tm_t+1, d_t)\rangle$, $\vec{d} \in \{0, t\}^t$. We obtain

$$p_t(tr+1, t\vec{m}+1, \vec{d}) = \frac{1}{|\mathscr{F}_t(tr+1)|} \prod_{1 \leq j \leq t} |\mathscr{F}_t(tm_j+1)|.$$

Using this relation together with (5), a simple induction shows that the probability $p_t(T)$ of a tree $T \in \mathscr{F}_t(tr+1)$, $r \in \mathbb{N}$, is given by $p_t(T) = |\mathscr{F}_t(tr+1)|^{-1}$. Thus, the class $\mathscr{F}_t$ of all t-ary trees together with the characteristics $Q_{j,t}^{(t)}(z)=T(z)-z$, $1\leqslant j\leqslant t$, is equal to the class of all t-ary trees, where all trees $T\in\mathscr{F}_t(tr+1)$, $r\in\mathbb{N}$, are equally likely.

Theorem 3. *Let $\mathscr{F}_t$ be the set of all t-ary trees and let f be the weight function defined in (10a). Assuming that all trees $T\in\mathscr{F}_t$ with $(tr+1)$, $r\in\mathbb{N}$, nodes are equally likely, the average weight $E[w_f^{(t)}(tr+1)]$ is given by*

$$E[w_f^{(t)}(tr+1)]$$
$$=(tr+1)\binom{tr+1}{r}^{-1}\sum_{0\leqslant l\leqslant N}a_l\sum_{1\leqslant\mu\leqslant r}\frac{1}{(r-\mu)t+1}\binom{(r-\mu)t+1}{r-\mu}\binom{\mu t}{\mu}\binom{r-\mu}{l}.$$

□

In this case, Corollary 1 leads to an algebraic equation. Choosing $(N, a_0, a_1) = (1, 1, t)$ and $(N, a_0, a_1, a_2) = (2, 1, t^2+2t, 2t^2)$, Theorem 3 yields the following result.

Corollary 3. *Assume that all t-ary trees with $(tr+1)$, $r\in\mathbb{N}$, nodes are equally likely. The average total path length $\underline{p}(tr+1)$ and the average free search cost measure $\underline{\mathrm{FCOST}}(tr+1)$ are given by*

(i) $\quad \underline{p}(tr+1)=(tr+1)\sum_{\mu\geqslant 1}\binom{(r-\mu)t+1}{r-\mu}\binom{\mu t}{\mu}\Big/\binom{tr+1}{r},$

(ii) $\quad \underline{\mathrm{FCOST}}(tr+1)=t(tr+1)\sum_{\mu\geqslant 1}\mu\binom{(r-\mu)t+1}{r-\mu}\binom{\mu t}{\mu}\Big/\binom{tr+1}{r}.$ □

These closed form expressions are well-known and have been derived in ([8; p. 590], [3]) by a different method. Using the results given in ([4], [5]), the asymptotic behaviour of $\underline{p}(tr+1)$ and $\underline{\mathrm{FCOST}}(tr+1)$ is given by

$$\underline{p}(tr+1)=\sqrt{\frac{\pi}{2(t-1)}}(tr+1)^{3/2}+o(r^{3/2})$$

and

$$\underline{\mathrm{FCOST}}(tr+1)=\frac{1}{2}\sqrt{\frac{\pi}{2(t-1)}}(tr+1)^{5/2}+o(r^{3/2}).$$

Note that $\underline{\mathrm{FCOST}}(tr+1)\sim\frac{1}{2}(tr+1)\underline{p}(tr+1)$.

3.3. Binary search trees

Consider the class $\mathscr{F}_2$ of all 2-ary (extended binary) trees with the characteristics $Q_{j,2}^{(2)}(z)=z^3/(1-z^2)$, $j \in \{1,2\}$. An inspection of (3b) and Corollary 1 shows that $q_j^{(2)}(2m_j+1,2)=\langle z^{2m_j+1}; Q_{j,2}^{(2)}(z)\rangle=1$, $j \in \{1,2\}$, and $Q_2(z)=z^3/(1-z^2)^2$. Therefore, $q(2r+1,2)=r$, $r \in \mathbb{N}$. Inserting these expressions for $q_j^{(2)}(2m_j+1,2)$ and $q(2r+1,2)$ into (1), we obtain the probability $p_2(2r+1,(2m_1+1,2m_2+1),\vec{d})$ that a tree $T \in \mathscr{F}_2(2r+1)$, $r \in \mathbb{N}$, is of type $\langle (2r+1,2); (2m_1+1,d_1), (2m_2+1,d_2)\rangle$, $\vec{d} \in \{0,2\}^2$. We find

$$p_2(2r+1,(2m_1+1,2m_2+1),\vec{d})=r^{-1}.$$

Using this relation, formula (5) shows that the probability $p_2(T)$ of a tree $T \in \mathscr{F}_2(2r+1)$ is given by

$$p_2(T)= \prod_{T' \in \mathrm{SUB}(T)} |T'|^{-1},$$

where $|T|$ denotes the number of internal nodes of a tree $T \in \mathscr{F}_2$. Now, it is well-known ([9; p. 671]) that this is the probability of a binary search tree if each of the possible orderings of the keys is an equally likely sequence of insertions for building the tree. Thus, the class $\mathscr{F}_2$ of all 2-ary trees together with the characteristics $Q_{j,2}^{(2)}(z)=z^3/(1-z^2)$, $j \in \{1,2\}$, is equal to the class of all binary search trees.

Computing the average weight $E[w_f^{(2)}(2r+1)]$ for these trees, we obtain the following result.

Theorem 4. *Let $\mathscr{F}_2$ be the set of all binary search trees and let f be the weight function defined in (10a). The average weight $E[w_f^{(2)}(2r+1)]$ is given by*

$$E[w_f^{(2)}(2r+1)]=2a_0 r+2 \sum_{1 \leq l \leq N} a_l \left[(-1)^{l+1}l(r+1)\{H_{r+1}-1\}-\binom{r}{l} \right.$$
$$\left. +l \sum_{1 \leq i < l} \frac{(-1)^{i+1}}{l-i}\binom{r+1}{l-i+1} \right],$$

where H_r is the r-th harmonic number. □

Here, Corollary 1 leads to an Euler differential equation. Choosing $(N, a_0, a_1) = (1, 1, 2)$ and $(N, a_0, a_1, a_2)=(2, 1, 8, 8)$, Theorem 4 implies the following expressions for $p(2r+1)$ and $\mathrm{FCOST}(2r+1)$.

Corollary 4. *The average total path length* $\underline{p}(2r+1)$ *and the average free search cost measure* $\underline{\text{FCOST}}(2r+1)$ *of a binary search tree* $T \in \mathscr{F}_2(2r+1)$ *is given by*

(i) $\quad \underline{p}(2r+1) = 4(r+1)\{H_{r+1} - 1\} - 2r,$

(ii) $\quad \underline{\text{FCOST}}(2r+1) = 4(r+1)(2r+5)\{H_{r+1} - 1\} - 12r(r+1).\quad\square$

The results appearing in this corollary are proved in ([9; p. 427]) and ([3]). Assuming that each of the r keys are equally likely, the average number of comparisons in a successful (unsuccessful) search is just equal to $\underline{p}(2r+1)/(2r)$ (($\frac{1}{2}\underline{p}(2r+1) + r)/(r+1)$). Since $H_r = \ln(r) + \gamma + \dfrac{1}{2r} - \dfrac{1}{12r^2} + O\!\left(\dfrac{1}{r^4}\right)$ (see [8]), the asymptotic behaviour of $\underline{p}(2r+1)$ and $\underline{\text{FCOST}}(2r+1)$ is described by

$$\underline{p}(2r+1) = 4(r+1)\ln(r) + r(4\gamma - 1) + 4\gamma + 2 + O\!\left(\frac{1}{r}\right),$$

and

$$\underline{\text{FCOST}}(2r+1) = 4(r+1)(2r+5)\ln(r) + 4r^2(2\gamma - 5) + 28r(\gamma - 1)$$
$$+ \frac{20}{3}(3\gamma + 2) + O\!\left(\frac{1}{r}\right),$$

where $\gamma = 0.577215\ldots$ is Euler's constant. Note that $\underline{\text{FCOST}}(2r+1) \sim 2r\underline{p}(2r+1)$.

3.4. Patricia trees

Consider the class $\mathscr{F}_2$ of all 2-ary trees with the characteristics $Q_{j,2}^{(2)}(z) = z^{-1}(\exp(z^2) - 1) - z$, $j \in \{1, 2\}$. Formula (3b) and Corollary 1 yield $q_j^{(2)}(2m_j + 1, 2) = \langle z^{2m_j+1}; Q_{j,2}^{(2)}(z)\rangle = (m_j + 1)!^{-1}$, $j \in \{1, 2\}$, and $Q_2(z) = z^{-1}(\exp(z^2) - 1)^2$. Hence, $q(2r+1, 2) = \langle z^{2r+1}; Q_2(z)\rangle = (2^{r+1} - 2)/(r+1)!$, $r \in \mathbb{N}$. Inserting these expressions for $q_j^{(2)}(2m_j+1, 2)$ and $q(2r+1, 2)$ into (1), we find the probability $p_2(2r+1, (2m_1+1, 2m_2+1), \vec{d})$ that a tree $T \in \mathscr{F}_2(2r+1)$, $r \in \mathbb{N}$, is of type $\langle(2r+1, 2); (2m_1+1, d_1), (2m_2+1, d_2)\rangle$, $\vec{d} \in \{0, 2\}^2$. We obtain

$$p_2(2r+1, (2m_1+1, 2m_2+1), \vec{d}) = \frac{1}{2} \frac{(r+1)!}{(m_1+1)!(m_2+1)!(2^r - 1)}.$$

Now, using (5), a simple induction shows that the probability $p_2(T)$ of a tree

$T \in \mathcal{F}_2(2r+1)$ is given by

$$p_2(T) = (|T|+1)! \, 2^{-|T|} \prod_{T' \in \text{SUB}(T)} (2^{|T'|} - 1)^{-1},$$

where $|T|$ denotes the number of internal nodes of a tree $T \in \mathcal{F}_2$. This is the probability of a Patricia tree ([9; p. 686]). Thus, the class of all 2-ary trees together with the characteristics $Q_{j,2}^{(2)}(z) = z^{-1}(\exp(z^2) - 1) - z$, $j \in \{1, 2\}$, is equal to the class of Patricia trees.

Theorem 5. *Let $\mathcal{F}_2$ be the set of all Patricia trees and let f be the weight function defined in (10a). The average weight $E[w_f^{(2)}(2r+1)]$ is given by*

$$E[w_f^{(2)}(2r+1)] = \sum_{0 \leq l \leq N} a_l \sum_{j \geq 1} (-1)^{j+1} \binom{r+1}{l+j+1} \frac{1}{2^{l+j}-1}$$

$$\times \left[1 - 2^{l+j+1} + \sum_{0 \leq i \leq l} 2^{l-i} \binom{i+j}{j} \right]. \quad \square$$

In this case, Corollary 1 leads to a functional equation. Choosing $(N, a_0, a_1) = (1, 1, 2)$ and $(N, a_0, a_1, a_2) = (2, 1, 8, 8)$, Theorem 5 implies the following expressions for $\underline{p}(2r+1)$ and $\underline{\text{FCOST}(2r+1)}$.

Corollary 5. *The average total path length $\underline{p}(2r+1)$ and the average free search cost measure $\underline{\text{FCOST}(2r+1)}$ of a Patricia tree $T \in \mathcal{F}_2(2r+1)$ is given by*

(i) $\quad \underline{p}(2r+1) = -2r + 2 \sum_{j \geq 0} (-1)^j \binom{r+1}{j+2} \frac{j+2}{2^{j+1}-1},$

(ii) $\quad \underline{\text{FCOST}}(2r+1) = -8r(r+1) + 2 \sum_{j \geq 0} (-1)^j \binom{r+1}{j+2} \frac{(j+2)(2r+2j+3)}{2^{j+1}-1}. \quad \square$

Using (12b) and part (i) of this corollary, the corresponding result concerning the average external path length $p_L(2r+1)$ of a Patricia tree $T \in \mathcal{F}_2(2r+1)$ appears in ([9; p. 498]). The average number of bit inspections made in a random successful search is just equal to $p_L(2r+1)$ divided by $(r+1)$.

In a similar way as in the case of t-ary digital search trees, the asymptotic behaviour of $\underline{p}(2r+1)$ and $\underline{\text{FCOST}}(2r+1)$ can easily be derived by the methods given in ([9; pp. 131–134]). There is proved that the series $U_r := \sum_{j \geq 0} \binom{r}{j+2} (-1)^j \times \frac{1}{2^{j+1}-1}$ has an asymptotic representation of the form

$$U_r = rld(r) + r\left[\frac{\gamma-1}{\ln(t)} - \frac{1}{2} + f_{-1}(2,r)\right]$$

$$+ 2 - \frac{1}{2\ln(2)} - \frac{1}{2}f_1(2,r) + O\left(\frac{1}{r}\right),$$

where $\gamma = 0.577215\ldots$ is Euler's constant and $f_s(t,r)$ is the function defined in (20). Since

$$\underline{p}(2r+1) = -2r + 2(r+1)\{U_{r+1} - U_r\}$$

and

$$\underline{\text{FCOST}}(2r+1) = (r+1)[-8r + 2(4r+1)\{U_{r+1} - U_r\} - 4r\{U_r - U_{r-1}\}],$$

we can apply the above approximation and obtain by an elementary computation

$$\underline{p}(2r+1) = 2(r+1)ld(r) + r\left[\frac{2\gamma}{\ln(2)} - 3 - 2f_0(2,r)\right] + O(1)$$

and

$$\underline{\text{FCOST}}(2r+1) = 2r(2r+3)ld(r) - r^2\left[10 - \frac{4\gamma}{\ln(2)} + 4f_0(2,r)\right] + O(r).$$

Note that $\underline{\text{FCOST}}(2r+1) \sim 2r\underline{p}(2r+1)$.

References

[1] Abramowitz, M., Stegun, I.A.: *Handbook of Mathematical Functions*, Dover Publications, New York, 1970.

[2] Comtet, L.: *Advanced Combinatorics*, D. Reidel Publishing Company, Dordrecht-Holland/Boston U.S.A., 1974.

[3] Flajolet, Ph., Ottmann, Th., Wood, D.: Search Trees and Bubble Memories, R.A.I.R.O. *19* (2), 137–164, (1985).

[4] Kemp, R.: *On a General Weight of Trees*, Lecture Notes in Computer Science *166* (M. Fontet, K. Mehlhorn, eds.), Springer-Verlag, Berlin/Heidelberg/New York/Tokyo, 109–120, 1984.

[5] Kemp, R.: Additive Weights of Trees, preprint, University of Frankfurt, 1984 (submitted for publication).

[6] Kemp, R.: *Free Cost Measures of Trees*, Lecture Notes in Computer Science *199* (L. Budach, ed.), Springer-Verlag, Berlin/Heidelberg/New York/Tokyo, 175–190, 1985.

[7] Kemp, R.: Additive Weights of Non-Regularly Distributed Trees, preprint, University of Frankfurt, 1985.

[8] Knuth, D. E.: *The Art of Computer Programming*, vol. 1, 2nd ed., Addison-Wesley, Reading, Mass., 1973.

[9] Knuth, D. E.: *The Art of Computer Programming*, vol. 3, Addison-Wesley, Reading, Mass., 1973.

[10] Meir, A., Moon, J. W.: On the Altitude of Nodes in Random Trees, *Can. Math. 30*, 997–1015, (1978).

A TREE ENUMERATION PROBLEM INVOLVING THE ASYMPTOTICS OF THE "DIAGONALS" OF A POWER SERIES

Peter KIRSCHENHOFER

Technical University of Vienna
A-1040 Vienna, Austria

1. Introduction

In this paper we will be concerned with the asymptotic behaviour of some enumerative double sequences $(a_{n,m})$ for $n, m \to \infty$ under certain side conditions. Although the special instance of our problem presented here arises from a question on random trees, our interest is mainly focussed on giving methodological insight into a technique that hopefully will be useful for the solution of other problems of this kind in asymptotics, too. For this reason let us start with some remarks of a more technical nature:

A classical approach establishing the asymptotic behaviour of an ordinary sequence (a_n) for $n \to \infty$ can be described as follows (compare the diagram):

$$(a_n)_{n \geq 0} \longrightarrow A(z) = \sum_{n \geq 0} a_n z^n$$
$$\downarrow$$

Asymptotic behaviour of $\qquad$ Local expansion about
(a_n) for $n \to \infty$ $\longleftarrow$ singularities

Starting from $(a_n)_{n \geq 0}$ we look for a suitable generating function (G.F.), e.g. the ordinary generating function $A(z) = \sum_{n \geq 0} a_n z^n$ and translate the information available on (a_n) into an explicit expression or implicit information (functional equation,...) on $A(z)$. If we are lucky enough to localize the singularities of $A(z)$ in the complex plane and to establish the local behaviour of $A(z)$ at least about the singularities nearest to the origin, we can apply certain "Translation Lemmas" to gain the asymptotics of (a_n) itself. As an example we cite the following result due to FLAJOLET and ODLYZKO ([4]):

1.1. Translation lemma: *Let $A(z)$ be analytic in the domain $z \neq 1$, $|z| < \delta$, $|\mathrm{Arg}(1-z)| < \theta$ with $\delta > 1$, $\pi/2 < \theta < \pi$. Assuming that inside the intersection of a neighbourhood of 1 with this domain $A(z)$ allows an expansion*

$$A(z) = C \log(1-z) + K + \sum_{i=1}^{l} b_i (1-z)^{v_i} + O(|1-z|^v),$$

where C, K, b_i, v_i, v are constants with $0 < v_1 < v_2 < \ldots < v_l < v < 1$, the Taylor coefficients a_n of $A(z)$ have the following asymptotic behaviour for $n \to \infty$:

$$a_n = [z^n] A(z)$$
$$= -Cn^{-1} + \sum_{i=1}^{l} b_i \Gamma(-v_i)^{-1} n^{-v_i-1} + O(n^{-v-1}).$$

Since in a large amount of enumerative problems we are concerned with sequences depending on more than one parameter, it is quite natural to ask for multivariate analogues of the above described procedure. As far as known to the author this problem is widely unsolved even for special instances (compare e.g. the remarks in BENDER's survey [1]). In the present article we want to report on some progress in a technique that can be used in the case of searching for the asymptotics of $(a_{n,m})$ where $n = rk$ and $m = pk$ tend to infinity at a fixed ratio $p/r = \rho$. If the indices of $(a_{n,\rho n})_{\rho n \in \mathbb{N}}$ are marked in two-dimensional lattice, they lie on a "diagonal" line of the first quadrant: For this reason the name "diagonal" is sometimes used for the subsequence $(a_{n,\rho n})_{\rho n \in \mathbb{N}}$ of $(a_{n,m})_{n,m \geq 0}$.

Returning to our initial comments, the analytic problem we are concerned with may also be described in the following way:

Find the analytic behaviour of the G.F.

$$B_\rho(x) = \sum_{k \geq 0} a_{rk, pk} x^{rk} \quad (\rho = p/r)$$

of a diagonal of the double sequence $(a_{n,m})$ by making use of known analytic properties of the double G.F.

$$A(z, u) = \sum_{n, m \geq 0} a_{n, m} z^n u^m.$$

The following Proposition, that has been proved by the author in [8], is one of the key tools (the special case $\rho = p/r = 1$ is due to HAUTUS and KLARNER [5]):

1.2. Proposition: *Suppose* $A(z,u) = \sum_{n,m \geq 0} a_{n,m} z^n u^m$ *converges for* $|z| < R_1$, $|u| < R_2$ *in* C^2. *Then* $B_\rho(x) = \sum_{k \geq 0} a_{rk, pk} x^{rk}$ *(where* $\rho = p/r$, $p, r \in \mathbb{N}$, $\gcd(p,r) = 1$*) allows for* $|x| < R_1 R_2^{p/r}$ *the representation*

$$B_\rho(x) = \frac{1}{2\pi i} \int_{C=C(x)} A(x/s^p, s^r) \, ds/s,$$

where $C = C(x)$ *is a positive oriented simple closed curve in the annulus* $(|x|/R_1)^{1/p} < |s| < (R_2)^{1/r}$ *or any homotopic curve surrounding those singularities in s that tend to 0 for x tending to 0.*

From the technical point of view we have to solve two main problems when applying this Proposition:

a) to *locate the singularities of the integrand* in order to choose an appropriate contour of integration C, and

b) to *find local expansions of the integrand* regarded as a function in s for fixed x about its singularities, where these expansions must be uniform in x in a domain as needed for application of a "Translation Lemma" afterwards.

If we have explicit knowledge of $A(z,u)$ or at least relatively simple implicit function relations that allow well-suited local expansions, point b) turns out to be manageable, although by some effort of calculations. Point a) appears to be of a more delicate nature: in the papers cited above, the location of singularities of $A(x/s^p, s^r)$ was in that sense easy, that $A(z, u)$ was in fact a simple expression in $y(zu^j)$, where $y(z)$ is a function with well known singularities. This is no longer the case in the instances we will treat below: Nevertheless it will turn out that by a (quite natural) restriction of the range of the ratio ρ the singularities in question are separated in a way that allows to choose an appropriate contour of integration C for the application of our Proposition.

In the following section we describe the concrete problem on random trees we will be concerned with, cite the main results and, for comparison, the results derived in foregoing papers.

2. The average stack size for traversing planted plane trees

In the following we consider certain families of planted plane trees, namely the family $\mathscr{B}$ of (extended) *binary trees*, the family $\mathscr{T}$ of (extended) *t-ary trees*, the family $\mathscr{M}$ of (extended) *Motzkin trees* and the family $\mathscr{P}$ of *general planted plane trees*. In the first three instances we distinguish between internal nodes (○) and leaves (□), whereas in the last case all nodes are regarded equal.

Using a by now standard notation (compare e.g. [3]) the families in consideration may be described by the symbolic equations

$$\mathcal{B} = \square + \underset{\mathcal{B} \quad \mathcal{B}}{\wedge}, \tag{2.1}$$

$$\mathcal{T} = \square + \underset{\underset{t \text{ times}}{T \; T \; \ldots \; T}}{\wedge}, \tag{2.2}$$

$$\mathcal{M} = \square + \underset{\mathcal{M}}{\vert} + \underset{\mathcal{M} \quad \mathcal{M}}{\wedge} \quad \text{and} \tag{2.3}$$

$$\mathcal{P} = \bigcirc + \underset{\mathcal{P}}{\vert} + \underset{\mathcal{P} \quad \mathcal{P}}{\wedge} + \underset{\mathcal{P} \; \mathcal{P} \; \mathcal{P}}{\wedge} + \ldots, \tag{2.4}$$

which can be translated immediately into functional equations for the G.F.s $y(z)$ of trees of specified size (i.e. number of nodes ○) in a given family:

$$y_{\mathcal{B}}(z) = 1 + z y_{\mathcal{B}}^2(z), \tag{2.5}$$

$$y_{\mathcal{T}}(z) = 1 + z y_{\mathcal{T}}^t(z), \tag{2.6}$$

$$y_{\mathcal{M}}(z) = 1 + z y_{\mathcal{M}}(z) + z y_{\mathcal{M}}^2(z) \quad \text{and} \tag{2.7}$$

$$y_{\mathcal{P}}(z) = z + z y_{\mathcal{P}}(z) + z y_{\mathcal{P}}^2(z) + z y_{\mathcal{P}}^3(z) + \ldots = \frac{z}{1 - y_{\mathcal{P}}(z)}. \tag{2.8}$$

Solving the quadratic equations and expanding, 2.5 and 2.8 lead to the well-known Catalan numbers ($[z^n]y(z)$ denotes the coefficient of z^n in $y(z)$):

$$[z^n] y_{\mathcal{B}}(z) = \frac{1}{n+1} \binom{2n}{n}, \tag{2.9}$$

$$[z^n] y_{\mathcal{P}}(z) = \frac{1}{n} \binom{2n-2}{n-1}. \tag{2.10}$$

Equation 2.6 may be treated by Lagrange's inversion formula to derive

$$[z^n]y_{\mathcal{F}}(z)=\frac{1}{(t-1)n+1}\binom{tn}{n}. \qquad (2.11)$$

The coefficients of

$$y_{\mathcal{M}}(z)=\frac{1-z-\sqrt{(1-z)^2-4z}}{2z} \qquad (2.12)$$

are more complicated, but a standard Darboux-type argument (compare e.g. [1]) leads to the asymptotic form

$$[z^n]y_{\mathcal{M}}(z)=\frac{1}{2}\sqrt{\frac{4+3\sqrt{2}}{\pi}}(3+2\sqrt{2})^n n^{-3/2}(1+O(1/n)), \quad n\to\infty. \qquad (2.13)$$

Families of planted plane trees occur as a special kind of *data structures in Computer Science*. One of the simplest recursive algorithms is the so-called *preorder traversal* of the planar tree, where to traverse in preorder means to "visit" (i.e. to do something with) the root of the tree followed by traversing the subtrees of the root from left to right. As is easily seen the running time of the traversal procedure is linear in the size of the input tree. Of more interest is the analysis of the *stack size* (or equivalently the *recursion depth*) during the performance of the algorithm. At each stage the stack size is equivalent to the height of the concerned node of the tree, i.e. its distance from the root. To determine the average stack size for all trees of size n (regarded equally likely) thus means to determine their average height. The corresponding results are due to

DE BRUIJN, KNUTH and RICE ([2])

$$\bar{h}_{\mathcal{P}}(n)\sim\sqrt{\pi n} \qquad (2.14)$$

and to FLAJOLET and ODLYZKO ([4])

$$\bar{h}_{\mathcal{B}}(n)\sim 2\sqrt{\pi n}, \qquad (2.15)$$

$$\bar{h}_{\mathcal{F}}(n)\sim\sqrt{2\pi\frac{t}{t-1}n} \quad \text{and} \qquad (2.16)$$

$$\bar{h}_{\mathcal{M}}(n)\sim\sqrt{\frac{(3+2\sqrt{2})\pi}{\sqrt{2}}n}. \qquad (2.17)$$

In order to get more detailed information about the development of the stack during the performance of the algorithm, it is convenient to consider the leaves of a given tree enumerated from left to right and to ask for the *average height of the j-th leaf of a tree of size n* (which means to determine the average value of the relative maxima of the stack size during preorder traversal). Results in that direction are due to KEMP [6], MOON [11], RUSKEY [12] and the author:

If, for convenience, $\alpha(n,j)$ denotes the average height of the $j+1$-st leaf of a tree of size n, the author has proved in [9] the explicit formula

$$\alpha_{\mathcal{B}}(n,j)=\frac{2(2j+1)(2(n-j)+1)}{n+2}\frac{\binom{2j}{j}\binom{2(n-j)}{n-j}}{\binom{2n}{n}}+1, \qquad (2.18)$$

so that

$$\alpha_{\mathcal{B}}(n,\rho n)\sim\frac{8}{\sqrt{\pi}}\sqrt{\rho(1-\rho)}\,n^{\frac{1}{2}}, \qquad (2.19)$$

and in [10]

$$\alpha_{\mathcal{T}}(n,\rho(t-1)n)\sim\frac{8}{\sqrt{2\pi}}\sqrt{\frac{t}{t-1}}\sqrt{\rho(1-\rho)}\,n^{\frac{1}{2}}. \qquad (2.20)$$

The main result of this paper is:

2.21. Theorem. *The average height of the ρn-th leaf of a Motzkin tree with n internal nodes fulfills for $0<\rho<1/\sqrt{2}$*

$$\alpha_{\mathcal{M}}(n,\rho n)\sim\frac{4\sqrt[4]{2}(1+\sqrt{2})}{\sqrt{\pi}}\sqrt{\rho(1/\sqrt{2}-\rho)}\,n^{\frac{1}{2}},\quad n\to\infty.$$

The average height of the ρn-th leaf of a general planted plane tree with n nodes in total fulfills for $0<\rho<1/2$

$$\alpha_{\mathcal{P}}(n,\rho n)\sim\frac{8}{\sqrt{\pi}}\sqrt{\rho(\tfrac{1}{2}-\rho)}\,n^{\frac{1}{2}},\quad n\to\infty.$$

The analysis for Motzkin trees and general planted plane trees turns out to be much more complicated than for the previous cases in the sense mentioned in section 1: If

$$H(z,u)=\sum_{n\geq0}z^n\sum_{j\geq0}u^j\sum_{\substack{\text{All trees of}\\ \text{size }n}}\text{height of }j\text{-th leaf}(+1)\text{ of tree} \qquad (2.22)$$

denotes the double G.F. involved in our problem (the $+1$ is usually added in the case $\mathscr{P}$), it has been shown by the author in [7] and [8] that

$$H_{\mathscr{B}}(z,u) = \left[\frac{y_{\mathscr{B}}(z) - u y_{\mathscr{B}}(zu)}{1-u}\right]^2 - \frac{y_{\mathscr{B}}(z) - u y_{\mathscr{B}}(zu)}{1-u}, \qquad (2.23)$$

$$H_{\mathscr{T}}(z,u) = \left[\frac{y_{\mathscr{T}}(z) - u y_{\mathscr{T}}(zu^{i-1})}{1-u}\right]^2 - \frac{y_{\mathscr{T}}(z) - u y_{\mathscr{T}}(zu^{i-1})}{1-u}, \qquad (2.24)$$

$$H_{\mathscr{M}}(z,u) = \left[\frac{y_{\mathscr{M}}(z) - y_{\mathscr{M}}(z,u)}{1-u}\right]^2 - \frac{y_{\mathscr{M}}(z) - y_{\mathscr{M}}(z,u)}{1-u}, \qquad (2.25)$$

where

$$y_{\mathscr{M}}(z,u) = \frac{1 - z - \sqrt{(1-z)^2 - 4zu}}{2z}$$

is the G.F. of Motzkin trees with internal nodes labelled by z and leaves labelled by u, and

$$H_{\mathscr{P}}(z,u) = \frac{u}{z}\left[\frac{y_{\mathscr{P}}(z) - y_{\mathscr{P}}(z,u)}{1-u}\right]^2 \qquad (2.26)$$

where

$$y_{\mathscr{P}}(z,u) = \frac{1 - z(1-u) - \sqrt{(1+z(1-u))^2 - 4z}}{2}$$

is the G.F. of general planted plane trees with all nodes labelled by z and leaves labelled by u.

For this reason the location of singularities of $H(x/s^p, s^r)$ is simply derived for $\mathscr{B}$ and $\mathscr{T}$, but this is no longer the case for $\mathscr{M}$ and $\mathscr{P}$. In the following sections we want to demonstrate the solution of the problem for the family $\mathscr{M}$ in some detail.

3. The contour integral representation of the "diagonal" generating function

In each of the right-hand sides of 2.23 to 2.26 the coefficient $[z^n u^{pn}]$ of the linear term is simple and $O(q^{-n} n^{-3/2})$, where q is the main singularity of $y(z)$, and thus negligible compared to the main term which is of order $q^{-n} n^{-1}$. Hence it suffices to study the quadratic term, which in the following will be denoted by $A(z, u)$. (We omit the index $\mathcal{M}$ in $A(z, u)$.)

Let

$$h_\rho(x) = \sum_{k \geq 0} x^{rk} [z^{rk} u^{pk}] A(z, u), \text{ where } \rho = p/r, \ p, r \in \mathbf{N}, \ \gcd(p, r) = 1, \quad (3.1)$$

be the G.F. of the "diagonal" in question. Applying Proposition 1.2 we have a complex contour integral representation

$$h_\rho(x) = \frac{1}{2\pi i} \int_{C = C(x)} A(x/s^p, s^r) \, ds/s, \quad (3.2)$$

as long as it is possible to choose a contour $C = C(x)$ that separates in s-plane the singularities of the integrand tending to 0 for x tending to 0 from the other singularities.

For $|z| < q_\mathcal{M} = 3 - 2\sqrt{2}$, $|u| < 1$, $A(z, u)$ converges absolutely, so that by Proposition 1.2 equation 3.2 is valid with $C = C(x)$ a circle in the annulus $|x|/q_\mathcal{M} < |s|^p < 1$.

Let now $|x| \geq q_\mathcal{M}$. As is easily seen, the singularities caused by $u = 1$ are removable, so that it remains to treat the singularities originating from $\sqrt{(1-z)^2 - 4z}$ and $\sqrt{(1-z)^2 - 4zu}$. Both square-roots give rise to some cuts in s-plane (fixed x), which occur for, say,

$$\frac{4z}{(1-z)^2} = \lambda \geq 1 \quad \text{and} \quad (3.3)$$

$$\frac{4zu}{(1-z)^2} = \mu \geq 1, \quad \text{where} \quad (3.4)$$

$$z = x/s^p \quad \text{and} \quad u = s^r. \quad (3.5)$$

Let us assume that one of the cuts 3.3 has a point in common with one of the cuts 3.4. Since x is fixed, that means common values of z and u fulfilling 3.3 and 3.4 in this point. Thus we have

$$u = \mu/\lambda \in \mathbf{R}^+.$$

Furthermore 3.3 forces z to be

$$z = \frac{\lambda + 2 \pm \sqrt{(\lambda+2)^2 - \lambda^2}}{\lambda} \in \mathbf{R}^+, \quad (3.6)$$

so that

$$x^r = z^r u^p \in \mathbf{R}^+.$$

In other words: *as long as we exclude r-th root of unity arguments for x*, the cuts of 3.3 are separated from those of 3.4.

Let us now consider each kind of cuts for itself:

3.3 will not cause any troubles for arbitrary $|x|$: Considering 3.6 with $\lambda \geqslant 1$, $z = x/s^p$, it turns out that the corresponding values of s are situated on the lines connecting the origin with the p-th roots of $x/q_\mathcal{M}$, namely

$$s^p = \tau x, \quad \tau \in [3 - 2\sqrt{2}, 3 + 2\sqrt{2}] - \{1\}. \tag{3.7}$$

The more delicate instance is 3.4. Let us assume that two of the cuts originating from 3.4 have a common point in s-plane. Since for fixed x the equality of s-values forces the corresponding z's and u's to be equal, the s-values in question must also correspond to the same value of μ. Thus we have to look for multiple solutions in s of 3.4 for fixed x and μ. It follows by ordinary calculus that no such multiple solutions can occur as long as

$$|x| < f(\rho) = \rho^{2\rho}(1-\rho)^{1-\rho}(1+\rho)^{-1-\rho} \quad (\rho = p/r). \tag{3.8}$$

Since we are interested in defining $h_\rho(x)$ for some $|x| > q_\mathcal{M}$, i.e. to find an analytic continuation outside the circle of convergence of the corresponding series, it comes out that 3.2 does the job as long as $f(\rho) > q_\mathcal{M}$, which is the case for

$$0 < \rho < 1/\sqrt{2}.$$

Let us call a contour $C = C(x)$ as used in 3.2 admissible for x (i.e. it separates the singularities of the integrand in the described manner). Since all cuts depend continuously on s, a contour $C = C(x_0)$ admissible for x_0 is also admissible for x in some neighbourhood $U(x_0)$, in other words

$$h_\rho(x) = \frac{1}{2\pi i} \int_{C(x_0)} A(x/s^p, s^r) \, ds/s \quad \text{for } x \in U(x_0).$$

Thus the contour integral defines an analytic function in $U(x_0)$, and since the same argument holds for any $x \in \mathbf{C}$ for which an admissible contour exists, we have proved:

3.9. Proposition. *Let G_ρ be the region*

$$G_\rho = \{x \in \mathbf{C} : |x| < q_\mathcal{M} = 3 - 2\sqrt{2}\} \cup \{x \in \mathbf{C} : q_\mathcal{M} \leq |x| < f(\rho), \operatorname{Arg} x^r \neq 0\},$$

with $f(\rho)$ from 3.8 and let $C = C(x)$ denote an admissible contour as described above. Then

$$h_\rho(x) = \frac{1}{2\pi i} \int_{C=C(x)} A(x/s^\rho, s^r) \, ds/s$$

is analytic in G_ρ. Furthermore $f(\rho) > q_\mathcal{M}$ for $0 < \rho < 1/\sqrt{2}$.

In order to apply a Translation Lemma we have to establish the local behaviour of $h_\rho(x)$ near the points $x_j = q_\mathcal{M} \omega_j$, where ω_j ($0 \leq j \leq r-1$) are the r-th roots of unity. Before proceeding in that direction let us spend some words on the range $0 < \rho < 1/\sqrt{2}$ that appears in the last Proposition:

The G.F. of the sums of numbers of leaves of Motzkin trees of size n is

$$\frac{\partial y_\mathcal{M}}{\partial u}(z, 1) = \frac{1}{1 - z - 2zy_\mathcal{M}(z)}. \tag{3.10}$$

By a straightforward expansion about the singularity $q_\mathcal{M}$ and application of Darboux's method (compare e.g. [1]), we have

$$[z^n] \frac{\partial y_\mathcal{M}}{\partial u}(z, 1) \sim \frac{1}{2 q_\mathcal{M} b \sqrt{\pi}} q_\mathcal{M}^{-n} n^{-\frac{1}{2}} \quad \text{with} \quad b^2 = 4 + 3\sqrt{2},$$

so that the average number of leaves of a Motzkin tree of size n is asymptotic to

$$\frac{1}{\sqrt{2}} n, \quad n \to \infty. \tag{3.11}$$

A more detailed analysis shows that the number of Motzkin trees with n internal nodes and ρn leaves grows exponentially like $f(\rho)^{-n}$, $f(\rho)$ from 3.8. Since $f(\rho)$ has its minimum in $\rho = 1/\sqrt{2}$, the number of Motzkin trees of size n having a ρn-th leaf for a value $\rho > 1/\sqrt{2}$ is exponentially small compared with the total number of such trees. In other words: The boundary $1/\sqrt{2}$ appearing in the analytic approach from above is a quite natural one.

4. Local expansions and asymptotics

In this section it is our aim to find the local expansions of $h_\rho(x)$ near $x_j = q_\mathcal{M} \omega_j$, where ω_j are the r-th roots of unity, resp. an analytic equivalent. Recalling the proof of Proposition 3.9, the singularities x_j occur from the coincidence of singularities of type 3.3 with singularities of type 3.4 for $|x| \geqslant q_\mathcal{M}$, Arg $x^r = 0$. In fact it follows from the discussion there, that for $|x| < f(\rho)$ no other cuts will intersect nontrivially. Hence, if we fix in s-plane a point sufficiently close to $(x/q_\mathcal{M})^{1/p}$ on each of the cuts 3.7 and split up the contour $C(x)$ into the part $C_1(x)$ connecting these points on the different cuts and the part $C_2(x) = \bigcup_{0 \leqslant l \leqslant p-1} C_{2,l}(x)$, where $C_{2,l}(x)$ runs along the top of the l-th cut, then the integral along $C_1(x)$ is analytic in a (full) neighbourhood of x_j ($0 \leqslant j \leqslant r-1$). In other words: Only the integral along the small part $C_2(x)$ of $C(x)$ running along the top of the cuts 3.7 will contribute to the nonanalytic part of $h_\rho(x)$ in x_j.

For further approximation it is convenient to transform the integral $h_{\rho,l}(x)$ along $C_{2,l}(x)$ in such a way that the contour of integration runs along an interval of the positive real axis. Substituting

$$s = (x/q_\mathcal{M})^{1/p}_l t^{1/p}, \quad t^{1/p} \in \mathbb{R}^+,$$

(the index denotes the l-th value of the p-th root) we have

$$h_{\rho,l}(x) = \frac{1}{2\pi i} \int_\Gamma A(q_\mathcal{M}/t, w_l t^{1/p}) \, dt/(pt) \tag{4.1}$$

where $w_l = ((x/q_\mathcal{M})^{1/p}_l)^r$ and Γ is a contour running along the cut $[-\infty, 1]$ from some δ ($0 < \delta < 1$) to 1 and back on the other side of the cut from 1 to δ. (δ is chosen sufficiently close to 1 afterwards.)

With the abbreviation

$$I_\rho(w) = \frac{1}{2\pi i} \int_\Gamma A(q_\mathcal{M}/t, w t^{1/p}) \, dt/t, \tag{4.2}$$

we therefore have

$$h_\rho(x) = \sum_{k \geqslant 0} q_\mathcal{M}^{-rk} x^{rk} [w^{pk}] I_\rho(w) + g_\rho(x) \tag{4.3}$$

where $g_\rho(x)$ is analytic around the x_j's.

Thus we may restrict our attention to the local behaviour of $I_\rho(w)$ near its singularity $w=1$. Observing that analytic parts of the integrand cancel when being integrated along Γ, we have

$$I_\rho(w) = -\frac{1}{2\pi i} \int_\Gamma \left[\frac{(1-z)^2}{2z^2(1-u)^2} \sqrt{1-\frac{4z}{(1-z)^2}} \sqrt{1-\frac{4zu}{(1-z)^2}} \right]_{\substack{z=q_\mathcal{M}/t \\ u=wt^{1/\rho}}} dt/t. \tag{4.4}$$

Expanding for t near 1, i.e. z near $q_\mathcal{M}$, we have with $w_1 = (1-w)/w$:

$$\left[\frac{(1-z)^2}{2z^2} \right]_{z=q_\mathcal{M}/t} = 2(3+2\sqrt{2}) + (1-t)\psi_1(t),$$

$$\left[1 - \frac{4z}{(1-z)^2} \right]^{\frac{1}{2}}_{z=q_\mathcal{M}/t} = \pm i \sqrt[4]{2}(1-t)^{\frac{1}{2}}(1+(1-t)\psi_2(t)),$$

$$\left[1 - \frac{4zu}{(1-z)^2} \right]^{\frac{1}{2}}_{\substack{z=q_\mathcal{M}/t, \\ u=wt^{1/\rho}}} = (w_1 + (1/\rho - \sqrt{2})(1-t))^{\frac{1}{2}}$$

$$\times (1 + (1-wt^{1/\rho - \sqrt{2}})\psi_3(wt^{1/\rho - \sqrt{2}})),$$

$$[(1-u)^2]_{u=wt^{1/\rho}} = (w_1 + 1/\rho(1-t))^2 (1 + (1-wt^{1/\rho})\psi_4(wt^{1/\rho})),$$

where $\psi_{1,2,3,4}$ are analytic in a neighbourhood of 1.

For w in a sector $|1-w| < \mu$, $|\text{Arg}(1-w)| < \theta$ with $\mu > 0$ and $\pi/2 < \theta < \pi$ chosen appropriately, we therefore get by a straightforward calculation

$$I_\rho(w) = \frac{2(3+2\sqrt{2})\sqrt[4]{2}}{\pi} \int_\delta^1 \frac{(1-t)^{\frac{1}{2}}(w_1 + (1/\rho - \sqrt{2})(1-t))^{\frac{1}{2}}}{(w_1 + 1/\rho(1-t))^2} dt \tag{4.5}$$

$$+ K'_\rho + O(|1-w|\log|1-w|)$$

with a constant K'_ρ.

The integral on the right-hand side is analytic in $w_1 \in \mathbf{C} -]-\infty, 0]$ and can be determined explicitly for $w_1 \in \mathbf{R}^+$ to be

$$-(1/\rho - \sqrt{2})^{\frac{1}{2}} \rho^2 \log w_1 + \varphi(w_1), \quad \varphi \text{ analytic in } U(0). \tag{4.6}$$

Thus 7.6 represents the integral in its whole domain of analyticity and we derive

4.7. Proposition: *There exist $\mu > 0$ and θ with $\pi/2 < \theta < \pi$ such that for w in the sector $w \neq 1$, $|1-w| < \mu$, $|\text{Arg}(1-w)| < \theta$ the following expansion holds*:

$$I_\rho(w) = -\frac{2(3+2\sqrt{2})\sqrt[4]{2}}{\pi}(1/\rho - \sqrt{2})^{\frac{1}{2}}\rho^2 \log(1-w)$$

$$+ K_\rho + O(|1-w|\log|1-w|)$$

(K_ρ a constant).

Applying the Translation Lemma and rewriting the result in terms of coefficients of $h_\rho(x)$, compare 7.3, we have

$$[z^n u^{\rho n}] A(z,u) \sim \frac{2\sqrt{2}(3+2\sqrt{2})}{\pi} \rho^{\frac{1}{2}}(1\sqrt{2}-\rho)^{\frac{1}{2}}(3+2\sqrt{2})^n n^{-1} \quad (4.8)$$

and Theorem 2.21, part "M", follows.

We want to point out that by a more detailed, but straightforward, computation it is also possible to determine the second order $(O(1)-)$-term in the asymptotic expansion. Since the calculations do not give more methodological insight, we omit them here.

The instance of *general planted plane trees* can be treated along the same lines as the case from above. It turns out that for a tree in $\mathscr{P}$ with n nodes in total, the natural restriction for ρ when considering the ρn-th leaf is $0 < \rho < 1/2$.

References

[1] Bender, E. A.: Asymptotic methods in enumeration, SIAM Review **16** (1974), 485–515.

[2] deBruijn, N. G., D. E. Knuth and S. O. Rice: The average height of planted plane trees, in: *Graph Theory and Computing* (R. C. Read, ed.), 15–22, Academic Press, New York 1972.

[3] Flajolet, Ph.: Analyse d'algorithmes de manipulation d'arbres et de fichiers, Cahiers de BURO 34-35 (1981), 1–209.

[4] Flajolet, Ph. and A. Odlyzko: The average height of binary trees and other simple trees, J. Comput. System Sci. **25** (1982), 171–213.

[5] Hautus, M. L. J. and D. A. Klarner: The diagonal of a double power series, Duke Math. J. **38** (1971), 229–235.

[6] Kemp, R.: On the average oscillation of a stack, Combinatorica **2** (1982), 157–176.

[7] Kirschenhofer, P.: On the average shape of simply generated families of trees, J. Graph Theory **7** (1983), 311–323.

[8] Kirschenhofer, P.: On the height of leaves in binary trees, J. Comb. Inf. Syst. Sci. **8** (1983), 44–60.

[9] Kirschenhofer, P.: Some new results on the average height of binary trees, Ars Comb. **16A** (1983), 255–260.

[10] Kirschenhofer, P.: Asymptotische Untersuchungen zur durchschnittlichen Gestalt gewisser Graphenklassen, in: *Zahlentheoretische Analysis* (E. Hlawka, ed.), Springer Lecture Notes in Math. **1114**, 40–54, Springer, Berlin 1985.

[11] Moon, J. W.: On level numbers of t-ary trees, SIAM J. Alg. Discr. Meth. **4** (1983), 8–13.

[12] Ruskey, F.: On the average shape of binary trees, SIAM J. Alg. Discr. Meth. **1** (1980), 43–50.

ON MATCHINGS AND HAMILTONIAN CYCLES IN SUBGRAPHS OF RANDOM GRAPHS

Tomasz ŁUCZAK

Institute of Mathematics, Adam Mickiewicz University,
Poznań, Poland

For a natural number k, let $C_k(G)$ denote the maximal subgraph of G with minimal degree $\geq k$ and let M_k be the property that a graph has $\left\lfloor \dfrac{k}{2} \right\rfloor$ disjoint hamiltonian cycles plus a disjoint perfect matching if k is odd. We show that for the random graph $K_{n,M}$ with n labelled vertices and $M(n) = \dfrac{n \log n}{2(k+1)} + \dfrac{k}{2} n \log \log n + c_n n$ edges we have

$$\lim_{n \to \infty} P(C_k(K_{n,M}) \text{ has } M_k) = \begin{cases} 0 & \text{if } c_n \to -\infty \\ \exp\left(-\dfrac{e^{-2c(k+1)}}{(k!)^{k+1}(k+1)!}\right) & \text{if } c_n \to c \\ 1 & \text{if } c_n \to \infty. \end{cases}$$

Introduction

Let $K_{n,M}$ denote a graph chosen at random from the family of all graphs with n labelled vertices and M edges where each of $\left(\binom{n}{2} \atop M\right)$ possible graphs is equally likely and let us suppose that A is certain graph property. The most natural question concerning A is about the limit of the probability that $K_{n,M}$ has this property for a given function $M(n)$ when n tends to infinity. But one can state also a little more general problem: how to characterize the maximal subgraph of $K_{n,M}$ with the property A? There are many results of the first type for most of graph properties but considerably few concerning the second one (except the case when A is the property that a graph is connected). In this paper we shall study the situation when the property A is a generalization of the property that a graph is hamiltonian.

For a fixed natural number k let M_k be the property that a graph has $\lfloor k/2 \rfloor$ disjoint hamiltonian cycles plus a disjoint perfect matching if k is odd (to avoid complications with graphs which has odd number of vertices we shall say that such a graph has the property M_{2l+1} if it has l disjoint hamiltonian cycles and, moreover, there exists a vertex v of G not adjacent to vertices of degree $\leq 2(l+1)$ such that there is a disjoint matching in G containing all its vertices except v). The threshold function for M_k was found by Bollobás and Frieze [3].

Theorem 1. [3] *If $k \geq 1$ and*

$$M(n) = \frac{n \log n}{2} + \frac{(k-1)}{2} n \log \log n + c_n n,$$

then

$$\lim_{n \to \infty} P(K_{n,M} \text{ has } M_k) = \lim_{n \to \infty} P(\delta(K_{n,M}) \geq k)$$

$$= \begin{cases} 0 & \text{if } c_n \to -\infty, \ c_n > -\log \log n \\ \exp\left(-\frac{e^{-2c}}{(k-1)!}\right) & \text{if } c_n \to c \\ 1 & \text{if } c_n \to \infty \end{cases}$$

where $\delta(K_{n,M})$ denote the minimal degree of $K_{n,M}$. (We must assume that $c_n > -\log \log n$ to exclude cases when $K_{n,M}$ is very sparse. Here and below this condition can be replaced by weaker ones, for example, by the requirement that $M(n)/n \to \infty$.)

Let us observe that vertices of degree $\leq k-1$ cannot belong to any subgraph with the property M_k in $K_{n,M}$ so it is natural to ask when the subgraph $L_k(K_{n,M})$ induced by all vertices of $K_{n,M}$ of degree $\geq k$ has M_k. The answer is the following.

Theorem 2. *Let $k \geq 2$ and*

$$M(n) = \frac{n \log n}{4} + \frac{k}{2} n \log \log n + c_n n.$$

Then

$$\lim_{n\to\infty} P\big(L_k(K_{n,M}) \text{ has } M_k\big) = \lim_{n\to\infty} P\big(\delta(L_k(K_{n,M})) \geq k\big)$$

$$= \begin{cases} 0 & \text{if } c_n \to -\infty, c_n > -\log\log n \\ \exp\left(-\dfrac{e^{-4c}}{(k-1)!(k-2)!}\right) & \text{if } c_n \to c \\ 1 & \text{if } c_n \to \infty. \end{cases}$$

Since we are still limited by the crucial condition that minimum degree of a graph with the property M_k must be at least k, let us consider the subgraphs H of $K_{n,M}$ for which the inequality $\delta(H) \geq k$ holds.

Let $C_k(G)$ be the k-core of a graph G, i.e. the maximal subgraph of G with the minimal degree at least k (one can easily see that there is exactly one such maximal subgraph, so k-core is well defined). Furthermore, by B_k we denote the property that a graph contains no vertices which are adjacent to at least $k+1$ vertices of degree exactly k. (In fact, in the definition of M_k for odd k and for graphs with the odd number of vertices (see p. 172) we can replace the condition that deleted vertex v is not adjacent to any vertex of degree $\leq k+2$ by the assumption that $G-\{v\}$ has B_k if G has that property. The author chose the first formulation of this definition finding it simpler although it might look strange at the first sight.) We shall show that the probability that $C_k(K_{n,M})$ has the property M_k is asymptotically the same as the probability that $C_k(K_{n,M})$ has the property B_k.

Theorem 3. *If $k \geq 1$ and*

$$M(n) = \frac{n\log n}{2(k+1)} + \frac{k}{2} n\log\log n + c_n n$$

then

$$\lim_{n\to\infty} P\big(C_k(K_{n,M}) \text{ has } M_k\big)$$

$$= \lim_{n\to\infty} P\big(C_k(K_{n,M}) \text{ has } B_k\big)$$

$$= \lim_{n\to\infty} P\big(K_{n,M} \text{ has } B_k\big)$$

$$= \begin{cases} 0 & \text{if } c_n \to -\infty, c_n > -\log\log n \\ \exp\left(-\dfrac{e^{-2c(k+1)}}{(k!)^{k+1}(k+1)!}\right) & \text{if } c_n \to c \\ 1 & \text{if } c_n \to \infty. \end{cases}$$

In the case $k=1$ this theorem was proved by Bollobás and Frieze [3]. We shall show it for $k \geq 2$ using the colouring method of Fenner and Frieze. This method describes a simple colouring procedure which shows that, under some additional conditions, if a random graph with n vertices is such that almost surely each subset S of its vertices with s elements, $s \leq \alpha n$, where α is a constant, has at least $2s$ neighbours which do not belong to S, i.e.

$$|N(S)| \geq 2|S|, \qquad (*)$$

then this random graph is a.s. hamiltonian.

Unfortunately, the condition (*) does not hold for the subgraph $C_k(K_{n,M})$ if $M(n) \leq \frac{1}{2} n \log n$ because of the existence of groups of incident vertices of a small degree. However, we can replace each such group by a single vertex of degree 2 in such a way that for obtained graph $C_k(K_{n,M})^*$ the condition (*) is fulfilled and the existence of the hamiltonian cycle in $C_k(K_{n,M})^*$ implies that $C_k(K_{n,M})$ is also hamiltonian.

Now, we describe the idea of the proof of Theorem 3 for $k \geq 2$. Let $\mathscr{A}_k$ be a family of graphs with n vertices and M edges and let $G \in \mathscr{A}_k$. Let us consider the following Algorithm looking for disjoint hamiltonian cycles (and, maybe, one perfect matching) in $C_k(G)$:

Algorithm

1. **if** k is odd **then** $l := 0$
2. **if** k is even **then** $l := 1$
3. $H := C_k(G)$
4. **if** $\delta(H) = 0$ **then** stop
5. find H^*
6. **if** H^* has a hamiltonian cycle **then** find the hamiltonian cycle related to that in H **else** stop
7. **if** $l = 0$ **then**
   begin
   $H := H'$ where H' is obtained from H by removing of all edges of one from two perfect matchings found in step 6 (in the case when the number of vertices of $C_k(G)$ is odd we remove all edges of the matching which does not contain vertex v (see definition p. 172))
   $l := l + 1$
   go to 5
   end

8. $H = H''$ where H'' is obtained from H by removing all edges of the hamiltonian cycle found in step 6
9. **go to 4**

We can define the family $\mathcal{A}_k$ in such a way that in command 6 for the graph H^* the condition (*) holds, so we can use colouring arguments to prove that this algorithm a.s. goes to the next step. Hence, the probability that it stops before finding $[k/2]$ disjoint hamiltonian cycles (and a disjoint perfect matching when k is odd) tends to zero, or equivalently

$$\lim_{n \to \infty} P(C_k(K_{n,M}) \text{ has not } M_k \text{ but } K_{n,M} \in \mathcal{A}_k) = 0.$$

But we can choose the family $\mathcal{A}_k$ in such a way that

$$\lim_{n \to \infty} P(K_{n,M} \in \mathcal{A}_k) = \lim_{n \to \infty} P(K_{n,M} \text{ has } B_k)$$

and

$$\lim_{n \to \infty} P(C_k(K_{n,M}) \text{ has } M_k \text{ but } K_{n,M} \text{ has not } B_k) = 0.$$

It is easy to show that these facts imply that

$$\lim_{n \to \infty} P(C_k(K_{n,M}) \text{ has } M_k) = \lim_{n \to \infty} P(K_{n,M} \text{ has } B_k).$$

The second equality of Theorem 3 follows from the fact that every group of $k+1$ vertices of degree k with a common neighbour is in $K_{n,M}$ a.s. far from the vertices of small degree, so the number of such structures is the same for both $K_{n,M}$ and $C_k(K_{n,M})$. When $c_n \to c$ then this number is asymptotically Poisson distributed with the expectation

$$\frac{e^{-2c(k+1)}}{(k!)^{k+1}(k+1)!},$$

which implies the last equality of Theorem 3.

Theorem 2 follows easily from Theorem 3; it requires only finding the threshold function for the event that $\delta(L_k(K_{n,M})) \geq k$ or, equivalently, that $C_k(K_{n,M}) = L_k(K_{n,M})$. This equality holds a.s. if in $K_{n,M}$ there are no vertices of degree k adjacent to vertices of degree $k-1$, the threshold function for the existence of such structures can be found by a standard way.

The structure of this paper is as follows. Firstly, we describe the construction of the 'contraction graph' G^*. Then we prove our main result using the algorithm presented above.

Notation

For a graph G, by $V(G)$ we denote the set of its vertices and by $E(G)$ its set of edges.

For $v \in V(G)$, $d_G(v)$ is the degree of v and $\delta(G)$ and $\Delta(G)$ denote the minimal and the maximal degrees of G.

For $S \subset V(G)$ let $N_G(S) = \{v \in V(G) \setminus S : \text{there exists } w \in S \text{ such that } \{v, w\} \in E(G)\}$ and $G[S] = (S, E_S)$ where $E_S = \{e \in E(G) : e \subset S\}$.

Now, let $k \geq 0$ be a natural number. Then $V_k(G) = \{v \in V(G) : d_G(v) \geq k\}$, $L_k(G) = G[V_k]$ and $D_k(G) = \{v \in V(G) : d_G(v) = k\}$.

Finally, by $C_k(G)$ we denote the k-core of G, i.e. the maximal subgraph H of G with property $\delta(H) \geq k$ and $CV_k(G) = V(C_k(G))$.

Definition of the j-contraction of a graph

Let G be a graph and let $W^* \subset W \subset V(G)$. A path $P(v_1, \ldots, v_m)$ in G is (W, W^*)-path if

(i) $v_1, v_m \in W$,

(ii) for every i, $2 \leq i \leq m-1$ either $v_i \in W$ or $v_{i-1}, v_{i+1} \in W$,

(iii) there exist some neighbours of vertices v_1 and v_m which do not belong to $V(P)$,

(iv) there are no paths in G for which conditions (i)–(iii) are fulfilled and which have more vertices from the set W^* than the path P,

(v) the path P is the maximal path with these properties.

For the (W, W^*)-path $P(v_1, \ldots, v_m)$ the P-contraction of G is the graph H obtained from G by choosing two vertices (say, v_0 and v_{m+1}) from the neighbours of vertices v_1 and v_m which do not belong to $V(P)$ and replacing the path P by a vertex W with exactly two neighbours v_0 and v_{m+1}. Since there is one-to-one correspondence between the vertices of H different from W and the vertices of the graph G which do not belong to the path P so we shall identify them. Consequently, by W_c^*, W_c (or, simply by W^*, W) we denote the set of those vertices of H which are related to those vertices of G which are in W^* or W but do not belong to the path P. In a similar way, we shall identify all edges of H with respective edges of G.

Now, let the (W, W^*)-contraction of G be the graph K, obtained by a series of P-contractions such that K contains no more (W, W^*)-paths. Finally, for a fixed natural number j, the (W_j, W_j^*)-contraction of G, where $W_j^* = \{v \in V(G) : d_G(v) = j\}$ and $W_j = \{v \in V(G) : d_G(v) \leq 200j\}$, will be called the j-contraction of G. Let us observe that if the j-contraction of G is hamiltonian then there exists a hamiltonian cycle in G which contains all contracted (W_j, W_j^*)-paths of G. This property is crucial for our further consideration.

Proof of Theorem 3. Let $\mathcal{A}_k$ be the family of graphs with n vertices and M edges such that for every graph G from $\mathcal{A}_k$ we have

(i) $\Delta(G) < 3 \log n$,

(ii) there are at most $\dfrac{n}{(\log n)^3}$ vertices of degree $\leq 200k$,

(iii) no two vertices of degree $\leq 200k$ belong to a cycle with length $\leq 3k$,

(iv) there are no $k+2$ vertices of degree $\leq 200k$ such that every two of them are within the distance $5k$,

(v) there are no vertices incident to at least $k+1$ vertices of degree exactly k,

(vi) every set of s vertices, $s \leq \dfrac{(4k+5)n}{(\log n)^2}$, spans $\leq 10s$ edges,

(vii) for every set S of s vertices, if $n/(\log n)^2 \leq s \leq n/(k+4)$ then $|N_G(S)| \geq (k+2)|S|$.

We shall now show that for $G \in \mathcal{A}_k$ the k-core $C_k(G)$ has the property M_k almost surely, i.e.

$$\lim_{n \to \infty} P(C_k(K_{n,M}) \text{ has not } M_k \text{ and } K_{n,M} \in \mathcal{A}_k) = 0. \tag{1}$$

To do this let us consider the Algorithm given in the Introduction where H^* in the sixth command denotes $(k - 2t' - t'')$-contraction of H, where $t' = 0, 1, \ldots, \lfloor k/2 \rfloor$, and $t'' = 0, 1$ are the numbers of hamiltonian cycles and perfect matchings found by the Algorithm in previous steps. It is enough to prove that this algorithm a.s. does not stop working in command 6, so we must show only that for every r, $0 \leq 2r \leq k-2$, we have

$$\lim_{n \to \infty} P(K_{n,M} \in \mathcal{A}_k, \text{ the Algorithm finds } r \text{ disjoint hamiltonian cycles in } C_k(K_{n,M}) \text{ and } C_k^r(K_{n,M}) \text{ is not hamiltonian})$$
$$= 0 \tag{2}$$

where $C_k^r(K_{n,M})$ is k-core of $K_{n,M}$ without edges which belong to r hamiltonian cycles found by the Algorithm up till now (we shall consider here only the case when k is even, the proof for k odd requires only some modifications).

In the proof of (2) we shall use the colouring method of Fenner and Frieze. Let $G \in \mathscr{A}_k$, G' be the graph obtained from G by removing cycles related to r hamiltonian cycles in $C_k(G)$, say $HC_1, \ldots, HC_r$, found by the Algorithm in previous steps, and let H be the $(k-2r)$-contraction of $C_k^r(G)$. Furthermore, let P^* be the lexicographically first maximal path in H and P be the path related to it in G.

Now let us colour red $R = \lfloor (\log n)^2 \rfloor$ edges of G in such a way that

(i) no edge of cycles $HC_1, \ldots, HC_r$ is red,

(ii) no edge of the path P is red,

(iii) no edge incident to a vertex of degree $\leq 200k$ or to a vertex adjacent to one of degree $\leq 200k$ is red,

(iv) the set of red edges is independent, i.e. there are no pairs of them with common vertex.

All other edges of G we colour blue.

Now let B be the blue subgraph of H, i.e. the subgraph of H such that $e \in E(B)$ iff e is coloured blue in G. We shall use the following simple fact about B.

Lemma 1. *Let vertex v of B belong at the same time also to G. Then*

$$d_B(v) \geq d_G(v) - 3k - 1 . \tag{3}$$

Moreover, for every $w \in V(B)$ we have

$$d_B(w) \geq 2 . \tag{4}$$

Proof. Let $v \in V(B) \cap V(G)$. In the procedure of obtaining B from G this vertex may lose:

(i) at most $k+1$ neighbours in passing from G to $C_k(G)$ since there are no more than $k+1$ vertices of degree $\leq 200k$ within the distance $2k$ from v,

(ii) $2r \leq k-2$ neighbours due to removing r cycles $HC_1, HC_2, \ldots, HC_r$,

(iii) at most one neighbour v' if $\{v, v'\}$ is red,

(iv) some neighbours in the process of obtaining B from coloured $C_k^r(G)$ due to removing some edges incident to vertices of contracted paths. Since no two vertices of degree $\leq 200k$ belong to a cycle with length $\leq 3k$ we can lose at most two edges contracting one path. But there are at most $k+1$ vertices of degree $\leq 200k$ within the distance $3k$ from v and each contracted path contains at least two such vertices, so the maximal number of neighbours we can lose in this way is $k+1$.

Since in these four steps we remove at most $3k+1$ edges incident with v so the proof of the first part of Lemma 1 is completed.

Now let $w_0 \in V(B)$. If w_0 is a vertex obtained from contraction of a path then its degree is 2 so (4) holds. Moreover, if $d_G(w_0) \geq 3(k+1)$ then (4) is implied by (3), so it is enough to consider the case when $w_0 \in V(G)$ and $d_G(w_0) \leq 200k$.

Let $d_G(w_0) \leq 200k$ and let us remove a hamiltonian cycle from $C_k(G)$ using the Algorithm. Then $d_{C_k^1(G)}(w_0) = d_{C_k(G)} - 2$ and there are at most $k-2$ vertices of degree $\leq 200k-2$ which are within the distance 2 from w_0 in $C_k^1(G)$. Indeed, if not than in $C_k(G)$ there are either k or $k-1$ vertices of degree $\leq 200k$ within the distance 2 from w_0; then, either w_0 belongs to a contracted path or is adjacent to some vertex of such a path. In both cases it loses at least two from near vertices of degree $\leq 200k$ due to removing edges of the hamiltonian cycle. Thus in such a case w_0 can have at most $k-2$ vertices of small degree within distance 2 in $C_k^1(G)$.

Using similar arguments one can deduce that for $r \leq k-1$ there are at most $k-2r$ vertices of small degree within the distance 2 from w_0 in $C_k^r(G)$ and w_0 has the degree at least $k-2r$ in this graph. Since there is at most one edge between w_0 and each contracted path, and there are at most $\frac{1}{2}(k-2r)$ such paths, so

$$d_B(w_0) \geq d_{C_k^r(G)}(w_0) - \frac{1}{2}(k-2r)$$
$$\geq \frac{1}{2}(k-2r).$$

Hence, (4) holds unless $k-2r=2$, $d_{C_k^r(G)}(w_0) = 2$ and there is an edge in $C_k^r(G)$ between w_0 and a vertex of a path which is contracted in B. To complete the proof it is enough to show that such a situation is impossible. Indeed, one can check that in such a case there are $k+1$ vertices $w_0, w_1, \ldots, w_k$ of degree k in $C_k(G)$ adjacent to one vertex so, since $G \in \mathcal{A}_k$, each of these vertices has degree k also in G. Using again the fact that $G \in \mathcal{A}_k$ we obtain that there are no such configurations in it.

Now we can prove the crucial lemma concerning B.

Lemma 2. *For every set $S \subset V(B)$, $|S| \leq n/(k+4)$, we have*

$$|N_B(S)| \geq 2|S|. \tag{5}$$

Furthermore, B is connected.

Proof. Let us split the proof of the first part of the Lemma into two cases, according to the size of the set S.

Case 1. Let $S \subset V(B)$, $n/(\log n)^2 \leqslant |S| \leqslant n/(k+4)$ and let $S' \subset S$ denote the set of vertices of G which are related to the vertices from S. Since there are no more than $n/(\log n)^3$ vertices of degree $\leqslant 200k$ and there are no groups of $k+2$ such vertices which are near each other, we can deduce that

$$|V(G) \setminus CV_k(G)| \leqslant \frac{n}{(\log n)^3}$$

and

$$|S \setminus S'| \leqslant \frac{n}{2(\log n)^3}.$$

Hence, from properties of G we have

$$|N_G(S')| \geqslant (k+2)|S'|$$

so

$$|N_{G'}(S')| \geqslant (k-2r+2)|S'| \geqslant 4|S'|$$

and finally

$$|N_B(S)| \geqslant |N_B(S')| - |S \setminus S'|$$
$$\geqslant |N_{G'}(S')| - |S \setminus S'| - \frac{n}{(\log n)^3}$$
$$\geqslant 3|S'| > 2|S|.$$

Case 2. Let $S \subset V(B)$, $|S| \leqslant n/(\log n)^2$ and let $S_1 = \{v \in S : d_B(v) \leqslant 190k\}$, $S_2 = S \setminus S_1$.

Since no two vertices of S_1 are incident then from Lemma 1 we obtain immediately that

$$|N_B(S_1)| \geqslant 2|S_1|. \tag{6}$$

Now let us suppose that (4) fails for the set S, i.e.

$$|N_B(S)| < 2|S|. \tag{7}$$

From the equality

$$|N_B(S)| = |N_B(S_2)\setminus(S_1 \cup N_B(S_1))| + |N_B(S_1)\setminus S_2|$$

and (6) and (7), we obtain

$$|N_B(S_2)\setminus(S_1 \cup N_B(S_1))| < 3|S_2|.$$

Since no $k+2$ vertices from S_1 are within distance 5 of each other, then

$$|N_B(S_2)| < (k+4)|S_2|.$$

Hence, from Lemma 1 we have

$$|N_G(S_2)| < (4k+5)|S_2| \leq \frac{(4k+5)n}{(\log n)^2}.$$

But $G \in \mathcal{A}_k$, so the graph induced in G by the set $T = S_2 \cup N_G(S_2)$, $|T| < (4k+6)n/(\log n)^2$, has no more than $10|T| < 10(4k+6)|S_2| \leq 90k|S_2|$ edges. On the other hand, since every vertex of S_2 has degree $\geq 190k$, the graph $G(T)$ has at least $95k|S_2|$ edges. Contradiction. So our assumption (7) is false and (5) holds also for $|S| \leq n/(\log n)^2$.

The proof of the second part of Lemma 2 is now obvious. Indeed, it is enough to prove that for every set $S \subset V(B)$, $|S| \leq \frac{1}{2}|V(B)| \leq \frac{1}{2}n$ we have

$$|N_B(S)| > 0. \tag{8}$$

From the first part of Lemma 2, (8) holds for every set $S \subset V(B)$, $|S| \leq 3n/(k+4)$. Now let $T \subset S \subset V(B)$, $3n/(k+4) < |S| \leq \frac{1}{2}n$, $|T| = n/\lfloor k+4 \rfloor$ and let T' denote the set of all vertices from T which belong also to G (i.e. we exclude all vertices obtained by contracting paths). Then we have

$$|N_G(T')| \geq (k+2)|T'|. \tag{9}$$

But there are at most $n/(\log n)^3$ vertices of degree $\leq 100k$ in G and $\Delta(G) < 3\log n$, so passing from G to B we do not lose more than $3n/\log n$ vertices. So, from (9), we obtain

$$|N_B(T)| \geq |N_B(T')| - |T\setminus T'|$$
$$\geq (k+1.5)|T'| \geq (k+1)|T|. \tag{10}$$

Clearly, (10) implies (8) so the graph B is connected.

In the proof of the equality (2) we shall need also the following lemma (see [1], p. 186) which is an easy consequence of the Pósa well known result ([5], p. 360).

Lemma 3. *Let l and u be natural numbers and G be a graph such that the longest path in G has length l and there are no cycles of length $l+1$ in G. Let us assume also that for every $S \subset V(G)$, $|S| \leq u$, we have*

$$|N_G(S)| \geq 2|S|.$$

Then, there exists a set $F \subset \{\{v, w\} : v, w \in V(G)\}$ which has the following three properties:

(i) $\quad F \cap E(G) = \emptyset$,

(ii) $\quad |F| \geq \dfrac{u^2}{2}$,

(iii) $\quad$ *for every pair of vertices $\{v, w\} \in F$ the graph $(V(G), E(G) \cup \{v, w\})$ has a cycle of length $l+1$.*

Now we are able to prove (2). Let us count the number of blue-red colourings of such graphs $G \in \mathscr{A}_k$ that the Algorithm finds in $C_k(G)$ r disjoint hamiltonian cycles and there are no hamiltonian cycles in $C_k^r(G)^*$. To construct one of these graphs we must first choose all its blue edges (we can do it in no more than $\left(\binom{n}{2} \atop M-R\right)$ ways). These edges determine the graph B, i.e. the blue part of the graph $C_k^r(G)^*$. Due to our assumption, B is not hamiltonian so let us denote the length of the longest path P^* in B by l. From Lemmas 2 and 3 there are at least $n^2/(2(k+4)^2)$ pairs of vertices of B such that adding one of them to B creates a cycle of length $l+1$ and, since B is connected, a path longer than P^*. So, since all edges of the maximal path in $C_k^r(G)^*$ must be blue, we cannot add to B a red edge in such places. Let us consider only "red prohibited" places $\{v, w\}$ such that neither v nor w is a vertex obtained by the contraction of a path of the graph G. Since the number of vertices of degree $\leq 200k$ in G is smaller than $n/(\log n)^3$ the number of such pairs $\{v, w\}$ is still greater than $n^2/(2(k+4)^2) - n^2/(\log n)^3 > n^2/(80k^2)$. But then both v and w are vertices of G and it is clear that we cannot add to the blue part of G the red edge $\{v, w\}$ because we would create a new blue-red path in $C_k^r(G)^*$ longer than the longest blue one.

Then, there are at least $n^2/(80k^2)$ places in G prohibited for red edges, so we can choose R red edges in no more than $\left(\binom{n}{2} - n^2/(80k^2) \atop R\right)$ ways. Hence, the upper bound for the number of blue-red colourings of our graphs is

$$\binom{\binom{n}{2}}{M-R}\binom{\binom{n}{2}-n^2/(80k^2)}{R}. \tag{11}$$

To obtain the equality (2) we must divide (11) by the number of all graphs with n vertices and M edges multiplied by the number of colourings for one of the graphs from $\mathscr{A}_k$. So let $G \in \mathscr{A}_k$ be fixed and let us have a look at the definition of the colouring. We cannot colour the red edges of r cycles, $HC_1, \ldots, HC_r$, of the path P and the edges incident to vertices of degree $\leq 200k$. Let M_0 denote the number of all such edges in G. It is easy to see that since $G \in \mathscr{A}_k$ we have $M_0 \leq rn + n + n \leq kn$. The condition of independence of the set of red edges does not change the number of blue-red colourings very much. Indeed, if we choose at random a set of R red edges for which conditions (i)–(iii) hold, then the probability that such a set is not independent is not greater than

$$n\binom{3\log n}{2}\binom{M-M_0-2}{R-2}\binom{M-M_0}{R}^{-1}$$
$$= O(n^{-1}(\log n)^2 R^2) = o(1).$$

So we can choose red edges in G in at least

$$(1-o(1))\binom{M-M_0}{R} \geq \frac{1}{2}\binom{M-kn}{R}$$

ways. Finally, we have

$$P(C_k^r(K_{n,M})^* \text{ is not hamiltonian but } K_{n,M} \in \mathscr{A}_k$$
$$\text{and the Algorithm found } r \text{ disjoint}$$
$$\text{hamiltonian cycles in } C_k(K_{n,M}))$$

$$\leq 2\binom{\binom{n}{2}}{M-R}\binom{\binom{n}{2}-n^2/(80k^2)}{R}\binom{M-kn}{R}^{-1}\binom{\binom{n}{2}}{M}^{-1}$$

$$= o(1).$$

So we proved that the equality (2) and then the equality (1) is true.

To complete the proof of Theorem 3 we need more information about the family $\mathscr{A}_k$. Let B_k be the property that a graph contains no vertices adjacent to at least $k+1$ vertices with degree exactly k. Then, if $M(n) = n \log n/(2(k+1)) + k/2n \log \log n + c_n n$ where $c_n > -\log \log n$, we have

$$\lim_{n \to \infty} P(K_{n,M} \in \mathscr{A}_k) = \lim_{n \to \infty} P(K_{n,M} \text{ has } B_k). \tag{12}$$

To see that one has to examine all conditions of the definition of the family $\mathscr{A}_k$. Using the first moment method, it is easy to see that conditions (i), (iii), (iv), (vi) and (vii) hold for almost all graphs with n vertices and M edges, and one can check that the Chebyshev inequality implies that for $K_{n,M}$ (ii) also holds a.s. (The calculations are rather simple but tedious and therefore omitted. For details see, for example, [2], [3] or [4] where similar facts are proved.) Hence, the probability that a graph G belongs to $\mathscr{A}_k$ is determined by the condition (v), which is nothing else but the property that G has the property B_k.

It can easily be verified that no graph with vertex adjacent to $k+1$ vertices of degree k has the property M_k. One can also check (see Introduction) that the probability that $C_k(K_{n,M})$ has B_k is asymptotically the same as the probability that $K_{n,M}$ has B_k, i.e.

$$\lim_{n \to \infty} P(K_{n,M} \text{ has } B_k)$$
$$= \lim_{n \to \infty} P(C_k(K_{n,M}) \text{ has } B_k)$$

for every function $M(n)$. Hence

$$\lim_{n \to \infty} P(C_k(K_{n,M}) \text{ has } M_k \text{ but } K_{n,M} \text{ has not } B_k)$$
$$= \lim_{n \to \infty} P(C_k(K_{n,M}) \text{ has } M_k \text{ but } C_k(K_{n,M}) \text{ has not } B_k)$$
$$= 0. \tag{13}$$

Finally, from (1), (12) and (13) we obtain the following series of equalities

$$P(C_k(K_{n,M}) \text{ has } M_k)$$
$$= P(C_k(K_{n,M}) \text{ has } M_k \text{ and } K_{n,M} \in A_k)$$

$$+ P\bigl(C_k(K_{n,M}) \text{ has } M_k \text{ but } K_{n,M} \notin \mathscr{A}_k\bigr)$$
$$= P(K_{n,M} \in \mathscr{A}_k)$$
$$+ P\bigl(C_k(K_{n,M}) \text{ has not } M_k \text{ but } K_{n,M} \in \mathscr{A}_k\bigr)$$
$$+ P\bigl(C_k(K_{n,M}) \text{ has } M_k \text{ but } K_{n,M} \text{ has not } B_k\bigr) + o(1)$$
$$= P(K_{n,M} \text{ has } B_k) + o(1).$$

This completes the proof of Theorem 3.

ACKNOWLEDGEMENTS

I would like to thank Alan Frieze and Andrzej Ruciński for their helpful comments and discussions.

References

[1] B. Bollobás, *Random Graphs*, Academic Press, London, 1985.

[2] B. Bollobás, The evolution of sparse graphs. In *Graph Theory and Combinatorics*, Cambridge, 1984, 35–37.

[3] B. Bollobás, A. Frieze, On matchings and hamiltonian cycles in random graphs, *Ann. Disc. Math.* 28 (1985), 23–46.

[4] A. Frieze, On large matchings and cycles in sparse random graphs, *Disc. Math.* 59 (1986), 243-256.

[5] L. Pósa, Hamiltonian circuit in random graphs, *Disc. Math.* 14 (1976), 359–364.

GENERAL PERCOLATION AND ORIENTED MATROIDS

Colin McDIARMID

Institute of Economics and Statistics
Oxford

Certain simple conditions involving clutters and equivalence relations are useful in the analysis of applied probability problems concerning random graphs and directed graphs, percolation, reliability, epidemics and random networks. We shall see here that oriented matroids form a rich source of examples satisfying the various conditions, and provide a natural framework within which to understand certain relationships amongst the conditions.

§ 1. Introduction

Consider two possible experiments on a simple (undirected) graph G. In the former we *d*elete the edges of G at random, that is independently and with probability $\frac{1}{2}$. In the latter we *o*rient the edges of G at random, that is independently and so that for each edge the two directions are equally likely.

Given two vertices s and t let $P_d(s \to t)$ and $P_o(s \to t)$ denote the corresponding probabilities that there is a (directed) path from s to t. Then ([14])

(1.1) $P_d(s \to t) = P_o(s \to t)$.

Now suppose that the graph G is plane, with a given embedding in the plane. Then with notation as above

(1.2) $P_d(\text{exists cycle}) = P_o(\text{exists clockwise cycle})$.

Results such as these and analogous results concerning shortest paths or cycles may be analysed in terms of clutters and equivalence relations satisfying appropriate conditions. These conditions are labelled (C), (C*), (C*1), (C*2) and theorems involving them are useful in the analysis of applied probability problems concerning random graph and directed graphs, percolation, reliability, epidemics and random networks (see [5, 7–11, 13–16]).

The condition (C) says simply that no two equivalent elements are in the same set in the clutter. Condition (C*) is equivalent to the blocking clutter satisfying condition (C). Condition (C*1) is a strengthening of condition (C*), and condition (C*2) is a further strengthening. Details are given in section 2 below.

It was remarked in [15] that although condition (C*1) is precisely what is needed to yield the relevant theorems for first-passage times ((2.6)–(2.8) of [15]), it is not attractive and general examples of clutters satisfying condition (C*1) seem also to satisfy the stronger and more elegant condition (C*2). Further, for these examples a clutter satisfies the conditions if and only if the blocking clutter does. The examples arise from paths and cuts in directed graphs. They will be seen to be 'ports' of oriented matroids, and by generalising to this level we shall gain in section 3 below a good explanation of the relationships noted above.

In section 4 we discuss joint distribution of first-passage times. The basic theorem on first-passage clutter percolation, theorem (2.6) of [15], concerns a clutter satisfying conditions (C) and (C*1). It shows for example that in a random network the marginal distributions of the edge-lengths determine the distribution of the random first-passage time from a given vertex s to another vertex t. This result is a special case of theorem (3.1) of [15] which says that the marginals actually determine the joint distribution of all the first-passage times from s. It is of interest to extend the clutter theorem so that it yields this joint distribution theorem on random networks. This requires us (unfortunately!) to introduce a new condition (C*3) on a family of clutters.

Consider further the extension of the clutter theorem. The basic theorem shows that if the 'edge-lengths' are regarded as capacities then the marginals determine also the distribution of MAXFLOW (s, t), the maximum value of a flow from s to t (theorem (3.3) of [15]). However, we have no analogous result concerning joint distributions for flows. We can gain further understanding of this partial symmetry between first-passage times and flows by applying the extended clutter theorem to oriented matroids, and using a recent theorem of Seymour.

Finally in section 5 we introduce 'splits' of oriented matroids, and show that they satisfy conditions (C) and (C*). This generalises the result (1.2) on clockwise cycles. Also it shows that all the interesting known classes of clutters satisfying conditions (C) and (C*) arise from oriented matroids. More such classes are always welcome!

Most of these results were first announced at the XIth International Symposium on Mathematical Programming in Bonn in 1982.

§ 2. Preliminaries

Let $\mathscr{C}$ be a clutter on a (finite) set I, that is a collection of subsets of I with no one containing another. Let $\sim$ be an equivalence relation on I. The following are the four conditions on the pair $\mathscr{C}$ and $\sim$ to which we referred above.

(C)   If $a \sim b$, $a \neq b$, $C \in \mathscr{C}$ then $\{a, b\} \not\subseteq C$.

(C*) If $a \sim b$, $A, B \in \mathscr{C}$, $a \in A \backslash B$, $b \in B \backslash A$ then there is a set $C \in \mathscr{C}$ with $C \subseteq (A \cup B) \backslash \{a, b\}$.

(C*1) With premises as in condition (C*), some convex combination of the incidence vectors $\mathbf{1}_C$ for C in $\mathscr{C}$ is less than or equal to some convex combination of the incidence vectors of $A \backslash \{a\}$ and $B \backslash \{b\}$.

(C*2) With premises as in condition (C*), there are sets $C, D \in \mathscr{C}$ such that the sum of the incidence vectors of C and D is less than or equal to the sum of the incidence vectors of $A \backslash \{a\}$ and $B \backslash \{b\}$.

The *blocking clutter* or *blocker* $\mathscr{C}^*$ of the clutter $\mathscr{C}$ is the collection of minimal sets that intersect each set in $\mathscr{C}$. It is well known that $(\mathscr{C}^*)^* = \mathscr{C}$. The clutter $\mathscr{C}$ satisfies condition (C) if and only if its blocker $\mathscr{C}^*$ satisfies condition (C*) [14]. (We have rather loosely dropped reference to the equivalence relation $\sim$, and we shall continue to do this whenever no confusion can arise.) Clearly condition (C*1) is a strengthening of condition (C*), and condition (C*2) is a further strengthening.

If $\mathscr{C}$ is a clutter on I and A and B are disjoint subsets of I then the *clutter minor* $\mathscr{C} \backslash A / B$ or $\mathscr{C} / B \backslash A$ is the collection of all minimal sets of the form $C \backslash B$ for $C \in \mathscr{C}$, $C \cap A = \emptyset$ (see [19, 20, 14]). Note that $(\mathscr{C} \backslash A / B)^* = \mathscr{C}^* / A \backslash B$. It is straightforward to check that each of the conditions (C), (C*), (C*1), (C*2) is closed under taking clutter minors.

We let $\tilde{I}$ denote the set of equivalence classes $[i]$ of I under the given equivalence relation $\sim$. For $S \subseteq I$ let $\tilde{S} = \{[i] : i \in S\}$, so that $\tilde{S} \subseteq \tilde{I}$. Given a clutter $\mathscr{C}$ on I we let $\tilde{\mathscr{C}}$ be the clutter on $\tilde{I}$ of minimal sets of the form $\tilde{C}$ for $C \in \mathscr{C}$, and call $\tilde{\mathscr{C}}$ the *underlying clutter* of $\mathscr{C}$. Finally, if $\mathscr{C}$ is a clutter on I and $i \in I$, the *port* of i in $\mathscr{C}$ is the clutter

$$P_i(\mathscr{C}) = \{C \backslash \{i\} : i \in C \in \mathscr{C}\}.$$

For an introduction to the theory of oriented matroids see [2, 3, 6]. We give here only a few definitions.

Suppose that the equivalence relation $\sim$ partitions I into pairs. We call I a *paired set*. For each $x \in I$ we denote the other member of its pair by $-x$; and for each $S \subseteq I$ we let

$$-S = \{-x : x \in S\}.$$

We call S *pair-free* if $S \cap (-S) = \emptyset$.

Note that a clutter $\mathscr{C}$ on I satisfies condition (C) if and only if each set C in $\mathscr{C}$

is pair-free. The clutter $\mathscr{C}$ is *symmetric* if for each $C \in \mathscr{C}$ we have $-C \in \mathscr{C}$. An *oriented matroid* on a paired set I is a symmetric clutter $\mathscr{M} \neq \{\emptyset\}$ which satisfies both conditions (C) and (C*). Each port of an oriented matroid also satisfies these conditions (see (II) of [3]). If $\mathscr{M}$ is an oriented matroid on I then the underlying clutter $\widetilde{\mathscr{M}}$ on $\widetilde{I}$ is (the collection of circuits of) a matroid on $\widetilde{I}$, the *underlying matroid* of $\mathscr{M}$ and $\mathscr{M}$ is an *orientation* of this matroid.

(2.1) **Example** Let G be a graph. Split each edge e into a pair of oppositely directed edges, yielding the directed graph $D(G)$ with edge-set the naturally paired set I. The cycles and cuts of G yield natural dual oriented matroids $\mathscr{M}$ and $\mathscr{M}^\perp$ on I, which are orientations of the cycle and cocycle matroid of G (see example 3.3 of [3]).

Now let s and t be distinct vertices of G and suppose that in $D(G)$ there is an edge i from t to s. The port of i in $\mathscr{M}$ is the clutter of minimal edge-sets of $s-t$ paths in $D(G)$, and the port of $-i$ in $\mathscr{M}^\perp$ is the clutter of minimal $s-t$ cuts in $D(G)$. The clutters form a blocking pair, and both satisfy conditions (C) and (C*2) ([15]).

§ 3. Conditions (C*1), (C*2) and blocking clutters

In general a clutter may satisfy conditions (C) and (C*1) but not (C*2). Also, a clutter may satisfy conditions (C) and (C*2) but its blocker need not satisfy even condition (C*1).

(3.1) **Example** (i) Let $I = \{1, 2, \ldots, 8\}$, let $7 \sim 8$ (and let no other distinct elements be equivalent), and let $\mathscr{C}$ be the clutter on I with members $\{1, 2, 3\}$, $\{1, 5, 6\}$, $\{2, 4, 6\}$, $\{3, 4, 5\}$, $\{1, 4, 6, 7\}$ and $\{2, 3, 5, 8\}$. Then $\mathscr{C}$ satisfies conditions (C) and (C*1) but not (C*2).

(ii) Let $I = \{1, 2, \ldots, 6\}$, let $5 \sim 6$, and let $\mathscr{C}$ be the clutter on I with members $\{1, 3\}$, $\{1, 4\}$, $\{2, 3\}$, $\{2, 4\}$, $\{1, 2, 6\}$ and $\{3, 4, 5\}$. Then $\mathscr{C}$ satisfies both conditions (C) and (C*2). However, the blocking clutter $\mathscr{C}^*$ has members $\{1, 2, 3\}$, $\{1, 2, 4\}$, $\{1, 3, 4\}$, $\{2, 3, 4\}$, $\{1, 2, 5\}$ and $\{3, 4, 6\}$, and does not satisfy condition (C*1).

We have already noted that the general examples of clutters from [15] do not exhibit such untidy behaviour. Also we have noted that these examples are in fact ports of oriented matroids. The theorem below shows that for such clutters life is tidy: if one satisfies condition (C*1) then both it and its blocker must satisfy the stronger condition (C*2).

(3.2) **Theorem** *Let the clutter $\mathscr{C}$ be a port of an oriented connected matroid $\mathscr{M}$. Then $\mathscr{C}$ satisfies both conditions (C) and (C*); the blocking clutter is a port of the dual oriented matroid; and the following statements are equivalent.*

(i) *$\mathscr{C}$ satisfies condition (C*1).*

(ii) *$\mathscr{C}$ and its blocker satisfy condition (C*2).*

(iii) *$\mathscr{M}$ is binary.*

(iv) *$\mathscr{C}$ has no clutter minor isomorphic to the clutter $\mathscr{C}_0 = \{\{a, c\}, \{b, d\}, \{c, d\}\}$ on the set $\{a, b, c, d\}$, where $a \sim b$ and no other distinct elements are equivalent.*

In order to prove this theorem let us first note two lemmas.

(3.3) **Lemma** *Let $\mathscr{C}$ and $\mathscr{D}$ be dual oriented matroids on a paired set I, and let $i \in I$. Then the ports $P_i(\mathscr{C})$ and $P_{-i}(\mathscr{D})$ are blocking clutters.*

This first lemma is a direct translation into 'blocking' terms of theorem 13 of [6], or a less direct translation of part of theorem 2.2 of [3]. It proves the first part of theorem (3.2).

There is also a converse result. Thus if $\mathscr{C}$ and $\mathscr{D}$ are pair-free symmetric clutters on I not equal to $\{\emptyset\}$, then $\mathscr{C}$ and $\mathscr{D}$ are dual oriented matroids if and only if for each $i \in I$ the ports $P_i(\mathscr{C})$ and $P_{-i}(\mathscr{D})$ are blocking. This may be proved along the lines of the proof of theorem (2.2) in [3].

The second lemma is straightforward to check, and its proof is omitted.

(3.4) **Lemma** *Consider the uniform matroid $\mathscr{U}_{2,4}$ of rank 2 on a set of four elements. Each port of each orientation of $\mathscr{U}_{2,4}$ is isomorphic to the clutter $\mathscr{C}_0$ (with their respective equivalence relations).*

Proof of theorem (3.2) Suppose that the underlying matroid $\tilde{\mathscr{M}}$ is not binary. Let e be the element $\{i, -i\}$ of $\tilde{I}$. Since $\tilde{\mathscr{M}}$ is connected it has a matroid minor using e which is isomorphic to the uniform matroid $\mathscr{U}_{2,4}$ ([1]). Hence some oriented matroid minor $\mathscr{N}$ of $\mathscr{M}$ is an orientation of $\mathscr{U}_{2,4}$ using i; and so by lemma (3.4) the port $P_i(\mathscr{N})$ is isomorphic to the clutter $\mathscr{C}_0$ (with the appropriate equivalence relations). But a port of an oriented matroid minor of $\mathscr{M}$ is the corresponding clutter minor of the original port $\mathscr{C}$, and so the clutter $\mathscr{C}$ has a clutter minor isomorphic to $\mathscr{C}_0$. But condition (C*1) is closed under clutter minors, and clearly $\mathscr{C}_0$ does not satisfy condition (C*1). Hence the clutter $\mathscr{C}$ does not satisfy condition (C*1).

So far we have established that (i) implies (iv) and (iv) implies (iii). Suppose now that $\mathscr{M}$ (or $\tilde{\mathscr{M}}$) is binary. Since the dual of a binary matroid is binary it will suffice for us to show that $\mathscr{C}$ satisfies condition (C*2). This will follow easily from results in [3], but we shall spell out the proof since it will also yield proposition (4.5) below.

First note that $\mathscr{M}$ arises from a unimodular subspace. Let $A, B \in \mathscr{C}$, $x \in A\backslash B$, $-x \in B\backslash A$. Then $A' = A \cup \{i\}$ and $B' = B \cup \{i\}$ are in $\mathscr{M}$. Let the vector y be the sum of the incidence vectors on I of the sets A' and B'; and let the vector z be obtained by setting to zero each co-ordinate j such that both the j and $-j$ co-ordinates of y are 1. Then z may be expressed as a sum of incidence vectors of sets in $\mathscr{M}$. Exactly two of these sets say C', D' must contain i. Let $C = C'\backslash\{i\}$ and $D = D'\backslash\{i\}$. Then $C, D \in \mathscr{C}$ and

$$1_C + 1_D \leqslant z \leqslant 1_{A\backslash\{x\}} + 1_{B\backslash\{-x\}}.$$

This shows that the clutter $\mathscr{C}$ satisfies condition (C*2), as required.

Further equivalent statements may be added to theorem (3.2) using results of Seymour [18], [19].

§ 4. Joint distributions of first-passage times

In this section we shall introduce a new condition (C*3) on families of clutters, and use it to extend the first-passage clutter theorem of [15] so that it yields results concerning joint distributions of first-passage times. We shall see that an extension of the idea of a port of an oriented matroid yields examples satisfying this new condition. We shall thus gain further understanding of the partial symmetry discussed in the introduction between first-passage times and flows in random networks, and see that the theorem on joint distributions of first-passage times seems really to be a 'network result'.

Let $\mathsf{C} = (\mathscr{C}_j : j \in J)$ be a family of clutters on the set I. We say that C satisfies condition (C) if each clutter $\mathscr{C}_j$ does. The following condition is a refinement of condition (C*2).

(C*3) If $a \sim b$, $j, k \in J$, $A \in \mathscr{C}_j$, $B \in \mathscr{C}_k$, $a \in A\backslash B$, and $b \in B\backslash A$, then there exist $\hat{A} \in \mathscr{C}_j$, $\hat{B} \in \mathscr{C}_k$ with

$$1_{\hat{A}} + 1_{\hat{B}} \leqslant 1_{A\backslash\{a\}} + 1_{B\backslash\{b\}}.$$

(4.1) **Example** Let D be a directed graph with vertex set V. Let $\sim$ be an equivalence relation on the edge set I of D such that if $i \sim j$ then $i = j$ or i and j are a pair of opposite edges. Let s be a fixed vertex and for each vertex $v \in V\backslash\{s\}$ let $\mathscr{C}_v$ be the clutter of minimal edge-sets of $s-v$ paths. Then the family $\mathsf{C} = (\mathscr{C}_v : v \in$

$\in V\setminus\{s\})$ satisfies conditions (C) and (C*3). This is easily proved, and will also follow from proposition (4.5) below on oriented matroids.

A family $X=(X_i : i \in I)$ of non-negative random variables satisfies the *independence-equality condition* with respect to an equivalence relation $\sim$ on I if the subfamilies corresponding to the equivalence classes are independent of each other, and if X_i and X_j have the same marginal distribution whenever $i \sim j$. A digraph D together with an equivalence relation as in example (4.1) and a family $X=(X_i : i \in I)$ of non-negative random variables satisfying the independence-equality condition form a *random network* D_X.

Denote by R^I_+ the set of non-negative real-valued vectors $x=(x_i : i \in I)$ indexed by I. Given $x \in R^I_+$ and a clutter $\mathscr{C}$ on I the corresponding *first-passage time* FPT $(\mathscr{C}, x)$ is the minimum value over the sets C in $\mathscr{C}$ of the quantity $x(C) = \sum\{x_i : i \in C\}$.

The following theorem is our extension of (the equality part of) the basic first-passage clutter theorem, theorem (2.6) of [15].

(4.2) **Theorem** *Let the family* $\mathbf{C}=(\mathscr{C}_j : j \in J)$ *of clutters on I satisfy both conditions* (C) *and* (C*3), *and let the family* $X=(X_i : i \in I)$ *of non-negative random variables satisfy the independence-equality condition. Then the marginal distributions of the X_i determine the joint distribution of the first-passage times* FPT $(\mathscr{C}_j, X)$ *for* $j \in J$.

When this theorem is applied to example (4.1) we obtain immediately the theorem on random networks discussed earlier; that is, we have

(4.3) **Corollary** [15] *In a random network with a specified vertex s the marginal distributions of the edge-lengths determine the joint distribution of the first-passage times from s.*

Proof of theorem (4.2) For each $j \in J$ let $t_j \geq 0$ and let $S_j = \{x \in R^I_+ : \text{FPT}(\mathscr{C}_j, x) \leq t_j\}$. By theorem (2.1) of [15] we must show that the set $S = \cap \{S_j : j \in J\}$ satisfies the following two conditions (D) and (D*). Given a vector x in R^I_+ and an index $a \in I$ we let x^a denote any vector y in R^I_+ with $y_i = x_i$ for $i \neq a$.

(D) $a \sim b$, $a \neq b$, $x \in S \Rightarrow$ all $x^a \in S$ or all $x^b \in S$

(D*) $a \sim b$, $a \neq b$, $x \in R^I_+$, some $x^a \in S$, some $x^b \in S \Rightarrow x \in S$.

Let us show first that the set S satisfies condition (D). Let $a \sim b$, $a \neq b$, $x \in S$ but suppose that $y \notin S$ for $y =$ some x^a and also some $x^b \notin S$. For each $j \in J$ denote FPT$(\mathscr{C}_j, x)$ by f_j.

Since $y \notin S$ there is an index $j \in J$ with $\mathrm{FPT}(\mathscr{C}_j, y) > t_j$. Thus there is a set $A \in \mathscr{C}_j$ with $a \in A$ and $x(A) = f_j \leqslant t_j$. Similarly there exists $k \in J$ and $B \in \mathscr{C}_k$ with $b \in B$ and $x(B) = f_k$. By condition (C*3) there exist $\widehat{A} \in \mathscr{C}_j$, $\widehat{B} \in \mathscr{C}_k$ such that $a \notin \widehat{A}$ and

$$x(\widehat{A}) + x(\widehat{B}) \leqslant x(A \setminus \{a\}) + x(B \setminus \{b\}) \leqslant f_j + f_k.$$

But $f_j \leqslant x(\widehat{A})$ and $f_k \leqslant x(\widehat{B})$ and so in particular $x(\widehat{A}) = f_j$. Hence

$$y(\widehat{A}) = x(\widehat{A}) = f_j \leqslant t_j,$$

which contradicts the statement that $\mathrm{FPT}(\mathscr{C}_j, y) > t_j$. We have now shown that the set S satisfies condition (D).

It remains to show that the set S satisfies condition (D*). But each clutter $\mathscr{C}_j$ satisfies conditions (C) and (C*1) and so by proposition 2.5(c) of [15] each set S_j satisfies condition (D*). Hence also $S = \cap \{S_j : j \in J\}$ satisfies condition (D*).

Remark The careful reader will note that for the proof to work it would be sufficient to have a condition on the family C of clutters that is based on condition (C*1) rather than on (C*2). Thus the theorem still holds if the family C satisfies the following weaker but less attractive condition.

(4.4) With the same premises as in condition (C*3), the sum of some convex combination of incidence vectors of sets in $\mathscr{C}_j$ and some convex combination of incidence vectors of sets in $\mathscr{C}_k$ is at most the sum $\mathbf{1}_{A \setminus \{a\}} + \mathbf{1}_{B \setminus \{b\}}$.

Given an oriented matroid $\mathscr{M}$ on a paired set I, and a pair-free subset J of I such that $|X \cap J| \leqslant 1$ for each $X \in \mathscr{M}$ we may form a natural extension of the idea of a port. For each $j \in J$ let

$$\mathscr{C}_j = \{X \setminus \{j\} : j \in X \in \mathscr{M}, X \cap (-J) = \emptyset\}.$$

We call the family of clutters $(\mathscr{C}_j : j \in J)$ the *port-family* $\mathbf{C}_J$ of J in $\mathscr{M}$. The following result may be proved easily using the last part of the proof of theorem (3.3).

(4.5) **Proposition** *Let $\mathscr{M}$ be an oriented binary matroid on a paired set I, and let J be a pair-free subset of I such that $|X \cap J| \leqslant 1$ for each $X \in \mathscr{M}$. Then the port-family $\mathbf{C}_J$ satisfies conditions (C) and (C*3).*

Suppose that s is a fixed vertex in a directed graph D, as in example (4.1). Construct a directed graph $\widehat{D}$ by adding to D a set J of edges, one from v to s for each vertex $v \neq s$. Let G be an undirected graph such that $\widehat{D}$ is contained in

$D(G)$ (in an obvious sense). If we consider the oriented binary matroid $\mathcal{M}$ arising from the cycles in G we see that the result in example (4.1) is indeed a special case of proposition (4.5).

It may be of interest to observe that if J is a pair-free subset of I then the condition that $|X \cap J| \leq 1$ for each X in $\mathcal{M}$ is equivalent to the condition that $Y \cap J = \emptyset$ or $(-Y) \cap J = \emptyset$ for each set Y in the dual oriented matroid $\mathcal{M}^\perp$.

We shall now see that corollary (4.3) on joint distributions of first-passage times is essentially a 'network result' rather than a more general 'oriented matroid result'. For a recent theorem of Seymour [20] shows that if in proposition (4.5) the underlying matroid of $\mathcal{M}$ is 4-connected and not graphic, and if J is a pair-free subset of I with $|J \cap X| \leq 1$ for each X in $\mathcal{M}$, then we must have $|J| \leq 2$. Thus for example if we wish to obtain a general result by applying theorem (4.2) and proposition (4.5) to the cuts of a graph we must take $|J| \leq 2$. We may in fact obtain the following odd little result, by taking $J = \{(r, s), (s, t)\}$.

(4.6) **Corollary** *Let r, s, t be distinct vertices in a random network. Then the marginal distributions of the capacities determine the joint distribution of* MAXFLOW $(r, \{s, t\})$ *and* MAXFLOW $(\{r, s\}, t)$.

§ 5. Splits of oriented matroids

In this last section we introduce another generalisation of a port of an oriented matroid, namely a 'split' of an oriented matroid. We show that a split satisfies conditions (C) and (C*) and that we can thus generalise the 'clockwise cycles' result (1.2). We now see that all our general examples of clutters satisfying conditions (C) and (C*) arise from oriented matroids either as ports or splits.

Let $\mathcal{M}$ be an oriented matroid on a paired set I. Given a linear order on the pairs we obtain a natural partial order on I. Suppose that we are also given a pair-free subset J of I. Then the corresponding split is the subset $\mathcal{N}$ of $\mathcal{M}$ constructed as follows: for each $X \in \mathcal{M}$, put X in $\mathcal{N}$ if and only if the first element in X is in J.

(5.1) **Proposition** *A split of an oriented matroid satisfies conditions* (C) *and* (C*).

Proof Let $X_1, X_2 \in \mathcal{N}$ and $x \in X_1$, $-x \in X_2$. Let y be the first element in $X_1 \cup X_2$. Then $y \in J$ and $-y \notin X_1 \cup X_2$. Hence, by property (II) of [3], there exists $X_3 \in \mathcal{M}$ with

$$y \in X_3 \subseteq (X_1 \cup X_2) \setminus \{x, -x\}.$$

But y is the first element in X_3, so $X_3 \in \mathcal{N}$. Thus condition (C*) holds, and of course condition (C) holds.

Let G be a plane simple graph, with a given embedding. Let $D(G)$ be the corresponding directed graph obtained by replacing each edge of G by a pair of opposite edges, and embedded in the plane in the obvious way. Let us see that the clutter $\mathscr{C}$ of clockwise cycles on the edge set I of $D(G)$ is a split of the oriented matroid $\mathscr{M}$ on I corresponding to the cycle matroid of G.

Construct a pair-free subset J of I by repeatedly applying the following rules. As edges of G are deleted we number them $1, 2, \ldots$.

(i) Delete any edges of G not in any cycle.

(ii) Let C be the outer cycle of any block in G, oriented clockwise. Delete the edges in C from G and add to J the corresponding edges of $D(G)$.

To see that the corresponding split does indeed yield $\mathscr{C}$, consider any $X \in \mathscr{M}$, with first element x. Let C be the cycle in G which was the 'outer cycle' when x was numbered. Then no edge in $\tilde{X}$ lies outside C, and so X is clockwise if and only if $x \in J$.

We may now see that a split of a binary oriented matroid need not satisfy condition (C*1). For example, if G consists of three parallel edges the corresponding clockwise cycles clutter is isomorphic to the clutter $\mathscr{C}_0$ of theorem (3.2).

Lastly let us consider a related application of proposition (5.1). Let G now be any simple graph. Then, with notation as in (1.1) and (1.2),

(5.2) $\quad P_d(\text{exists cycle}) \leqslant P_o(\text{exists directed cycle})$,

and further if G contains a cycle then the inequality is strict. To see this note that the clutter of directed cycles of length >2 in $D(G)$ satisfies condition (C), and fails to satisfy condition (C*) if G has a cycle: now use theorem 4.1 of [14].

We may now refine this result. For by proposition (5.1) we see that it is possible to assign an orientation to each cycle of G in such a way that

(5.3) $\quad P_d(\text{exists cycle}) = P_o$ (exists correctly oriented directed cycle).

References

[1] R. E. Bixby, *l*-matrices and a characterisation of binary matroids, *Discrete Math.* **8** (1974), 139–145.

[2] R. G. Bland, A combinatorial abstraction of linear programming, *J. Comb. Th. B.* **23** (1977), 33–57.

[3] R. G. Bland and M. Las Vergnas, Orientability of matroids, *J. Comb. Th. B.* **24** (1978), 94–123.

[4] J. Edmonds and D. R. Fulkerson, Bottleneck extrema, *J. Comb. Th.* 8 (1970), 299–306.

[5] T. I. Fenner and A. M. Frieze, On the existence of Hamiltonian cycles in a class of random graphs, *Discrete Math.* 45 (1983), 301–305.

[6] J. Folkman and J. Lawrence, Oriented matroids, *J. Comb. Th. B.* 25 (1978), 199–236.

[7] G. R. Grimmett and W-C. S. Suen, The maximal flow through a directed graph with random capacities, *Stochastics* 8 (1982), 153–159.

[8] K. Kuulasmaa, The spatial general epidemic and locally dependent random graphs, *J. Appl. Prob.* 19 (1982), 745–758.

[9] K. Kuulasmaa, The product representation of a locally dependent graph, *Stochastic Processes Appl.*, to appear.

[10] K. Kuulasmaa, Locally dependent random graphs and their use in the study of epidemic models, preprint, 1984.

[11] K. Kuulasmaa and S. Zachary, On spatial general epidemics and bond percolation processes, *J. Appl. Prob.* 21 (1984), 911–914.

[12] A. Lehman, A solution of the Shannon switching game, *J. SIAM* 12 (1964), 687–725.

[13] C. J. H. McDiarmid, Clutter percolation and random graphs, *Math. Progr. Study* 13 (1980), 17–25.

[14] C. J. H. McDiarmid, General percolation and random graphs, *Adv. Appl. Prob.* 13 (1981), 40–60.

[15] C. J. H. McDiarmid, General first-passage percolation, *Adv. Appl. Prob.* 15 (1983), 149–161.

[16] C. J. H. McDiarmid, Maths Reviews MR 84c: 05077.

[17] G. J. Minty, On the axiomatic foundations of the theories of directed linear graphs, electrical networks, and network-programming, *J. Math. Mech.* 15 (1966), 485–520.

[18] P. D. Seymour, A forbidden minor characterisation of matroid ports, *Quart. J. Math. (Oxford)* (2) 27 (1976), 407–413.

[19] P. D. Seymour, The forbidden minors of binary clutters, *J. London Math. Soc.* (2) 12 (1976), 356–360.

[20] P. D. Seymour, Triples in matroid circuits, *Europ. J. Combinatorics* 7 (1986), 177–185

SOME ENUMERATIVE RESULTS ON SERIES-PARALLEL NETWORKS

John W. MOON

University of Stirling,
Stirling, Scotland

1. Introduction

Let A and B denote two connected graphs each having two distinguished nodes called *terminals*. If a terminal of A is identified with a terminal of B, the resulting graph G is said to be obtained by joining A and B in *series*; the terminals of G are the terminals of A and B that were not identified. If the two terminals of A are identified with the two terminals of B in pairs, the resulting graph H is said to be obtained by joining A and B in *parallel*; the terminals of H are the two nodes obtained by identifying the terminals of A and B in pairs. The class of *series-parallel networks* may be defined recursively as follows; (i) the graph consisting of a single edge joining two terminals is a series-parallel network, the *trivial* network; (ii) any graph formed by joining two not necessarily distinct series-parallel networks in series or in parallel is also a series-parallel network.

If a series-parallel network has one or more cut-nodes, then it is called a *series-network* or a σ-*network*; if not, it is called a *parallel-network* or a π-*network*. Notice that the trivial network is a π-network but not a σ-network according to these definitions. If a σ-network N has $k-1$ cut-nodes where $k \geq 2$, then it is formed by joining in series k σ-networks called the constituents of N. Let j denote the maximum number of paths joining the terminals of a π-network N no two of which have any nodes in common other than the terminals: if $j \geq 2$, then N is formed by joining in parallel j networks each of which is a σ-network or the trivial network and these j networks are called the constituents of N; if $j=1$, then N is the trivial network and is said to have one constituent, namely, the network itself. Two π-networks or two σ-networks are regarded as the same if and only if they have the same constituents; thus the ordering of the constituents and their terminals is not taken into account.

MacMahon [11] gave a relation for the generating function for the number of series-parallel networks with n unlabelled edges. Riordan and Shannon [16] discussed procedures for calculating these numbers and gave some inequalities and approximations. Knödel [10] derived a relation for the generating function for the number of series-parallel networks with n labelled edges (as well as rediscovering the corresponding relation for the unlabelled case). Carlitz and Riordan [2] (see also [17; pp. 139–143]) considered the generating function for the number of series-parallel networks with n edges r of which have distinct labels and they derived recurrence and congruence relations for these numbers.

There is an obvious one-to-one correspondence between the π-networks and the σ-networks with $n \geq 2$ edges. It so happens that there is a natural one-to-one correspondence between these sets and two other sets of combinatorial interest: the set of rooted trees with n endnodes in which each interior node is incident with at least two edges leading away from the root; and, the set of bracketings of a product of n symbols in which at least two factors are adjacent within each pair of brackets and in which the order of the factors within each pair of brackets is immaterial. As a consequence, certain equivalent results have been obtained independently by different authors in different contexts.

Cayley [3] gave a relation for the generating function for the number of such trees with n unlabelled endnodes that was the same as the relation given later by MacMahon [11] for the generating function for networks with unlabelled edges. Schröder [20] derived a relation for the generating function for the number of such bracketings of n distinguishable symbols that was the same as the relation given later by Knödel [10] for the generating function for networks with labelled edges. Comtet [4] (see also [5; p. 224]) pointed out the correspondence between the bracketings of n distinguishable symbols and the trees with n labelled endnodes and determined the asymptotic behaviour of the number of these objects. Riordan [18] discussed the correspondence between these trees, bracketings, and networks and, among other things, derived relations for the generating functions that enumerate these trees not only by the number of endnodes but also by the number of interior nodes or by the degree of the root; he considered both the labelled and the unlabelled cases. More recently, Foulds and Robinson [7] also have determined the asymptotic behaviour of the number of these trees with n labelled endnodes in the course of enumerating various classes of what they call phylogenetic trees.

Usually there is no restriction upon the number of edges that may join a given pair of nodes in a series-parallel network; but Pfaff, Laskar, and Hedetniemi [15] prohibit multiple edges in the networks they consider. Our object here is to establish certain enumerative results for four types of networks: those in which the edges are or are not labelled and in which multiple edges are or are not permitted. We determine the asymptotic behaviour of the number of networks of these four types (repeating the results of Comtet [4] and Foulds and Robinson [7] for the sake of completeness). We also determine the asymptotic behaviour of the expected number of interior nodes in these networks (or, equivalently, the expected number of series operations involved in constructing them) and the expected number of multiple edges when such edges are permitted.

2. Preliminaries

In the later sections we shall derive relations for the generating functions for various numbers. To determine the asymptotic behaviour of the coefficients in these generating functions we shall appeal to the following special case of a result of Darboux [6; p. 20] (see also [1; p. 498]).

Lemma. *Suppose the function $F(x)=\sum F_n x^n$ has a non-zero finite radius of convergence τ and that $x=\tau$ is the only singularity on the circle of convergence. If $F(x)$ has an expansion about $x=\tau$ of the form*

$$F(x)=f(x)(\tau-x)^{-\kappa}+g(x) \qquad (2.1)$$

when $f(x)$ and $g(x)$ are analytic for $|x|\leqslant\tau$, $f(\tau)\neq 0$, and $\kappa\neq 0,1,2,\ldots$, then

$$F_n \sim \frac{f(\tau)}{\Gamma(\kappa)}\tau^{-n-\kappa}n^{\kappa-1} \qquad (2.2)$$

as $n\to\infty$.

We shall frequently encounter products of the form $G(x)F'(x)$ where $F(x)$ satisfies the hypothesis of the lemma with $\kappa=\frac{1}{2}$ and $G(x)$ does also or else converges when $|x|<\tau+\varepsilon$ for some $\varepsilon>0$ and, in either case, $G(\tau)\neq 0$. In these circumstances, it follows readily from the lemma (see also [12; p. 1001]) that

$$C_n\{G(x)F'(x)\}\sim \tau^{-1}G(\tau)nF_n \qquad (2.3)$$

where, for notational convenience, we let $C_n\{T(x)\}$ denote the coefficient of x^n in the power series $T(x)$.

Suppose we are given a generating function $F(x)$ and a functional relation $W(x,z)=0$ that is satisfied when $z=F(x)$. Harary, Robinson, and Schwenk [8] have described a procedure for establishing that $F(x)$ satisfies the hypothesis of the lemma with $\kappa=\frac{1}{2}$. (See [12; p. 1000] for a result when the relation $W(x,F)=0$ can be expressed in the form $F=x\varphi(F)$ where φ is a power series with non-negative coefficients.) If the steps in the procedure can be carried through, then the values of τ and $F(\tau)$ are determined by the relations $W(\tau, F(\tau))=0$ and

$$W_z(\tau, F(\tau))=0. \tag{2.4}$$

It turns out that the procedure described in [8] can be carried through, where necessary, for the generating functions we shall encounter here. The steps involved in applying the procedure to the generating functions here are similar to the corresponding steps in the cases illustrated in [8], so we shall omit the details and mention only the relations that determine the values of the constants involved. We remark that asymptotic results on trees established by arguments of a nature similar to those to be used here may be found, for example, in [19], [12], [14], or [13].

3. Labelled networks without multiple edges

Let Q_n and R_n denote the number of π-networks and σ-networks, respectively, with n labelled edges; in this section multiple edges are not permitted. We begin by giving relations for the generating functions $Q=Q(x)=\sum Q_n x^n/n!$ and $R=R(x)=\sum R_n x^n/n!$ which we use to determine the asymptotic behaviour of Q_n and R_n.

Theorem 1.

$$Q(x)=R(x)+\log(1+x) \tag{3.1}$$

and

$$2Q(x)=e^{Q(x)}-1+\log(1+x). \tag{3.2}$$

Proof. Every σ-network is obtained by joining in series two or more π-networks; the σ-networks thus obtained have no multiple edges if and only if the constituent π-networks have no multiple edges. Since the ordering of the π-networks is immaterial, it follows that

$$R=\sum_{2}^{\infty} \frac{Q^k}{k!},$$

or that

$$1+Q+R=e^Q. \tag{3.3}$$

Since we are enumerating networks without multiple edges, the required

π-networks may be split into three classes: (i) the trivial network; (ii) networks obtained by joining in parallel the trivial network and one or more σ-networks; and (iii) networks obtained by joining in parallel two or more σ-networks. The π-networks thus obtained have no multiple edges if and only if the constituent σ-networks have no multiple edges (bearing in mind that there are no edges joining the terminals in the σ-networks). The ordering of the constituent networks is immaterial, and when we consider separately these three classes of π-networks we find that

$$Q = x + x \sum_{1}^{\infty} \frac{R^k}{k!} + \sum_{2}^{\infty} \frac{R^k}{k!},$$

or that

$$1 + Q + R = (1+x)e^R. \tag{3.4}$$

Relation (3.1) now follows from (3.3) and (3.4); and relation (3.2) follows from (3.1) and (3.3).

Corollary 1.1.

$$\frac{Q_n}{n!} \sim \frac{R_n}{n!} \sim A\alpha^{-n} n^{-3/2}$$

as $n \to \infty$, where $\alpha = 4/e - 1 = .4715\ldots$ and $A = \frac{1}{4}((4-e)/\pi)^{\frac{1}{2}} = .1596\ldots$.

Proof. We have seen that the generating function $Q(x)$ satisfies the functional relation $W(x, Q) = 0$, where

$$W(x, z) = \log(1+x) + e^z - 1 - 2z.$$

When we apply the procedure described in [8] to the function $Q(x)$ (or a slight modification of the argument used in [12; p. 1000]), we find that $Q(x)$ is analytic for $|x| \leq \alpha$, $x \neq \alpha = 4/e - 1$ and that in the neighbourhood of $x = \alpha$ it has an expansion of the form

$$Q(x) = Q(\alpha) - a(\alpha - x)^{\frac{1}{2}} + a_2(\alpha - x)^1 + \ldots. \tag{3.5}$$

The conditions that $W_z(\alpha, Q(\alpha)) = 0$ and $W(\alpha, Q(\alpha)) = 0$ imply that $Q(\alpha) = \ln 2$

and $\alpha = 4/e - 1$. Furthermore, relations (3.5) and (3.2) imply that

$$\tfrac{1}{2}a^2 = \lim_{x \to \alpha^-} (\ln 2 - Q(x)) Q'(x) = \lim_{x \to \alpha^-} \frac{\ln 2 - Q(x)}{2 - e^{Q(x)}} \frac{1}{1+x}$$

$$= \tfrac{1}{4} e \lim_{x \to \alpha^-} e^{-Q(x)} = \tfrac{1}{8} e,$$

so $a = \tfrac{1}{2}\sqrt{e}$. The required result now follows upon applying Darboux's result to the function $Q(x)$, since

$$\frac{Q_n}{n!} = \frac{R_n}{n!} - \frac{(-1)^n}{n}$$

by (3.1).

Let $V = V(x) = Q(x) + R(x)$. We observe for later use that it follows from the preceding results that $V(\alpha) = 1$ and that $(1 + V(x))^{-1} = e^{-Q(x)}$ is analytic for $|x| \leq \alpha$, $x \neq \alpha$ and that in the neighbourhood of $x = \alpha$ it has an expansion of the form

$$(1 - V(x))^{-1} = \tfrac{1}{2} + \tfrac{1}{4}\sqrt{e}(\alpha - x)^{\frac{1}{2}} + \ldots. \tag{3.6}$$

We remark that it is not difficult to derive some results on the number of constituents in networks. In particular, it can be shown that as the number of edges tends to infinity the expected number of constituents in a σ-network tends to $1 + 2\ln 2 = 2.3862\ldots$ and the expected number of constituents in a π-network tends to $3 - \tfrac{1}{4}e = 2.3204\ldots$. Furthermore, the probability that the terminals are joined in a π-network tends to $2 - \tfrac{1}{2}e = .6408\ldots$; consequently, if we did not count constituents consisting of a single edge in π-networks, the expected number of constituents over both types of networks would tend to 2.

Let $I(N_n)$ denote the number of *interior nodes* in the network N_n with n edges, i.e., the number of nodes in N_n other than the terminals. It is easy to see that $I(N_n)$ is the number of series operations involved in constructing N_n and $n - 1 - I(N_n)$ is the number of parallel operations involved. (We remark that $n - 1 - I(N_n)$ is also the number of independent cycles in the network N_n.) In order to determine the expected value of $I(N_n)$ we need to introduce an additional variable into the generating functions we have been considering.

Let

$$Q(x, y) = \sum Q_{ni} y^i \frac{x^n}{n!}$$

and
$$R(x,y) = \sum R_{ni} y^i \frac{x^n}{n!}$$

where Q_{ni} and R_{ni} denote the number of π-networks and σ-networks, respectively, with n labelled edges and with i interior nodes. If

$$I_Q = I_Q(x) = \sum I_Q(n) Q_n \frac{x^n}{n!}$$

and

$$I_R = I_R(x) = \sum I_R(n) Q_n \frac{x^n}{n!},$$

where $I_Q(n)$ and $I_R(n)$ denote the expected value of $I(N_n)$ over all π-networks and σ-networks N_n, respectively, then $I_Q = Q_y(x, 1)$ and $I_R = R_y(x, 1)$. It will also be convenient to let $V(x, y) = Q(x, y) + R(x, y)$ and $I_V = I_Q + I_R$. We shall continue to write V, Q, and R for the functions $V(x) = V(x, 1)$, $Q(x) = Q(x, 1)$, and $R(x) = R(x, 1)$.

Theorem 2.

$$(1 - V)I_V = VQ - R.$$

Proof. We need to obtain the analogues of relations (3.3) and (3.4) for the generating functions $Q(x, y)$ and $R(x, y)$. If we repeat the observations that led to relation (3.3) and bear in mind that in a σ-network with k constituents the number of interior nodes in the entire network is $k-1$ plus the number of interior nodes in the constituents, we find that

$$R(x, y) = \sum_{2}^{\infty} y^{k-1} \frac{Q^k(x, y)}{k!}$$

or that

$$1 + yV(x, y) = e^{yQ(x, y)}. \tag{3.7}$$

The number of interior nodes in a π-network is the number of interior nodes in its constituents, so essentially the same argument that led to relation (3.4) implies that

$$1 + V(x, y) = (1 + x) e^{R(x, y)}. \tag{3.8}$$

When we differentiate both sides of equations (3.7) and (3.8) with respect to y and then set $y=1$, we find that

$$I_V + V = (1+V)(I_Q + Q)$$

and

$$I_V = (1+V)I_R. \tag{3.9}$$

The required result now follows upon adding these last two equations together and simplifying.

Corollary 2.1.

$$I_Q(n) \sim I_R(n) \sim \lambda n$$

as $n \to \infty$, where

$$\lambda = \frac{2}{4-e} \ln(4/e) = 0.6027\ldots.$$

Proof. It follows from (3.3) and (3.2) that

$$(1-V)Q' = (2-e^Q)Q' = (1+x)^{-1},$$

so Theorem 2 implies that

$$I_V = (1+x)(VQ - R)Q'.$$

It is not difficult to see that we may apply formula (2.3) to this product and conclude that

$$C_n\{I_V\} \sim (1+\alpha)\alpha^{-1}(V(\alpha)Q(\alpha) - R(\alpha))n\frac{Q_n}{n!} = 2n\lambda\frac{Q_n}{n!} \tag{3.10}$$

as $n \to \infty$.

Furthermore,

$$I_R = (1+V)^{-1}I_V,$$

by (3.9). In view of relation (3.6) we may apply formula (2.3) to this product also and conclude that

$$I_R(n)\frac{R_n}{n!} \sim (1+V(\alpha))^{-1}C_n\{I_V\} \sim \lambda n \frac{Q_n}{n!}. \qquad (3.11)$$

This implies that $I_R(n) \sim \lambda n$ since $R_n \sim Q_n$; and the conclusion that $I_Q(n) \sim \lambda n$ follows from the fact that $I_Q = I_V - I_R$.

We remark that it can be shown that the variance of $I(N_n)$ equals $O(n^{3/2})$ for both π-networks and σ-networks N_n; hence the probability that $|I(N_n)/n - \lambda| > \varepsilon$ tends to zero as $n \to \infty$ for any fixed $\varepsilon > 0$.

4. Labelled networks with multiple edges permitted

Let $P = P(x) = \sum P_n x^n/n!$ and $S = S(x) = \sum S_n x^n/n!$ where P_n and S_n denote the number of π-networks and σ-networks, respectively, with n labelled edges; in this section multiple edges are permitted. There is an obvious one-to-one correspondence between these π-networks and σ-networks with n edges when $n \geq 2$, so $P(x) = x + S(x)$.

As we mentioned earlier, the numbers P_n have been considered by several different authors in different contexts. In particular, the following relation for the generating function $P(x)$ was given by Schröder [20] in the context of bracketings; by Knödel [10] in the context of networks; and by Foulds and Robinson [7] in the context of trees; see also [2], [4], and [18]. We deduce it as a consequence of Theorem 1.

Theorem 3.

$$2P(x) = e^{P(x)} - 1 + x.$$

Proof. The generating function $Q(x)$ enumerates those labelled π-networks that have no multiple edges. If we replace x in this generating function by $x + x^2/2! + \ldots = e^x - 1$, then it is not difficult to see that each labelled π-network in which multiple edges are permitted is obtained once and only once. (The factorials are present in the denominators because the ordering of the edges joining a given pair of nodes is immaterial.) Hence, $P(x) = Q(e^x - 1)$ and the required relation for $P(x)$ now follows from (3.2). Similarly, $S(x) = R(e^x - 1)$ so the obvious relation $P(x) = x + S(x)$ is implied by (3.1).

The following result was deduced from the relation in Theorem 3 by Comtet [4] and Foulds and Robinson [7]; it is also a special case of a result in [12; p. 1000].

Corollary 3.1.

$$\frac{P_n}{n!} \sim B\beta^{-n} n^{-3/2}$$

as $n \to \infty$, where $\beta = \ln(4/e) = .3862...$ and $B = \frac{1}{2}(\beta/\pi)^{\frac{1}{2}} = .1753...$.

Foulds and Robinson [7] established this by showing, by a somewhat more direct version of the procedure described in [8], that the function $P(x)$ is analytic for $|x| \leq \beta$, $x \neq \beta = \ln(4/e)$ and that in the neighbourhood of $x = \beta$ it has an expansion of the form

$$P(x) = \ln 2 - (\beta - x)^{\frac{1}{2}} + b_2(\beta - x)^1 + \dots. \tag{4.1}$$

The required conclusion then follows upon appealing to Darboux's result.

Notice that it follows from Corollaries 1.1 and 3.1 that the probability that a π-network or a σ-network with n labelled edges has no multiple edges is asymptotic to

$$\tfrac{1}{2}(e\alpha/\beta)^{\frac{1}{2}}(\beta/\alpha)^n = (.9107...)(.8192...)^n$$

as $n \to \infty$.

Let $U = U(x) = P(x) + S(x)$. We observe for later use that it follows from the preceding results that $U(\beta) = 1$ and that $(1 + U(x))^{-1} = e^{-P(x)}$ is analytic for $|x| \leq \beta$, $x \neq \beta$ and that in the neighbourhood of $x = \beta$ it has an expansion of the form

$$(1 + U(x))^{-1} = \tfrac{1}{2} + \tfrac{1}{2}(\beta - x)^{\frac{1}{2}} + \dots. \tag{4.2}$$

As in the case considered in the last section, it is not difficult to derive some results on the number of constituents in these networks. (See [18; pp. 11–13] for some material on the analogous problem for the family of trees mentioned earlier.) In particular, it can be shown that the expected number of constituents in π-networks and σ-networks tends to $1 + 2\ln 2 = 2.3862...$ as the number of edges tends to infinity. Furthermore, the probability that the terminals are joined in a π-network tends to $2 - \tfrac{1}{2}e = .6408...$, as in the case considered in the last section; and the expected number of edges joining the terminals of a π-network N_n equals $2nP_{n-1}/P_n$ for $n \geq 3$ and this quantity tends to $2\beta = .7725...$ as $n \to \infty$.

Let $M(N_n)$ denote the number of *multiple edges* in the network N_n; if there are $k \geq 2$ edges joining a given pair of nodes this counts as $k - 1$ multiple edges. We now give a relation for the generating functions

$$M_P = M_P(x) = \sum M_P(n) P_n \frac{x^n}{n!}$$

and

$$M_S = M_S(x) = \sum M_S(n) S_n \frac{x^n}{n!}$$

where $M_P(n)$ and $M_S(n)$ denote the expected value of $M(N_n)$ over all labelled π-networks and σ-networks N_n, respectively.

Theorem 4.

$$M_P(x) = M_S(x) + e^{-x} - 1 + x \tag{4.3}$$

and

$$M_P(x) = (e^{-x} - 1 + x) P'(x). \tag{4.4}$$

Proof. We observed earlier that $P(x) = Q(e^x - 1)$. More generally, it is not difficult to see that if we replace x by $(e^{xt} - 1)/t$ in $Q(x)$, then the coefficient of $t^k x^n/n!$ in the resulting expression is the number of π-networks N_n with k multiple edges. It follows, therefore, that

$$M_P(x) = \frac{\partial}{\partial t} (Q((e^{xt} - 1)/t))_{t=1}$$

and similarly, that

$$M_S(x) = \frac{\partial}{\partial t} (R((e^{xt} - 1)/t))_{t=1}.$$

If we replace x by $(e^{xt} - 1)/t$ in relation (3.1), differentiate both sides with respect to t and set $t = 1$, we obtain relation (4.3). And if we apply the same procedure to relation (3.2), we obtain the relation

$$2M_P(x) = M_P(x) e^{P(x)} + e^{-x} - 1 + x.$$

This implies relation (4.4) since $(2 - e^P)P' = 1$ by Theorem 3.

Corollary 4.1.

$$M_P(n) \sim M_S(n) \sim \chi n$$

as $n \to \infty$, where

$$\chi = (e^{-\beta} - 1 + \beta)\beta^{-1} = .1705\ldots.$$

Proof. Relation (4.3) implies that

$$M_P(n)P_n = M_S(n)S_n + (-1)^n$$

for $n \geq 2$ and we know that $P_n = S_n$ for $n \geq 2$. The required result now follows readily upon applying formula (2.3) to the product in (4.4).

It can be shown that the variance of $M(N_n)$ equals $O(n^{3/2})$ for both π-networks and σ-networks N_n; hence the probability that $|M(N_n)/n - \chi| > \varepsilon$ tends to zero as $n \to \infty$ for any fixed $\varepsilon > 0$.

For both π-networks and σ-networks N_n, it can be shown that as $n \to \infty$ the expected number of pairs of nodes joined by two or more edges is asymptotic to $(1 - (1+\beta)e^{-\beta})\beta^{-1}n = (.1499\ldots)n$ and the expected number of pairs of nodes joined by exactly k edges is asymptotic to $\beta^{k-1}e^{-\beta}/k!$ for each fixed positive integer k.

In the last section we considered the expected number of interior nodes in labelled networks without multiple edges; we now consider the corresponding problem for networks in which multiple edges are permitted. Let

$$I_P = I_P(x) = \sum I_P(n) P_n \frac{x^n}{n!}$$

and

$$I_S = I_S(x) = \sum I_S(n) S_n \frac{x^n}{n!}$$

where $I_P(n)$ and $I_S(n)$ denote the expected value of $I(N_n)$ over all labelled π-networks and σ-networks N_n, respectively.

Theorem 5.

$$I_S = (1+U)^{-1}(xP' - P).$$

Proof. It is not difficult to see that $I_P(x) = I_Q(e^x - 1)$ and $I_S(x) = I_R(e^x - 1)$. Hence it follows from (3.9) that

$$I_P + I_S = (1+U)I_S.$$

Now I_P and I_S are the generating functions for the expected number of series operations involved in π-networks and σ-networks, respectively. Because of the duality between non-trivial π-networks and σ-networks, it follows that I_S is also the generating function for the expected number of parallel operations involved in π-networks. But in any π-network N_n the number of series operations involved plus the number of parallel operations involved equals $n-1$. Consequently

$$xP' - P = I_P + I_S = (1+U)I_S,$$

as required. (We remark that this result can also be deduced from (3.9) and Theorems 2 and 3; but the present argument, based on the duality between π-networks and σ-networks, is perhaps a bit more direct.)

Corollary 5.1.

$$I_P(n) \sim I_S(n) \sim \tfrac{1}{2} n \text{ as } n \to \infty.$$

Proof. If we appeal to relation (4.2) and apply formula (2.3) to the expression for I_S, we find that

$$I_S(n) \frac{S_n}{n!} \sim \tfrac{1}{2} n \frac{P_n}{n!}.$$

The required result now follows immediately since $S_n = P_n$ for $n \geq 2$ and $I_P(n) + I_S(n) = n - 1$.

We remark that if we take into account more terms in the expansion of $(1+U)^{-1}(xP' - P)$, we find that

$$I_S(n) = \tfrac{1}{2} n - 1.9367\ldots + O(n^{-1})$$

as $n \to \infty$. Furthermore, it can be shown that the variance of $I(N_n)$ equals $O(n^{3/2})$ for both π-networks and σ-networks N_n, so the probability that $|I(N_n)/n - \tfrac{1}{2}| > \varepsilon$ for labelled networks in which multiple edges are permitted tends to zero as $n \to \infty$ for any fixed $\varepsilon > 0$.

The first few values of some of the quantities we have been considering in the past two sections are given in Table 1 for the case of π-networks.

Table 1.

Values for π-networks with labelled edges.

n	1	2	3	4	5	6
Q_n	1	0	3	7	90	676
$I_Q(n)Q_n$	0	0	3	14	195	2059
P_n	1	1	4	26	236	2752
$I_P(n)P_n$	0	0	3	32	410	6164
$M_P(n)P_n$	0	1	2	21	224	3075

5. Unlabelled networks without multiple edges

Let q_n and r_n denote the number of π-networks and σ-networks, respectively, with n unlabelled edges; in this section multiple edges are not permitted. We derive relations for the generating functions $q=q(x)=\sum q_n x^n$ and $r=r(x)=\sum r_n x^n$ from which we deduce the asymptotic behaviour of q_n and r_n.

Theorem 6.

$$q(x) = x - x^2 + r(x) \tag{5.1}$$

and

$$1 - x + x^2 + 2q(x) = \prod_{1}^{\infty} (1-x^j)^{-q_j} = \exp \sum_{1}^{\infty} q(x^k)/k. \tag{5.2}$$

Proof. It is not difficult to see that for $n \geq 1$ the coefficient of x^n in the expansion of $\Pi(1-x^j)^{-q_j}$ is the number of unordered collections with repetitions permitted of one or more π-networks with a total of n unlabelled edges; the number of such collections consisting of a single π-network is simply the number of π-networks N_n, and the number of such collections consisting of two or more π-networks equals the number of σ-networks N_n. It follows, therefore, that

$$1 + q(x) + r(x) = \prod_{1}^{\infty} (1-x^j)^{-q_j}. \tag{5.3}$$

Similarly, for $n \geq 1$ the coefficient of x^n in the expansion of $\Pi(1-x^j)^{-r_j}$ is the number of collections of one or more σ-networks with a total of n edges; the number of collections consisting of a single σ-network is simply the number of σ-networks N_n, and the number of such collections consisting of two or more σ-networks equals the number of π-networks N_n in which the terminals are not joined by an edge. Since the generating function for π-networks in which the terminals are joined by an edge is $x\prod(1-x^j)^{-q_j}$, it follows that

$$1+q(x)+r(x)=(1+x)\prod(1-x^j)^{-r_j}. \tag{5.4}$$

Now $q_1=r_2=1$ and $r_1=q_2=0$; consequently

$$\prod_{3}^{\infty}(1-x^j)^{-q_j}=\prod_{3}^{\infty}(1-x^j)^{-r_j},$$

in view of (5.3) and (5.4). The conclusion that $q_j=r_j$ for $j\geqslant 3$ now follows by induction and this implies relation (5.1). Relations (5.1) and (5.3) imply the first part of relation (5.2) and the final expression is obtained by taking the logarithm of the product and interchanging the order of summation.

Corollary 6.1.

$$q_n\sim G\gamma^{-n}n^{-3/2}$$

as $n\to\infty$, where $G=.1972\ldots$ and $\gamma=.3462\ldots$.

Proof. The generating function $q(x)$ satisfies the functional relation $W(x, q)=0$ where

$$W(x,z)=\exp\left(z+\sum_{2}^{\infty}q(x^k)/k\right)-2z-1+x-x^2.$$

When we apply the procedure described in [8] to the function $q(x)$, we find that $q(x)$ is analytic for $|x|\leqslant\gamma$, $x\neq\gamma$ and that in the neighbourhood of $x=\gamma$ it has an expansion of the form

$$q(x)=q(\gamma)-g(\gamma-x)^{\frac{1}{2}}+g_2(\gamma-x)^1+\ldots. \tag{5.5}$$

It then follows from Darboux's result that $q_n\sim G\gamma^{-n}n^{-3/2}$ where $G=\frac{1}{2}g(\gamma/\pi)^{1/2}$. The conditions $W(\gamma, q(\gamma))=0$ and $W_z(\gamma, q(\gamma))=0$ imply that

$$q(\gamma)=\tfrac{1}{2}(1+\gamma-\gamma^2), \tag{5.6}$$

and this and relation (5.2) imply that

$$\prod_{1}^{\infty}(1-\gamma^j)^{-q_j}=2.$$

Knowing the first few values of the numbers q_j, we can use this condition to estimate the value of γ and we find that $\gamma = .3462\ldots$.

To determine the value of g we observe that relation (5.2) implies that

$$(1-v(x))q'(x) = 1 - 2x + (1+v(x)) \sum_{2}^{\infty} x^{k-1} q'(x^k) \tag{5.7}$$

where

$$v(x) = q(x) + r(x) = -x + x^2 + 2q(x).$$

It follows from expansion (5.5) and equation (5.6) that

$$\lim_{x \to \gamma^-} (1-v(x))q'(x) = g^2.$$

Consequently,

$$g^2 = 1 - 2\gamma + (1+v(\gamma)) \sum_{2}^{\infty} \gamma^{k-1} q'(\gamma^k)$$

$$= 1 - 2\gamma + 2 \sum_{1}^{\infty} nq_n \gamma^{2n-1}(1-\gamma^n)^{-1}. \tag{5.8}$$

From this we find that $g = 1.1882\ldots$ and, hence, that $G = \frac{1}{2}g(\gamma/\pi)^{1/2} = .1972\ldots$ as required.

We observe for later use that it follows from the preceding results that the functions $(1+v(x))^{-1}$ and $(1-v(x))^{-1}$ are both analytic for $|x| \leq \gamma$, $x \neq \gamma$ and that in the neighbourhood of $x = \gamma$ they have expansions of the form

$$(1+v(x))^{-1} = \tfrac{1}{2} + \tfrac{1}{2}g(\gamma - x)^{\frac{1}{2}} + \ldots \tag{5.9}$$

and

$$(1-v(x))^{-1} = \tfrac{1}{2}g^{-1}(\gamma - x)^{-\frac{1}{2}} + k_0 + k_1(\gamma - x)^{\frac{1}{2}} + \ldots . \tag{5.10}$$

We also remark that it can be shown that the expected number of constituents in unlabelled π-networks N_n with no multiple edges tends to $2.3254\ldots$ as $n \to \infty$ and the corresponding limit for π-networks is $2.5979\ldots$. Hence the expected number of constituents over both types of networks tends to $2.4616\ldots$.

We now consider the generating functions

$$g(x, y) = \sum q_{ni} y^i x^n$$

and

$$r(x, y) = \sum r_{ni} y^i x^n$$

where q_{ni} and r_{ni} denote the number of π-networks and σ-networks, respectively, with n unlabelled edges and i interior nodes. Let

$$I_q = I_q(x) = \sum I_q(n) q_n x^n$$

and

$$I_r = I_r(x) = \sum I_r(n) r_n x^n$$

where $I_q(n)$ and $I_r(n)$ denote the expected value of $I(N_n)$ over all such π-networks and σ-networks N_n, respectively. Then, as before, $I_q = q_y(x, 1)$ and $I_r = r_y(x, 1)$. For convenience we shall let $v(x, y) = q(x, y) + r(x, y)$ and $I_v = I_q + I_r$.

Theorem 7.

$$2I_v(x) + v(x) = (1 + v(x)) \sum_1^\infty \{I_v(x^k) + q(x^k)\}.$$

Proof. We observed earlier, in proving Theorem 2, that if N_n is a σ-network with k constituents then $I(N_n)$ equals $k-1$ plus the number of interior nodes within the constituents, and if N_n is a π-network then $I(N_n)$ equals the number of interior nodes within the constituents. If we repeat the arguments that led to relations (5.3) and (5.4), keeping these observations in mind, we find that

$$1 + yv(x, y) = \prod (1 - y^{i+1} x^j)^{-q_{ji}}$$

$$= \exp \sum_1^\infty y^k q(x^k, y^k)/k \qquad (5.11)$$

and

$$1 + v(x, y) = (1 + x) \prod (1 - y^i x^j)^{-r_{ji}}$$

$$= (1 + x) \exp \sum_1^\infty r(x^k, y^k)/k. \qquad (5.12)$$

The extra factor y is present in the first part of relation (5.11) because when the product is expanded the exponent of y in terms corresponding to collections of k π-networks is k plus the number of interior nodes within the networks and this is one more than it should be in $v(x, y)$.

When we differentiate these relations with respect to y and set $y=1$, we find that

$$I_v(x)+v(x)=(1+v(x))\sum_1^\infty \{I_q(x^k)+q(x^k)\}$$

and

$$I_v(x)=(1+v(x))\sum_1^\infty I_r(x^k). \tag{5.13}$$

The required result now follows upon adding these last two equations together.

Corollary 7.1.

$$I_q(n)\sim I_r(n)\sim \psi n$$

as $n\to\infty$, where

$$\psi=\frac{1}{2\gamma g^2}(\gamma-\gamma^2+2\sum_2^\infty \{I_v(\gamma^k)+q(\gamma^k)\})=.6615\ldots.$$

Proof. Theorem 7 implies that

$$I_b=(1-v)^{-1}(vq-r+(1+v)w) \tag{5.14}$$

where

$$w=w(x)=\sum_2^\infty \{I_v(x^k)+q(x^k)\}.$$

Now the function $w(x)$ converges when $|x|\leqslant \gamma^{1/2}$ since $I_v(n)\leqslant n-1$, so it follows from the preceding results that the last factor on the right hand side of equation (5.14) is analytic when $|x|\leqslant \gamma$, $x\neq \gamma$; furthermore, in the neighbourhood of $x=\gamma$ it has an expansion of the form

$$vq-r+(1+v)w=K-h(\gamma-x)^{\frac{3}{2}}+\ldots \tag{5.15}$$

where, recalling (5.1) and the fact that $v(\gamma)=1$,

$$K = v(\gamma)q(\gamma) - r(\gamma) + (1+v(\gamma))w(\gamma)$$
$$= \gamma - \gamma^2 + 2w(\gamma) = 2\gamma g^2 \psi.$$

Consequently, in view of (5.14) and expansions (5.10) and (5.15), the function $I_v(x)$ itself is analytic when $|x| \leqslant \gamma$, $x \neq \gamma$; and in the neighbourhood of $x = \gamma$ it has an expansion of the form

$$I_v(x) = \gamma g \psi (\gamma - x)^{-\frac{1}{2}} + j_0 + j_1(\gamma - x)^{\frac{1}{2}} + \ldots. \tag{5.16}$$

We may therefore conclude, appealing to Darboux's result and Corollary 6.1, that

$$C_n\{I_v(x)\} \sim (\gamma/\pi)^{\frac{1}{2}} g \psi \gamma^{-n} n^{-\frac{1}{2}} \sim 2\psi n q_n = 2\psi n r_n \tag{5.17}$$

as $n \to \infty$.

Relation (5.13) may be rewritten as

$$\sum_1^\infty I_r(x^k) = (1+v(x))^{-1} I_v(x). \tag{5.18}$$

If we apply Darboux's result to the product $(1+v)^{-1} I_v$ and make use of relations (5.9) and (5.16), we find that

$$C_n\{(1+v(x))^{-1} I_v(x)\} \sim \tfrac{1}{2} C_n\{I_v(x)\} \sim \psi n r_n \tag{5.19}$$

as $n \to \infty$. Now the coefficient of x^n in $\sum_2^\infty I_r(x^k)$ is negligible compared with $\psi n r_n$ since the function converges for $|x| \leqslant \gamma^{1/2}$. Hence $I_r(n) \sim \psi n$ as $n \to \infty$, from (5.18) and (5.19); and the result that $I_q(n) \sim \psi n$ follows immediately from (5.17) and the fact that $I_q = I_v - I_r$.

6. Unlabelled networks with multiple edges permitted

Let $p = p(x) = \sum p_n x^n$ and $s = s(x) = \sum s_n x^n$ where p_n and s_n denote the number of π-networks and σ-networks, respectively, with n unlabelled edges; in this section multiple edges are permitted. As in the labelled case, there is an obvious one-to-one correspondence between the π-networks and the σ-networks with n edges when $n \geqslant 2$, so $p(x) = x + s(x)$.

If we appeal to the fact that every network is either a π-network or an arrangement in series of an unordered collection of two or more π-networks, then it is not difficult to see that the following analogue of relation (5.3) holds:

$$1+p(x)+s(x)=\prod(1-x^j)^{-p_j}. \tag{6.1}$$

Since $s(x)=p(x)-x$, this implies the following result given by Cayley [3] in the context of the trees described earlier and by MacMahon [11] and Knödel [10] in the context of networks.

Theorem 8.

$$1-x+2p(x)=\prod(1-x^j)^{-p_j}=\exp\sum_1^\infty p(x^k)/k.$$

Corollary 8.1.

$$p_n \sim D\delta^{-n}n^{-3/2}$$

as $n\to\infty$, where $D=.2063\ldots$ and $\delta=.2808\ldots$.

Proof. The generating function $p(x)$ satisfies the functional relation $W(x.p)=0$ where

$$W(x,z)=\exp\left(z+\sum_2^\infty p(x^k)/k\right)-2z-1+x.$$

When we apply the procedure described in [8] to the function $p(x)$, we find that $p(x)$ is analytic for $|x|\leqslant\delta$, $x\neq\delta$ and that in the neighbourhood of $x=\delta$ it has an expansion of the form

$$p(x)=p(\delta)-d(\delta-x)^{\frac{1}{2}}+d_2(\delta-x)^1+\ldots. \tag{6.2}$$

It then follows from Darboux's result that $p_n \sim D\delta^{-n}n^{-3/2}$ where $D=\frac{1}{2}d(\delta/\pi)^{1/2}$. The conditions $W(\delta,p(\delta))=0$ and $W_z(\delta,p(\delta))=0$ imply that

$$p(\delta)=\tfrac{1}{2}(1+\delta), \tag{6.3}$$

and this and Theorem 8 imply that

$$\prod_1^\infty (1-\delta^j)^{-p_j}=2.$$

From this condition we can estimate the value of δ and we find that $\delta=.2828\ldots$.

To determine the value of d we observe that Theorem 8 implies that

$$(1-u(x))p'(x)=1+(1+u(x))\sum_{2}^{\infty} x^{k-1}p'(x^k) \qquad (6.4)$$

where

$$u(x)=p(x)+s(x)=-x+2p(x).$$

It follows from expansion (6.2) and equation (6.3) that

$$\lim_{x\to\delta^-}(1-u(x))p'(x)=d^2.$$

Consequently,

$$d^2=1+(1+u(\delta))\sum_{2}^{\infty} \delta^{k-1}p'(\delta^k)$$

$$=1+2\sum_{1}^{\infty} np_n\delta^{2n-1}(1-\delta^n)^{-1}. \qquad (6.5)$$

From this we find that $d=1.3805\ldots$ and, hence, that $D=\tfrac{1}{2}d(\delta/\pi)^{1/2}=.2063\ldots$ as required.

Notice that it follows from Corollaries 6.1 and 8.1 that the probability that a π-network or a σ-network with n unlabelled edges has no multiple edges is asymptotic to

$$G/D(\delta/\pi)^n=(.9557\ldots)(.8109\ldots)^n$$

as $n\to\infty$.

We observe for later use that it follows from the preceding results that the functions $(1+u(x))^{-1}$ and $(1-u(x))^{-1}$ are both analytic for $|x|\leqslant\delta$, $x\neq\delta$ and that in the neighbourhood of $x=\delta$ they have expansions of the form

$$(1+u(x))^{-1}=\tfrac{1}{2}+\tfrac{1}{2}d(\delta-x)^{\frac{1}{2}}+\ldots \qquad (6.6)$$

and

$$(1-u(x))^{-1}=\tfrac{1}{2}d^{-1}(\delta-x)^{-\frac{1}{2}}+t_0+t_1(\delta-x)^{\frac{1}{2}}+\ldots. \qquad (6.7)$$

We also remark that it can be shown that the expected number of constituents in unlabelled networks N_n tends to 2.5161... as $n \to \infty$. Riordan [18; p. 10], in considering the trees described earlier, has in effect determined the number of unlabelled networks N_n with a given number of constituents for $n \leq 9$.

Recall (see, e.g., [5; p. 53]) that the Catalon numbers

$$c_n = \frac{1}{n}\binom{2n-2}{n-1}$$

satisfy the recurrence relation $c_n = \sum_{1}^{n-1} c_i c_{n-i}$ for $n \geq 2$ and that

$$\sum_{1}^{\infty} c_n x^n = \tfrac{1}{2}(1-(1-4x)^{\frac{1}{2}}).$$

Riordan and Shannon [16] gave some estimates for the numbers p_n. In particular, they showed that

$$p_n \leq c_n \tag{6.8}$$

for $n \geq 1$; this follows by induction on n upon observing that $p_n \leq \sum_{1}^{n-1} p_i p_{n-i}$ for $n \geq 2$. Inequality (6.8) is useful in bounding the error made in estimating the value of d^2 by the first few terms of the infinite series in equation (6.5), and a similar remark applies to the series in equation (5.8) since $q_n \leq p_n \leq c_n$.

Riordan and Shannon [16] also considered various computational procedures for determining the numbers $u_n = p_n + s_n$. One of their procedures involved the coefficients e_n of the generating function $e(x) = \sum e_n x^n$ defined by the relation

$$\Pi(1-x^j)^{p_j} = 1 - e(x).$$

It follows from equation (6.1) that

$$u(x) = e(x) + u(x)e(x). \tag{6.9}$$

Consequently,

$$u_n = e_n + \sum_{i=1}^{n-1} e_i u_{n-i}$$

for $n \geq 1$ or, since $e_1 = e_2 = 1$, that

$$u_n = e_n + u_{n-1} + u_{n-2} + \sum_{i=3}^{n-1} e_i u_{n-i} \qquad (6.10)$$

for $n \geq 3$.

They observed empirically that $4e_n$ is approximately equal to u_n when n is large. This led them to estimate the numbers u_n by the numbers m_n where $m_1 = u_1 = 1$, $m_2 = u_2 = 2$, and

$$3m_n = 4m_{n-1} + 4m_{n-2} + \sum_{i=3}^{n-1} m_i m_{n-i} \qquad (6.11)$$

for $n \geq 3$; this recurrence relation is obtained by replacing u_n and e_n in (6.10) by m_n and $\frac{1}{4}m_n$, respectively. From (6.9) they deduced that

$$m_n \sim L\omega^n n^{-3/2}$$

where L is about $3/7$ and ω is about 3.56.

Notice that equation (6.9) implies that $e(x) = u(x)(1 + u(x))^{-1}$, or that

$$e'(x) = (1 + u(x))^{-2} u'(x).$$

If we appeal to (6.6) and apply formula (2.3) to this product, we find that

$$ne_n \sim (1 + u(\delta))^{-2} nu_n = \tfrac{1}{4} nu_n$$

so it is indeed true that $u_n/e_n \to 4$ as $n \to \infty$.

Let $g(x, t) = \sum g_{ni} t^i x^n$ and $h(x, t) = \sum h_{ni} t^i x^n$ where g_{ni} and h_{ni} denote the number of π-networks and σ-networks, respectively, with n unlabelled edges i of which are multiple edges. (As before, if there are k edges joining a pair of nodes in a network N_n, then this contributes $k-1$ to the number $M(N_n)$ of multiple edges in N_n.) If

$$M_p = M_p(x) = \sum M_p(n) p_n x^n$$

and

$$M_s = M_s(x) = \sum M_s(n) s_n x^n,$$

where $M_p(n)$ and $M_s(n)$ denote the expected value of $M(N_n)$ over all unlabelled π-networks and σ-networks N_n, respectively, then $M_p = g_t(x, 1)$ and $M_s = h_t(x, 1)$. For notational convenience we let $M_u = M_p + M_s$.

Theorem 9.

$$\sum_{1}^{\infty} M_p(x^k) = x^2(1-x)^{-1} + \sum_{1}^{\infty} M_s(x^k) \tag{6.12}$$

and

$$2M_u(x) = (1+u(x))\{x^2(1-x)^{-1} + \sum_{1}^{\infty} M_u(x^k)\}. \tag{6.13}$$

Proof. Since every network is a π-network or an arrangement in series of an unordered collection of two or more π-networks and, in the latter case, the number of multiple edges in the entire network is simply the number of multiple edges within the constituents, it is not difficult to see that

$$1 + g(x,t) + h(x,t) = \prod (1 - t^i x^j)^{-g_{ij}}$$

$$= \exp \sum_{1}^{\infty} g(x^k, t^k)/k. \tag{6.14}$$

Similarly, we find that

$$1 + g(x,t) + h(x,t) = \left(1 + \frac{x}{1-xt}\right) \prod (1 - t^i x^j)^{-h_{ij}}$$

$$= \left(1 + \frac{x}{1-xt}\right) \exp \sum_{1}^{\infty} h(x^k, t^k)/k. \tag{6.15}$$

Notice that if a π-network contains b trivial constituents, then the number of multiple edges in the network equals the number of multiple edges within all the constituents plus $b-1$; the separate factor in the last two parts of relation (6.15) takes into account the contribution of the trivial constituents.

When we differentiate these relations with respect to t and set $t=1$, we find that

$$M_p(x) + M_s(x) = (1 + u(x)) \sum_{1}^{\infty} M_p(x^k) \tag{6.16}$$

and

$$M_p(x) + M_s(x) = (1 + u(x))\{x^2(1-x)^{-1} + \sum_{1}^{\infty} M_s(x^k)\}. \tag{6.17}$$

Relation (6.12) now follows upon equating the right hand sides of equations (6.16) and (6.17), and relation (6.13) follows upon adding these two equations together.

Corollary 9.1.

$$M_p(n) \sim M_s(n) \sim \zeta n$$

as $n \to \infty$, where

$$\zeta = \frac{1}{\delta d^2}\{\delta^2(1-\delta)^{-1} + \sum_{2}^{\infty} M_u(\delta^k)\} = .2212....$$

Proof. Let $\mu(n)$ denote the Möbius function defined as follows: $\mu(1)=1$, $\mu(n) = (-1)^k$ if n is a product of k distinct primes, and $\mu(n)=0$ otherwise. We recall (see [9; p. 235]) that

$$\sum_{d \mid n} \mu(d) = 0$$

for $n \geqslant 2$. It follows readily from this identity and relation (6.12) that

$$M_s(n) s_n = M_p(n) p_n + \mu(n) \tag{6.18}$$

for $n \geqslant 2$.

Equation (6.13) can be rewritten as

$$M_u = (1-u)^{-1}(1+u)\{x^2(1-x)^{-1} + \sum_{2}^{\infty} M_u(x^k)\}.$$

From this it may be deduced that

$$C_n\{M_u(x)\} \sim 2\zeta n p_n \tag{6.19}$$

as $n \to \infty$; the argument is very similar to the argument by which relation (5.17) was deduced from equation (5.14) and we therefore omit the details. The required result now follows immediately from (6.19) in view of relation (6.18) and the fact that $p_n = s_n$ where $n \geqslant 2$.

We remark that it can be shown by a similar argument that the expected number of pairs of nodes joined by multiple edges in both π-networks and σ-networks N_n tends to $(.1625...)n$ as $n \to \infty$.

Let

$$p(x, y) = \sum p_{ni} y^i x^n$$

and

$$s(x, y) = \sum s_{ni} y^i x^n$$

where p_{ni} and s_{ni} denote the number of π-networks and σ-networks, respectively, with n unlabelled edges and i interior nodes. If

$$I_p = I_p(x) = \sum I_p(n) p_n x^n$$

and

$$I_s = I_s(x) = \sum I_s(n) s_n x^n,$$

where $I_p(n)$ and $I_s(n)$ denote the expected value of $I(N_n)$ over all unlabelled π-networks and σ-networks N_n, respectively, then $I_p = p_y(x, 1)$ and $I_s = s_y(x, 1)$.

Theorem 10.

$$\sum_{1}^{\infty} I_s(x^k) = (1+u)^{-1}(xp' - p).$$

Proof. The fact that

$$1 + p(x, y) + s(x, y) = (1-x)^{-1} \prod (1 - y^i x^n)^{-s_{ni}}$$

$$= (1-x)^{-1} \exp \sum_{1}^{\infty} s(x^k, y^k)/k$$

follows by essentially the same argument that led to equation (5.12); the factor $(1-x)^{-1}$ instead of $(1+x)$ is present here because there is no restriction now on the number of edges that can join the terminals. When we differentiate this relation with respect to y and set $y = 1$, we find that

$$I_p + I_s = (1+u) \sum_{1}^{\infty} I_s(x^k).$$

This implies the required result since, as in the proof of Theorem 5, I_s is also the generating function for the expected number of parallel operations involved in π-networks and so

$$I_p + I_s = xp' - p.$$

Corollary 10.1.

$$I_p(n) \sim I_s(n) \sim \tfrac{1}{2}n, \quad as \ n \to \infty.$$

Proof. If we appeal to (6.6) and apply formula (2.3), we find that

$$C_n\{(1+u)^{-1}(xp'-p)\} \sim (1+u(\delta))^{-1} n p_n = \tfrac{1}{2} n p_n$$

as $n \to \infty$. The coefficient of x^n in $\sum_{2}^{\infty} I_s(x^k)$ is negligible compared with $\tfrac{1}{2} n p_n$ since the function converges for $|x| \leq \delta^{1/2}$. Hence it follows from Theorem 10 that

$$I_s(n) s_n \sim \tfrac{1}{2} n p_n$$

as $n \to \infty$. This implies the required result since $p_n = s_n$ for $n \geq 2$ and $I_p(n) + I_s(n) = n - 1$.

We conclude by giving in Table 2 the first few values of some of the quantities we have been considering in the past two sections for the case of π-networks.

Table 2.

Values for π-networks with unlabelled edges.

n	1	2	3	4	5	6	7	8
q_n	1	0	1	2	4	9	20	47
$I_q(n) q_n$	0	0	1	4	10	30	78	217
p_n	1	1	2	5	12	33	90	261
$I_p(n) p_n$	0	0	1	6	20	74	245	845
$M_p(n) p_n$	0	1	2	5	15	47	149	489

ACKNOWLEDGEMENTS

I am indebted to Professor A. Meir for some helpful explanations. The preparation of this paper was assisted by a grant from the Natural Sciences and Engineering Research Council of Canada.

References

[1] E. A. Bender, Asymptotic methods in enumeration, SIAM Review 16 (1974) 485–515.

[2] L. Carlitz and J. Riordan, The number of labelled two-terminal series-parallel networks, Duke Math. J. 23 (1956) 435–446.

[3] A. Cayley, On the theory of the analytical forms called trees, Phil. Mag. 13 (1857) 172–176.

[4] L. Comtet, Sur le quatrième problème et les nombres de Schröder, C. R. Acad. Sci. Paris 271 (1970) 913–916.

[5] L. Comtet, *Advanced Combinatorics*, Reidel, Dordrecht, 1974.

[6] G. Darboux, Mémoire sur l'approximation des fonctions de très grands nombres, et sur une classe étendu de développements en série, J. Math. Pures et Appliquées 4 (1878) 5–56.

[7] L. R. Foulds and R. W. Robinson, Enumeration of phylogenetic trees without points of degree two. Ars Comb. 17A (1984) 169–183.

[8] F. Harary, R. W. Robinson, and A. J. Schwenk, Twenty-step algorithm for determining the asymptotic number of trees of various species, J. Austral. Math. Soc. 20A (1975) 483–503.

[9] G. H. Hardy and E. M. Wright, *An Introduction to the Theory of Numbers*, 4th ed., Oxford, 1965.

[10] W. Knödel, Über Zerfällungen, Monatsch. Math. 55 (1951) 20–27.

[11] P. A. MacMahon, The combination of resistances, The Electrician 28 (1892) 601–602.

[12] A. Meir and J. W. Moon, On the altitude of nodes in random trees, Can. J. Math. 30 (1978) 997–1015.

[13] A. Meir, J. W. Moon, and J. Mycielski, Hereditarily finite sets and identity trees, J. Comb. Th. B35 (1983) 142–155.

[14] E. M. Palmer and A. J. Schwenk, On the number of trees in a random forest, J. Comb. Th. B27 (1979) 109–121.

[15] J. Pfaff, R. Laskar, and S. T. Hedetniemi, Linear algorithms for independent domination and total domination in series-parallel graphs, Congressus Numerantium 45 (1984) 71–82.

[16] J. Riordan and C. E. Shannon, The number of two-terminal series-parallel networks, J. Math. and Physics 21 (1942) 83–93.

[17] J. Riordan, *An Introduction to Combinatorial Analysis*, Wiley, New York, 1958.

[18] J. Riordan, The blossoming of Schröder's fourth problem, Acta Math. 137 (1976) 1–16.

[19] R. W. Robinson and A. J. Schwenk, The distribution of degrees in a large random tree, Discrete Math. 12 (1975) 359–372.

[20] E. Schröder, Vier combinatorische Probleme, Z. Math. Physik 15 (1970) 361–376.

UNSOLVED PROBLEMS IN THE THEORY OF RANDOM GRAPHS

Edgar M. PALMER

Michigan State University

Several problems for random graphs are suggested that involve the investigation of hamiltonicity, connectivity, independence number, reconstruction and properties of trees.

1. Introduction

Since Erdös and Rényi founded the theory of random graphs over twenty years ago ([ErR59] and [ErR60]), several hundred research papers have been written about the subject and it is still undergoing very active investigation. In this article we present a few recent results and some random unsolved problems. The notation, definitions, necessary background and most references are provided in the introductory book, *Graphical Evolution* [Pa85].

We use the usual two closely related probability models.

Model A: For each positive integer n, we have a number p with $0<p<1$. The sample space consists of all

$$2^{\binom{n}{2}}$$

labeled graphs G of order n. The probability of a graph G of order n with q edges is defined by

$$P(G) = p^q (1-p)^{\binom{n}{2}-q}, \qquad (1.1)$$

and hence p is sometimes called the "probability of an edge".

Model B: The sample space consists of all

$$\binom{\binom{n}{2}}{q}$$

labeled graphs of order n with q edges. Each of these graphs G is assigned the same probability, namely

$$P(G) = \binom{\binom{n}{2}}{q}^{-1}. \tag{1.2}$$

2. Hamiltonicity

In *Graphical Evolution* [Pa85] we discussed at some length the strength of various sufficient conditions for a graph to be hamiltonian. Here is an example. Bondy and Chvátal [BoC76] found the essential ingredient in a whole family of hamiltonian hypotheses that require not too many vertices of low degree. They defined the *closure* of the graph G of order n, denoted $c(G)$, to be the graph obtained from G by successively adding edges which join vertices u and v that satisfy Ore's condition:

$$\deg u + \deg v \geq n. \tag{2.1}$$

Their theorem is proved in much the same way as Ore's [O60].

Theorem 1.1. (Bondy and Chvátal). *A graph G of order $n \geq 3$ is hamiltonian if and only if its closure $c(G)$ is hamiltonian.*

Since complete graphs are hamiltonian, there is the following immediate consequence.

Corollary 1.1. *If the closure of a graph of order $n \geq 3$ is complete, the graph is hamiltonian.*

Note that no more than $O(n^4)$ steps are needed to close any graph and so the theorem can be applied in polynomial time. However it may be as difficult to determine hamiltonicity from the closure as from the graph itself. It would be helpful to know what the chances are that a random graph has a complete closure.

Recent results of Pósa [Po76], Koršunov [Ko76], and Komlós and Szemerédi [KoS83] have given us sharp thresholds for hamiltonicity. For example, let $\mathcal{H}$ be the set of labeled hamiltonian graphs of order n. It follows from the papers just referenced that if the probability of an edge is $p = c \log n/n$ with constant $c > 1$, then in Model A

$$\lim_{n \to \infty} P(\mathcal{H}) = 1. \tag{2.2}$$

We usually write $P(\mathcal{H}) \to 1$ and say "almost all graphs are hamiltonian".

Now we are ready to analyze the theorem and its corollary. The question to be answered is "what can we take for the probability of an edge so that almost all graphs satisfy the hypothesis?"

First consider Model A with fixed $p<1/2$. Since $p>(1+\varepsilon)\log n/n$, almost all graphs are hamiltonian. An immediate consequence of the Central Limit Theorem (see section 5.1 of [Pa85]) is that the maximum degree Δ of almost all graphs is bounded above as in the inequality

$$\Delta < \lfloor pn + \sqrt{2p(1-p)n\log n} \rfloor. \tag{2.3}$$

Hence $\Delta < n/2$ for all n sufficiently large. Therefore for almost all graphs

$$\deg u + \deg v < \frac{n}{2} + \frac{n}{2} = n \tag{2.4}$$

for *every* pair of vertices u, v. And so in Model A with fixed $p<1/2$, $c(G)=G$, i.e. almost all graphs are closed, and thus the theorem is almost never helpful.

On the other hand, for fixed $p>1/2$, almost all graphs have minimum degree $\delta>n/2$ and so Ore's condition is always satisfied. Then almost all graphs have a complete closure and the corollary is almost always effective.

The remaining problem is to determine the strength of the corollary when $p=1/2$. That is, in Model A with $p=1/2$, what is the limiting probability that the closure is complete? Otherwise put, find

$$P(c(G)=K_n) \to ? \tag{2.5}$$

Let $X(G)$ be the number of pairs of non-adjacent vertices of G that do not satisfy Ore's condition (these are *bad* pairs). We showed in [Pa85] that when $p=1/2$,

$$E(X) \sim n^2/8 \tag{2.6}$$

and used the second moment method to show that almost every graph has a bad pair of vertices. Similarly almost every graph has a good pair of vertices and the expected number of these is also about $n^2/8$. Consequently almost all graphs are *not* closed but the closure has at least $3n^2/8$ edges. This would seem to be sufficient to ensure a complete closure, and indeed this has been shown by John Gimbel, David Kurtz, Linda Lesniak, Edward R. Scheinerman and John C. Wierman in another article of these proceedings [GiKLSW87], where the authors provide a thorough analysis of the closure operation.

As we observed in [Pa85], the current record holder in the hamiltonian sweepstakes is the theorem of Chvátal and Erdös [ChE72]. This beautiful result owes much to the smooth driving of Louise Guy in the fast lane of route 195 from Pullman to Spokane.

Theorem 2.1. (Chvátal and Erdös). *If the connectivity of a graph of order $n \geq 3$ is at least as large as the independence number, i.e. $\kappa \geq \beta$, then the graph is hamiltonian.*

Note that the connectivity κ of any graph can be found in at most $O(n^{4.5})$ steps using the maximum flow algorithm (see [Ev79] p. 120). But the determination of the independence number β is NP-hard. Hence it may be difficult to establish the Chvátal-Erdös hypothesis for some graphs, but of course the greedy coloring algorithm gives a fairly good lower bound for β for almost all graphs.

However, it was easy to show in [Pa85] that in Model A with any fixed $p>0$, almost all graphs have $\kappa \geq \beta$. Hence the theorem is almost always applicable. On the other hand, if $p = c \log n/n$ with constant $c<1$, then almost all graphs have several isolated vertices (see exercise 3.1.1 of [Pa85]). Thus $\beta \geq 1$ and $\kappa = 0$. The unsolved problem is to close this gap and find a sharp threshold for the Chvátal-Erdös hypothesis. Here is the beginning of a solution.

The independence number has been investigated to some extent by Burtin [Bu73] and later by Gazmuri [Ga84]. However, these papers focus primarily on phases of random graph evolution that are too early or too late for our purpose here. Therefore we start by finding a simple upper bound for the independence number.

Theorem 2.2. *In* Model A $0<p<1$ *and* $r = [(2 \log n)/p]$, *almost all graphs have independence number $\beta < r$.*

Proof. Let $X_r(G)$ be the number of r-sets of independent vertices of G. Then the expectation is

$$E(X_r) = \binom{n}{r}(1-p)^{\binom{r}{2}}$$

$$\leq \frac{n}{r!}(n(1-p)^{r/2})^{r-1}$$

$$\leq \frac{n}{r!}(ne^{-pr/2})^{r-1}. \tag{2.7}$$

By hypothesis $r/2 \geqslant (\log n)/p$, therefore

$$E(X_r) \leqslant \frac{n}{r!}. \tag{2.8}$$

But $r \geqslant 2 \log n$, hence

$$n \leqslant e^{r/2}, \tag{2.9}$$

and so

$$E(X_r) \leqslant \frac{(\sqrt{e})^r}{r!} \to 0. \tag{2.10}$$

Note that in this theorem there are hardly any restrictions on p at all.

I would like to thank the editor, Zbigniew Palka, for calling my attention to the next theorem of Bollobás and Thomason. It has just appeared in [BoT85] page 78 and also [Bo85] page 269, and it improves my own estimate.

Theorem 2.3. *Let $\varepsilon > 0$ be fixed and let $\omega_n \to \infty$ but with $\omega_n = o(n/\log n)$. In Model A with*

$$p = \omega_n/n \tag{2.11}$$

and

$$r = \frac{\log \omega_n - \varepsilon}{\omega_n} n, \tag{2.12}$$

almost all graphs have independence number $\beta \geqslant r$.

Now we can compare β and κ. We always have

$$\kappa \leqslant \delta \leqslant pn \tag{2.13}$$

for almost all graphs. If $pn = \omega_n$, we can conclude from Theorem 2.3 that

$$n(\log \omega_n - \varepsilon)/\omega_n \leqslant \beta. \tag{2.14}$$

Then $\kappa \leqslant \beta$ for almost all graphs if

$$\omega_n \leqslant (n(\log \omega_n - \varepsilon))^{\frac{1}{2}}. \tag{2.15}$$

This inequality suggests that we take

$$\omega_n = (n(\log\sqrt{n})c_1)^{\frac{1}{2}} \tag{2.16}$$

with constant $c_1 > 0$. Obviously the condition on ω_n in Theorem 2.3 is satisfied and it is also easy to check that if $c_1 < 1$, then (2.15) holds as well.

Since $pn = \omega_n$, we can conclude that if

$$p = \sqrt{\frac{c\log n}{n}}, \tag{2.17}$$

with constant $c < 1/2$, then almost all graphs have $\kappa \leqslant \beta$.

Now suppose $p = \omega_n \log n/n$ with $\omega_n \to +\infty$ arbitrarily slowly, so that $p \to 0$. It follows from a result of Ivchenko [Iv73] that if $p \to 0$, almost all graphs have $\delta = \kappa$. Erdös and Rényi (see [ErR60] or Theorem 5.1.4 of [Pa85]) showed that if $p = \omega_n \log n/n$, then

$$\delta > (1-\varepsilon) pn \tag{2.18}$$

for almost all graphs. Now Theorem 2.2 can be applied and we find that

$$\beta < (2\log n)/p = 2n/\omega_n. \tag{2.19}$$

Then $\beta < \kappa$ for almost all graphs if

$$2n/\omega_n \leqslant (1-\varepsilon) pn = (1-\varepsilon)\omega_n \log n \tag{2.20}$$

or

$$\omega_n \geqslant \sqrt{\frac{2}{(1-\varepsilon)} \frac{n}{\log n}}. \tag{2.21}$$

It follows that if

$$p = \sqrt{\frac{c\log n}{n}}, \quad c > 2, \tag{2.22}$$

then almost all graphs have $\kappa > \beta$.

The unsolved problem here is to sharpen these results by further exploration of the relation between β and $\kappa(=\delta)$ in the gap between $c = 1/2$ and $c = 2$.

For example, what is

$$P(\kappa - \beta \geq s) \to ?\qquad(2.23)$$

Our results here show that almost all graphs are hamiltonian if

$$p = \sqrt{c \log n / n}, \quad c > 2.$$

There is always the challenge of finding another nice sufficient condition for hamiltonicity that would imply almost all graphs are hamiltonian for p substantially less than

$$\sqrt{c \log n / n}.$$

However, it should be noted that there is heavy competition in the form of the non-traditional approach of the algorithm. Based on an idea of Pósa [Po76], Angluin and Valiant [AnV79] devised an algorithm that almost always produces a hamiltonian cycle whenever $p = (c \log n)/n$ and c is sufficiently large. Furthermore they showed that the algorithm operates in expected time $O(n(\log n)^2)$, proving this approach to be speedy as well as wide in scope!

3. Local properties

The *neighborhood* of a vertex v is the subgraph induced by the vertices adjacent to v. A graph is *locally connected* if the neighborhood of every vertex of degree ≥ 2 is connected. In [ErPR83] we found a sharp threshold for local connectivity. For connected graphs this turned out to be identical to the threshold for the property that every edge belongs to a triangle (a complete graph of order 3). Here we ask what should be the probability of an edge so that every *vertex* belongs to a triangle? Or when does the neighborhood of every vertex have at least one edge?

To begin the solution, let $X(G)$ be the number of vertices of G that do *not* belong to triangles (these are *bad* vertices). Then the expectation is

$$E(X) = n \sum_{k=0}^{n-1} \binom{n-1}{k} p^k (1-p)^{n-1-k+\binom{k}{2}}.\qquad(3.1)$$

Erdös and Rényi [ErR60] showed that if $p = \omega_n \log n / n$ with $\omega_n \to +\infty$ slowly, almost all graphs have

$$(1-\varepsilon) pn < \delta \leq \Delta < (1+\varepsilon) pn.\qquad(3.2)$$

In the proof (see Appendix of [Pa85]) it is shown that

$$n\sum \binom{n-1}{k} p^k(1-p)^{n-1-k} = O(1/n^{a\omega_n}), \qquad (3.3)$$

where $0 < a < \varepsilon^2/3$ and the sum is over all k such that $|k - pn| \geq \varepsilon pn$.

Therefore the top and bottom of the sum in (3.1) have limit zero and we can write

$$E(X) = o(1) + n\sum \binom{n-1}{k} p^k(1-p)^{n-1-k+\binom{k}{2}} \qquad (3.4)$$

where the sum is over all k such that $|k - pn| < \varepsilon pn$. Since $k > (1-\varepsilon)pn$,

$$E(X) \leq o(1) + n(1-p)^{\binom{(1-\varepsilon)pn}{2}}. \qquad (3.5)$$

It follows that $E(X) \to 0$ if

$$\omega_n(1-\varepsilon)^2 p^2 n/2 \geq 1+\varepsilon, \qquad (3.6)$$

and so we need only choose

$$\omega_n^3 \geq cn/(\log n)^2 \qquad (3.7)$$

for constant $c > 2$.

On solving for p we find that we should set

$$p = (c\log n/n^2)^{1/3} \qquad (3.8)$$

with $c > 2$ to have $E(X) \to 0$.

Similarly $E(X) \to +\infty$ if $0 < c < 2$ because

$$E(X) \geq o(1) + n(1-p)^{\binom{(1+\varepsilon)pn}{2}}. \qquad (3.9)$$

Presumably for this case we also have $E(X)^2 \sim E(X^2)$, in which event, almost all graphs have a vertex that is not in a triangle. But there remain some details to be worked out. To refine the threshold, set

$$p = ((2+\varepsilon_n)\log n/n^2)^{1/3} \qquad (3.10)$$

and find ε_n, a function of n and a new variable x such that $\varepsilon_n \to 0$ for each x and $E(X) \sim e^{-x}$. Show that the distribution is Poisson in the limit. Evidently, A. Ruciński has proved exactly this (communication from M. Karoński).

In general one could study the evolution of neighborhoods and try to find thresholds for the appearance of specified subgraphs such as triangles or cycles as well as thresholds for properties of graphs. For example, when is every neighborhood non-planar?

4. Reconstruction

The famous reconstruction conjecture of Ulam asks if a graph of order $n \geqslant 3$ can be reconstructed from its vertex − deleted subgraphs. For more details see the survey article [BoH77] which contains many results that support this conjecture. V. Müller was the first to find that the conjecture was indeed true for random graphs.

Theorem 4.1. (Müller). *In Model A with $p=1/2$, almost all graphs are vertex-reconstructible.*

The proof probably works also for any fixed p. But what if $p \to 0$ slowly? If $p = c \log n/n$ with constant $c < 1$, then almost all graphs are disconnected and hence reconstructible. What if $c > 1$? Or, can we find $\omega_n \to \infty$ slowly so that with $p = \omega_n \log n/n$, almost all graphs are vertex-reconstructible? We have just learned that this problem has been solved recently by Béla Bollobás [Bo-U] who proved the following theorem.

Theorem 4.1.1. (Bollobás). *If $c > 5/2$ and*

$$c \log n/n \leqslant p \leqslant 1 - c \log n/n,$$

then almost all graphs can be vertex-reconstructed from any three vertex-deleted subgraphs.

Evidently the theorem remains true for $c > 1$ but the details are cumbersome. There remains the case in which $c = 1$, or $p \sim (\log n)/n$.

By improving an idea of Lovász [Lo72], Müller was also able to determine an analogous result for the edge-reconstruction problem [Mu77].

Theorem 4.2. (Müller). *In Model A with $p = c \log n/n$, if $c > 2/\log 2$, almost all graphs are edge-reconstructible.*

If G is not connected and has at least four edges and at least two non-trivial components, then G is edge-reconstructible [BoH77]. As observed in section 5.4 of [Pa85], with $c<1/2$, almost every graph has isolated edges as well as a giant component. Hence these are edge-reconstructible. It remains to show that for $1/2 \leq c \leq 2/\log 2$, almost all graphs are edge-reconstructible. The corresponding problem in Model B is to prove that almost all graphs are edge-reconstructible when the number q of edges lies in the gap

$$\frac{1}{4} n \log n \leq q \leq \frac{1}{\log 2} n \log n. \tag{4.1}$$

5. Trees

There are many open problems involving labeled and unlabeled trees. We mention just a few.

Maximum degree

For any tree T of order n, $\Delta(T)$ denotes the maximum degree of the vertices of T. J. W. Moon (see [M68] or [M70]) derived an asymptotic formula for the average value of $\Delta(T)$ over the set of n^{n-2} labeled trees of order n.

Theorem 5.1. (Moon). *In the uniform probability model for labeled trees of order n, the probability of each tree T is*

$$P(T) = 1/n^{n-2}. \tag{5.1}$$

Then the expected maximum degree $E(\Delta)$ satisfies

$$E(\Delta) \sim \frac{\log n}{\log \log n}. \tag{5.2}$$

The proof uses such well known elementary counting formulas as Clarke's expression (see [M70]) for the number of vertices of degree k:

$$\binom{n-2}{k-1}(n-1)^{n-k-1}. \tag{5.3}$$

Robinson and Schwenk [RoS75] made a detailed study of the degree distribution for unlabeled trees but their asymptotic formulas for the number of vertices of degree k are for fixed values of $k \leqslant 10$. Hence for unlabeled trees, no simple formula corresponding to (5.3) has been found and so the behavior of $E(\varDelta)$ in this case remains a mystery.

Central vertices

Cayley counted unlabeled trees of order $n \leqslant 13$ by finding both the number with a single central vertex and the number with a double center (see [BiLW76]). His tables were extended a bit to $n \leqslant 20$ by Riordan [Ri60], and we find that the probability that an unlabeled tree of order 20 selected uniformly at random has a single central vertex is .50439..., i.e. very nearly 1/2. The unsolved problem is to show that this probability tends to 1/2 as $n \to \infty$. But the generating functions used to count single centered trees are quite unwieldly and the usual asymptoting techniques do not apply. Evidently the labeled version of this problem has been dealt with successfully by G. Szekeres [Sz83] and the solution is definitely nontrivial.

Fixed vertices

Let t_n be the number of unlabeled trees of order n. For each such tree T, we define the random variable $X(T)$ to be the proportion of vertices fixed by the automorphism group of T. Using the uniform probability model, i.e. each tree T has

$$P(T) = 1/t_n, \tag{5.4}$$

we found in [HP79] that

$$E(X) \sim .6995.... \tag{5.5}$$

The ultimate refinement of this question was treated by Bailey [Ba82]. A labeled version of the problem remains unsolved. In this case each labeled tree T of order n has uniform probability defined by (5.1) and $X(T)$ is again the proportion of vertices of T that are fixed by the group of T. To estimate $E(X)$ one would have to count fixed vertices of labeled trees.

Group order

With the uniform probability model for labeled trees, each tree T has probability defined by (5.1). The random variable $X(T)$ is now the order of the automorphism group of T. Then $E(X)$ is easy to calculate in terms of the number t_n of unlabeled trees of order n:

$$E(X) = n! \, t_n / n^{n-2}. \tag{5.6}$$

But the corresponding unlabeled problem is unsolved. The difficulty seems to lie in the fact that the ordinary generating function for group orders diverges. The coefficient of x^n is at least $(n-1)!$, the order of the group of the tree of order n with a vertex of degree $n-1$. Therefore the usual techniques do not apply. We are able to circumvent this obstruction to some extent in [PaR-P] by limiting the sample space to trees of maximum degree d (see also Chapter 6 of [Pa85]).

I want to conclude with a few words of thanks to all of the people in Poznań who made the random graph conference of 1985 such a success. This includes organizers, participants, students, supporting workers and many more who helped to make the week so worthwhile and enjoyable. Their extra efforts were especially appreciated by those of us who were far from home.

References

[AnV79]   D. Angluin and L. G Valiant, Fast probabilistic algorithms for hamiltonian circuits and matchings, *J. Comput. System Sci.* 18 (1979) 155–193.

[Ba82]   C. K. Bailey, Distribution of points by degree and orbit size in a large random tree, *J. Graph Theory* 6 (1982) 283–293.

[BiLW76]   N. L. Biggs, E. K. Lloyd and R. J. Wilson, *Graph Theory 1736–1936*, Clarendon, Oxford (1976).

[BoC76]   J. A. Bondy and V. Chvátal, A method in graph theory, *Discrete Math.* 15 (1976) 111–136.

[BoH77]   J. A. Bondy and R. L. Hemminger, Graph reconstruction – a survey, *J. Graph Theory* 1 (1977) 227–268.

[Bo85]   B. Bollobás, *Random Graphs*, Academic, London (1985).

[Bo-U]   B. Bollobás, Almost every graph has reconstruction number three, *J. Graph Theory*, to appear.

[BoT85]   B. Bollobás and A. Thomason, Random graphs of small order, *Annals of Discrete Math.* 28 (1985) 47–97.

[Bu73]   Yu. D. Burtin, Asymptotic estimates of the diameter and independence and denomination numbers of a random graph, *Soviet. Math. Dokl.* 14 (1973) 497–501.

[ChE72]   V. Chvátal and P. Erdös, A note on hamiltonian circuits, *Discrete Math.* 2 (1972) 111–113.

[ErPR83]   P. Erdös, E. M. Palmer and R. W. Robinson, Local connectivity of a random graph, *J. Graph Theory* 7 (1983) 411–417.

[ErR59]   P. Erdös and A. Rényi, On random graphs I, *Publ. Math. Debrecen* 6 (1959) 290–297.

[ErR60] P. Erdös and A. Rényi, On the evolution of random graphs, *Magyar Tud. Akad. Mat. Kutato Int. Kozl.* 5 (1960) 17–61.

[Ev79] S. Even, *Graph Algorithms*, Computer Science Press, Maryland (1979).

[Ga84] P. G. Gazmuri, Independent sets in random sparse graphs, *Networks* 14 (1984) 367–377.

[GiKLSW87] J. Gimbel, D. Kurtz, L. Lesniak, E. R. Scheinerman and J. C. Wierman, Hamiltonian closure in random graphs, *Annals of Discr. Math.* 33 (1987) 59–67.

[HP79] F. Harary and E. M. Palmer, The probability that a point of a tree is fixed, *Math. Proc. Cambridge Philos. Soc.* 85 (1979) 407-415.

[Iv73] G. I. Ivčhenko, The strength of connectivity of a random graph, *Theor. Probability Appl.* 18 (1973) 396–403.

[KoS83] J. Komlós and E. Szemerédi, Limit distribution for the existence of hamiltonian cycles in a random graph, *Discrete Math.* 43 (1983) 55–63.

[Ko76] A. D. Koršunov, Solution of a problem of Erdös and Rényi on hamiltonian cycles in nonoriented graphs, *Soviet. Math. Doklady* 17 (1976) 760–764.

[Lo72] L. Lovász, A note on the line reconstruction problem, *J. Combin. Theory Ser. B* 13 (1972) 309–310.

[M68] J. W. Moon, On the maximum degree in a random tree, *Mich. Math. J.* 15 (1968) 429–432.

[M70] J. W. Moon, *Counting Labelled Trees*, Canad. Math. Congress, Montreal (1970).

[Mu76] V. Müller, Probabilistic reconstruction from subgraphs, *Comment. Math. Univ. Carolinae* 17 (1976) 709–719.

[Mu77] V. Müller, The edge reconstruction hypothesis is true for graphs with more than $n\log_2 n$ edges, *J. Combin. Theory Ser. B* 22 (1977) 281–283.

[Or60] O. Ore, Note on hamilton circuits, *Amer. Math. Monthly* 67 (1960) 55.

[Pa85] E. M. Palmer, *Graphical Evolution*, Wiley-Interscience, New York (1985).

[PaR-P] E. M. Palmer and R. W. Robinson, Asymptotic number of symmetries in locally restricted trees, in preparation.

[Po76] L. Pósa, Hamiltonian circuits in random graphs, *Discrete Math.* 14 (1976) 359–364.

[Ri60] J. Riordan, The enumeration of trees by height and diameter, *IBM J. Res. Develop.* 4 (1960) 473–478.

[RoS75] R. W. Robinson and A. J. Schwenk, The distribution of degrees in a large random tree, *Discrete Math.* 12 (1975) 359–372.

[Sz83] G. Szekeres, Distribution of labelled trees by diameter, *Combinatorial Mathematics* X (L. R. A. Casse, ed.). Lecture Notes in Math. 1036, Springer, Berlin (1983) 392–397.

SOME RECENT RESULTS ON THE REGISTER FUNCTION OF A BINARY TREE

Helmut PRODINGER

Technische Universität Wien,
A-1040 Wien, Austria

> *In particular I had to make use of curves which are continuous but which are so crinkly that they can not properly said to have a direction. I have already pointed out in my discussion of the Brownian motion that these curves had been more or less the stepchildren of mathematics and had been regarded as rather unnatural museum pieces, derived by the mathematician from abstract considerations, and with no true representation in physics. Here I found myself establishing an essential physical theory in which such curves played an indispensable role.*
>
> Norbert Wiener, *I am a mathematician*

1. Introduction

We propose here to show that a large class of enumeration problems concerning trees can now be solved rather easily and automatically. This is in contrast to the situation several years ago when even the simplest problems in this class caused serious problems.

To be more explicit, we deal with binary trees; let $\mathfrak{B}$ be the family of binary trees, then we have the formal equation

$$\mathfrak{B} = \Box + \overset{\circ}{\underset{\mathfrak{B}\;\mathfrak{B}}{\bigwedge}}$$

expressing the fact that a binary tree is either empty or consists of a root together with a left and right subtree, each one being itself a binary tree.

We define the size $|t|$ of a tree t to be the number of internal nodes of t; t_n denotes the number of trees of size n. We find immediately from the formal equation for the generating function $B(z)$ of trees:

$$B(z) := \sum_{n \geq 0} t_n z^n = \sum_{t \in \mathfrak{B}} z^{|t|} = 1 + zB^2(z)$$

$$= \frac{1 - \sqrt{1-4z}}{2z} = \sum_{n \geq 0} \frac{1}{n+1} \binom{2n}{n} z^n.$$

We deal with the *register function* $\text{reg}(t)$ of a tree t. This function is defined inductively as follows:

$$\text{reg}(\square) = 0$$

$$\text{reg}\left(\bigwedge_{t_1 \; t_2}\right) = \begin{cases} 1 + \text{reg}(t_1) & \text{if } \text{reg}(t_1) = \text{reg}(t_2) \\ \max\{\text{reg}(t_1), \text{reg}(t_2)\} & \text{otherwise}. \end{cases}$$

This function is of relevance in Computer Science; for this we refer to [1].

For completeness we now cite all papers dealing with the register function: [1–3, 5, 7–13].

The recursive definition can be most easily visualised by labelling the nodes of the tree in a bottom-up-sense; the value of the root is then the desired value $\text{reg}(t)$. (Figure 1.)

The marked nodes cause the register function to increase; we call them *critical*. If we forget about all other nodes, we obtain the *(ordered binary) forest of critical nodes*, in the example presented in Figure 2.

In the next sections we deal with several enumeration problems concerning the register function. The interest is not so much in these parameters itself, but in the methodology that we are going to point out in a few seconds. However, there are still unsolved problems, for instance, what is the average number of components of the forest just mentioned?

We are interested in *average values* of certain parameters, where all trees of size n are to be considered equally likely. Since these values are of an intrinsic complexity, we confine ourselves to the determination of *asymptotic equivalents*; it will turn out that these contain *periodic fluctuations*, even if we do not compute them in all the examples. Using appropriate generating functions, the average value is

$$\frac{[z^n] E(z)}{[z^n] B(z)}$$

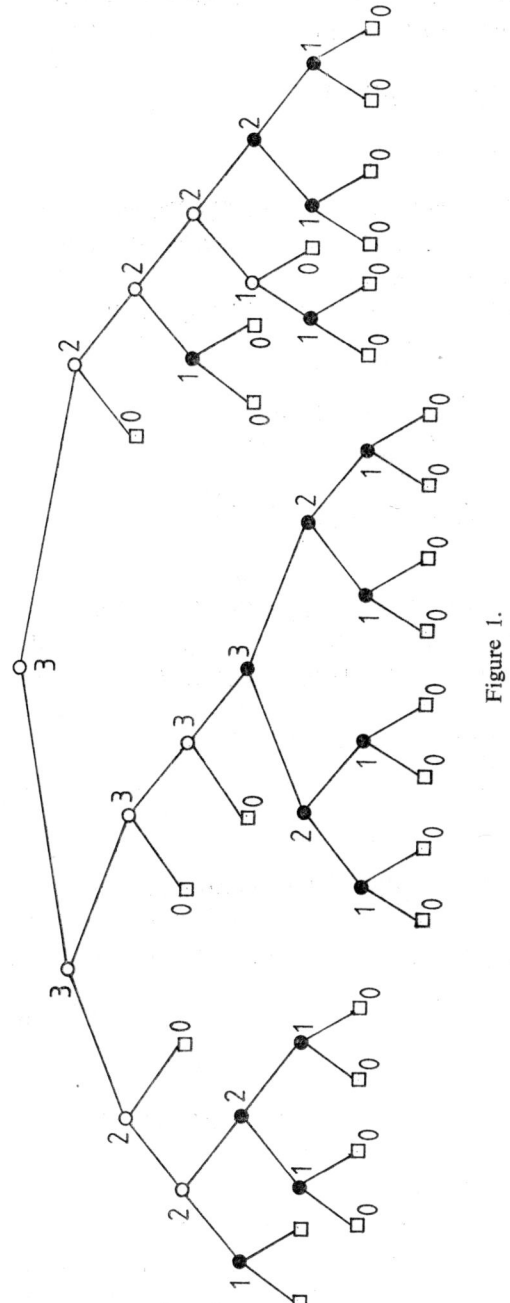

Figure 1.

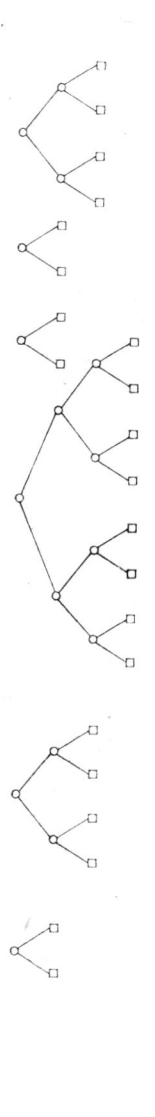

Figure 2.

where the notion $[z^n]f(z)$ refers to the coefficient of z^n in the Taylor expansion of the generating function $f(z)$. Such an $E(z)$ is often

$$E(z) = \sum_{p \geq 1} E_p(z) \quad \text{or} \quad E(z) = \sum_{p \geq 0} pE_p(z),$$

depending on what $E_p(z)$ counts.

So the first step is to find explicit expressions for the $E_p(z)$'s, whatever they might be, mostly by setting up recursions. Here, we use $R_p(z)$, the generating functions of trees with register function $= p$.

The next rule is to perform the change of variable $z = u/(1+u)^2$ in all generating functions. Here are some special functions:

$$B(z) = 1+u, \quad \sqrt{1-4z} = \frac{1-u}{1+u}, \quad \frac{dz}{du} = \frac{1-u}{(1+u)^3},$$

$$R_p(z) = \frac{1-u^2}{u} \frac{u^{2^p}}{1-u^{2^{p+1}}}, \quad S_p(z) := \sum_{j \geq p} R_j(z) = \frac{1-u^2}{u} \frac{u^{2^p}}{1-u^{2^p}}.$$

Remark that $S_p(z)$ is the generating function of trees with register function $\geq p$. The explicit values for $R_p(z)$ and $S_p(z)$ appear for the first time in [3] and [8]; an alternative (and easy) proof can be found in [12]. In Section 7 a further proof is sketched.

Then it usually turns out that $E(z)$ is a linear combination of

$$f(u) \sum_{p \geq 1} \omega_p u^p$$

where $f(u)$ is a rational function and ω_p are defined by some arithmetic properties. Each natural number n can be uniquely written as $n = 2^m(1+2j)$, $m, j \geq 0$; in a lot of cases ω_n is a function of m and j. The *corresponding Dirichlet series*

$$\sum_{p \geq 1} \omega_p p^{-s}$$

has then a closed form expression.

One is interested in a local expansion of $E(z)$ in a neighbourhood of its (unique) singularity at the radius of convergence at $z = 1/4$. This expansion can be nicely written in terms of $\varepsilon(z) = \sqrt{1-4z}$ for $\varepsilon \to 0$.

In this paper all expansions can be carried out to any desired degree of accuracy; henceforth we never worry about O-terms. In other words, if we have computed that a certain coefficient in a power series expansion is asymptotic to a certain quantity and we want to get the order of growth of the difference, we simply get it by computing one more term in the asymptotic expansion.

$z \to 1/4$ means $u \to 1$, or with $u = e^{-t}$ it means $t \to 0$. The desired local expansion is obtained in terms of t. However, we can easily rewrite it in terms of ε, since

$$t \sim 2\varepsilon + \dots.$$

The expansion of the rational function $f(u)$ is trivial.

For the sum one uses the *Mellin integral transform*. For details of this particularly useful feature we refer to a forthcoming book of Flajolet, Régnier and Sedgewick [6]

$$\mathfrak{M} V(t) := V^*(s) := \int_0^\infty t^{s-1} V(t) \, dt.$$

Since $V^*(at) = a^{-s} V^*(s)$, we see that

$$\mathfrak{M} \sum_p \omega_p e^{-tp} = \sum_p \omega_p p^{-s} \mathfrak{M} e^{-t},$$

so that the transform of the series in question is the product of the associated Dirichlet series and the classical gamma function $\Gamma(s)$. The Mellin inversion formula allows one to recover the original function:

$$V(t) = \frac{1}{2\pi i} \int_{c-i\infty}^{c+i\infty} V^*(s) t^{-s} ds,$$

where c is some appropriate real constant. If we shift the line of integration to the left, taking the residues into account, we thereby obtain an asymptotic series of the function $V(t)$ for $t \to 0$. Since the integrand contains as ingredients only things like $\Gamma(s)$, $\zeta(s)$, $(2^s - 1)^{-1}$, etc., the computation of the residues is particularly easy; one just needs some special values of the functions involved [14]. Observe that $(2^s - 1)^{-1}$ has poles at $\chi_k = (2k\pi i)/\log 2$, whence the periodic fluctuations already mentioned.

Once the local expansion is obtained, we use *translation lemmas* (see [4]); they allow to go from the local expansion of the generating function to the asymptotic behaviour of the coefficients. For this one has to have a repertoire of known expansion, like

$$[z^n] \frac{1}{\sqrt{1-4z}} \sim \frac{1}{\sqrt{\pi}} 4^n n^{-1/2}$$

etc. The last step is then to divide by

$$[z^n]B(z) \sim \frac{1}{\sqrt{\pi}} 4^n n^{-3/2}.$$

2. The register pathlength

Let us recall the (ordinary) pathlength of a binary tree: one considers all paths between the root and an internal node and counts the number of internal nodes on such a path. Summing all those numbers over all paths yields the *pathlength* of the tree.

If we count on each path only the *number of critical nodes*, we have defined in this way the *register path length* $\mathrm{rpl}(t)$ of the tree t.

In this section we are concerned with the average register pathlength, considering each tree of size n to be equally likely.

We have the following recursion: ($|t|$ stands for the number of nodes in the tree t.)

$$\mathrm{rpl}(\square) = 0$$

$$\mathrm{rpl}\left(\begin{array}{c}\mathring{\wedge}\\t_1\ t_2\end{array}\right) = \mathrm{rpl}(t_1) + \mathrm{rpl}(t_2)$$

$$+ \left\{ \left| \begin{array}{c}\mathring{\wedge}\\t_1\ t_2\end{array}\right| \quad \text{if the root is critical} \right\}.$$

We introduce the obvious generating function

$$P(z) = \sum_t \mathrm{rpl}(t) z^{|t|}$$

and obtain by a direct translation of the recursion for rpl the following equation for the corresponding generating functions:

$$P(z) = 2zB(z)P(z) + \sum_{p \geqslant 0} ([z^n] z R_p^2(z)) n z^n$$

or

$$P(z) = \frac{1}{\sqrt{1-4z}} z \frac{d}{dz} \sum_{p \geqslant 0} z R_p^2(z)$$

$$= \frac{u}{1-u^2} \frac{(1+u)^3}{1-u} \sum_{p\geqslant 1}\left[-\frac{1-u^2}{u^2}\frac{u^{2^p}}{(1-u^{2^p})^2}+\frac{(1-u)^2}{u^2}\frac{2^p u^{2^p}(1+u^{2^p})}{(1-u^{2^p})^3}\right]$$

$$= -\frac{(1+u)^3}{u(1-u)}A(u)+\frac{(1+u)^2}{u}B(u),$$

with

$$A(u)=\sum_{p\geqslant 1}\frac{u^{2^p}}{(1-u^{2^p})^2}$$

and

$$B(u)=\sum_{p\geqslant 1}\frac{2^p u^{2^p}(1+u^{2^p})}{(1-u^{2^p})^3}.$$

Now

$$A(u)=\sum_{p\geqslant 1}\frac{u^{2^p}}{(1-u^{2^p})^2}=-\frac{u}{(1-u)^2}+\sum_{p,\lambda\geqslant 0}\lambda u^{\lambda 2^p}$$

$$=-\frac{u}{(1-u)^2}+\sum_{n\geqslant 1}\psi(n)u^n,$$

with

$$\psi(n)=\sum_{n=\lambda 2^p}\lambda.$$

Now write $n=2^m(1+2j)$ in a unique way:

$$\psi(n)=\sum_{p=0}^{m}2^{m-p}(1+2j)=(1+2j)(2^{m+1}-1)$$

$$=2n-(1+2j),$$

so that

$$A(u)=\frac{u}{(1-u)^2}-F(u),$$

with

$$F(u)=\sum_{n\geqslant 1}(1+2j)u^n.$$

In a similar way,

$$B(u) = -\frac{u(1+u)}{(1-u)^3} + \sum_{p \geq 0} 2^p \sum_{\lambda \geq 0} \lambda^2 u^{\lambda 2^p}$$

$$= -\frac{u(1+u)}{(1-u)^3} + \sum_{n \geq 1} \theta(n) u^n,$$

with

$$\theta(n) = \sum_{n=\lambda 2^p} 2^p \lambda^2 = \sum_{p=0}^{m} [2^{m-p}(1+2j)]^2 2^p$$
$$= (1+2j)^2 2^m (2^{m+1} - 1) = 2n^2 - n(1+2j).$$

Hence

$$B(u) = \frac{u(1+u)}{(1-u)^3} - D(u),$$

with

$$D(u) = \sum_{n \geq 1} 2^m (1+2j)^2 u^n.$$

This gives us

$$P(z) = \frac{(1+u)^3}{u(1-u)} F(u) - \frac{(1+u)^2}{u} D(u).$$

Now let

$$C_p(u) = \sum_{n \geq 1} (1+2j) n^p u^n,$$

so that $C_0(u) = F(u)$, $C_1(u) = D(u)$.

The corresponding Dirichlet series $\hat{C}_p(s)$, which is obtained by replacing u^k by k^{-s} is then

$$\hat{C}_p(s) = \sum_{m, j \geq 0} [2^m(1+2j)]^{-s} (1+2j) [2^m(1+2j)]^p$$
$$= \sum_{m \geq 0} 2^{m(p-s)} \sum_{j \geq 0} (1+2j)^{p+1-s}$$

$$= 2 \frac{2^{s-(p+1)}-1}{2^{s-p}-1} \zeta(s-(p+1)).$$

Now we set, as proposed, $u=e^{-t}$ and find quite easily

$$\frac{(1+u)^3}{u(1-u)} \sim \frac{8}{t} \quad \text{and} \quad \frac{(1+u)^2}{u} \sim 4.$$

To find the local behaviour of $C_p(u)$ as t tends to zero, we Mellin transform it:

$$\mathfrak{M} C_p(e^{-t}) = \mathfrak{M} \sum_{n \geq 1} e^{-tn}(1+2j) n^p$$
$$= \Gamma(s) \hat{C}_p(s).$$

The Mellin inversion formula gives for some $c > p+2$:

$$C_p(e^{-t}) = \frac{1}{2\pi i} \int_{c-i\infty}^{c+i\infty} \Gamma(s) 2 \frac{2^{s-(p+1)}-1}{2^{s-p}-1} \zeta(s-(p+1)) t^{-s} ds.$$

Shifting the line of integration to the left and collecting residues we find

$$C_p(e^{-t}) \sim \tfrac{2}{3}(p+1)! t^{-p-2};$$

if $p=0$ we have to add the term (originating from the double pole)

$$\frac{\zeta(-1)}{\log 2} \log t.$$

In particular we have

$$C_0(e^{-t}) \sim \frac{2}{3} t^{-2} - \frac{1}{12} \frac{1}{\log 2} \log t,$$
$$C_1(e^{-t}) \sim \tfrac{4}{3} t^{-3}.$$

Hence

$$P(z) \sim -\frac{2}{3} \frac{1}{\log 2} \frac{\log t}{t}, \quad t \to 0$$

$$\sim -\frac{1}{6\log 2}\frac{1}{\sqrt{1-4z}}\log(1-4z), \quad z\to 1/4.$$

The coefficient of z^n in $P(z)$ is therefore asymptotic to

$$\frac{1}{6\log 2}\frac{4^n\log n}{\sqrt{\pi n}}.$$

We have to divide this quantity by $t_n\sim 4^n\pi^{-1/2}n^{-3/2}$ to obtain the average register pathlength:

Theorem 1. *The average register pathlength where all binary trees of size n are considered to be equally likely, is asymptotic to*

$$\tfrac{1}{6}n\log_2 n \quad as \quad n\to\infty.$$

The lower order terms involve periodic functions in $\log_4 n$ *which could be determined if desired.*

3. The leaves determining the register function

Let us consider the forest of critical nodes. We want to count the number of nodes on the highest level (this level is the register function). Intuitively speaking, if this number is small, deletion of just a few nodes would decrease the register function. It is somehow more natural to extend the definition of a critical node to the leaves. Counting in this way gives exactly twice the number described first.

Let $[z^n w^m] Q_p(z, w)$ count the number of trees of size n and register function p with m leaves "on the maximal level", furthermore

$$N_p(z):=\frac{\partial}{\partial w}Q_p(z,w)\Big|_{w=1}, \quad N(z):=\sum_{p\geq 0}N_p(z).$$

We easily obtain the recursion

$$Q_p(z,w)=2zQ_p(z,w)\sum_{j<p}R_j(z)+zQ_{p-1}^2(z,w), \quad p\geq 1,$$

$$Q_0(z,w)=w.$$

Hence

$$N_p(z) = 2zN_p(z)(B(z) - S_p(z)) + 2zN_{p-1}(z)R_{p-1}(z),$$

$$N_0(z) = 1,$$

or

$$N_p(z) = N_{p-1}(z) \frac{2zR_{p-1}(z)}{1 - 2z(B(z) - S_p(z))}$$

$$= N_{p-1}(z) \frac{2u^{2^p-1}}{1 + u^{2^p}}$$

$$= \prod_{j=1}^{p} \frac{2u^{2^j-1}}{1 + u^{2^j}}$$

$$= \frac{2^p u^{2^p-1}}{(1 - u^{2^{p+1}})/(1 - u^2)}$$

$$= \frac{1 - u^2}{u} \frac{2^p u^{2^p}}{1 - 2^{p+1}}.$$

$$N(z) = \frac{1 - u^2}{u} \sum_{p \geq 0} \frac{2^p u^{2^p}}{1 - u^{2^{p+1}}} \quad {}^{1)}$$

$$= \frac{1 - u^2}{u} \sum_{p \geq 0} \sum_{\lambda \geq 0} 2^p u^{2^p(1 + 2\lambda)}$$

$$= \frac{1 - u^2}{u} \sum_{n \geq 1} \chi(n) u^n,$$

with

$$\chi(n) = 2^m \text{ if } n = 2^m(1 + 2j).$$

Let

$$C(u) = \sum_{n \geq 1} \chi(n) u^n;$$

---

[1] Another way of reasoning is as follows: The register function p is the size of the largest "complete" binary tree; there is only one such tree in the forest. This complete tree has 2^p leaves, hence $N(z) = \sum_{p \geq 0} 2^p R_p(z)$.

the corresponding Dirichlet series $\hat{C}(s)$ is then

$$\hat{C}(s) = \sum_{m,j \geq 0} 2^m [2^m(1+2j)]^{-s}$$

$$= \frac{1}{2} \frac{2^s - 1}{2^{s-1} - 1} \zeta(s).$$

Therefore, as before,

$$C(e^{-t}) = \frac{1}{2\pi i} \int_{2-i\infty}^{2+i\infty} \Gamma(s) \frac{1}{2} \frac{2^s - 1}{2^{s-1} - 1} \zeta(s) t^{-s} ds.$$

We have a double pole at $s=1$ and simple poles at $s=1+(2k\pi i)/\log 2$, $k \in \mathbb{Z}\setminus\{0\}$. Since

$$\zeta(s) \sim \frac{1}{s-1} + \gamma, \quad \Gamma(s) \sim 1 - \gamma(s-1) \quad \text{as } s \to 1$$

we find

$$C(e^{-t}) \sim -\frac{1}{2\log 2} t^{-1} \log t + \frac{3}{4} t^{-1}$$

$$+ \frac{1}{2\log 2} \sum_{k \neq 0} \Gamma(1+\chi_k) \zeta(1+\chi_k) t^{-1-\chi_k}.$$

Since

$$\frac{1-u^2}{u} \sim 2t$$

we find

$$N(z) \sim -\log_2 t + \frac{3}{2} + \frac{1}{\log 2} \sum_{k \neq 0} \Gamma(1+\chi_k) \zeta(1+\chi_k) t^{-\chi_k}$$

$$\sim -\log_2 \varepsilon + \frac{1}{2} + \frac{1}{\log 2} \sum_{k \neq 0} \Gamma(1+\chi_k) \zeta(1+\chi_k) \varepsilon^{-\chi_k}.$$

For the coefficient of z^n in $N(z)$ we have the asymptotic equivalent

$$\frac{1}{2\log 2} \frac{4^n}{n} + \frac{1}{\log 2} \sum_{k \neq 0} \Gamma(1+\chi_k) \zeta(1+\chi_k) 4^n n^{-1+\chi_k/2} / \Gamma\left(\frac{\chi_k}{2}\right).$$

By the duplication formula for the gamma function we have

$$\Gamma(1+\chi_k)/\Gamma\left(\frac{\chi_k}{2}\right)=\frac{1}{\pi}\chi_k\Gamma\left(\frac{1+\chi_k}{2}\right).$$

We finally divide by $t_n \sim 4^n \pi^{-1/2} n^{-3/2}$ and have:

Theorem 2. *The average number of leaves on the maximum (register) level is asymptotic to*

$$\sqrt{n}\, G(\log_4 n), \quad n\to\infty$$

with a periodic function $G(x)$ with period 1 and the Fourier expansion

$$G(x)=\sum_{k\in Z} g_k e^{2k\pi i x}$$

where

$$g_0 = \frac{\sqrt{\pi}}{2\log 2}$$

and

$$g_k = \frac{1}{\sqrt{\pi}\log 2}\chi_k\Gamma\left(\frac{1+\chi_k}{2}\right)\zeta(1+\chi_k), \quad k\neq 0.$$

4. The average register value of a node

We now sum all register values obtained by the labelling procedure described in the introduction and divide by $2n+1$ which is the total number of nodes. The average of this quantity is then to be computed.

We consider a binary tree where each leaf is labelled (marked) and replace it by $\sum_{p\geq 0} p\mathfrak{R}_p$. In terms of generating functions this means

$$W(z) = \sum_{n\geq 0}(n+1)t_n z^n \sum_{p\geq 0} pR_p(z)$$

where the coefficient of z^n in $W(z)$ is the sum of all register values in all trees of size n.

$$W(z) = \frac{1}{\sqrt{1-4z}}\sum_{p\geq 0} pR_p(z).$$

Since $\sum_{p \geq 0} p R_p(z)$ appeared already in the computation of the average register function, and

$$\frac{1}{\sqrt{1-4z}} = \frac{1+u}{1-u} \sim \frac{2}{t},$$

we have (compare [5]) with a constant K

$$W(z) \sim \frac{2}{t}\left[t\log_2 t + Kt + \frac{1}{\log 2}\sum_{k \neq 0} 2\Gamma(\chi_k)\zeta(\chi_k)t^{-\chi_k} + 2\right]$$

$$\sim 2\log_2 \varepsilon + K' + \frac{4}{\log 2}\sum_{k \neq 0}\Gamma(\chi_k)\zeta(\chi_k)\varepsilon^{-\chi_k} + \frac{2}{\varepsilon}$$

and so

$$[z^n]W(z) \sim -\frac{1}{\log 2}\frac{4^n}{n} + \frac{4}{\log 2}\sum_{k \neq 0}\Gamma(\chi_k)\zeta(\chi_k)\frac{4^n n^{\chi_k/2 - 1}}{\Gamma\left(\frac{\chi_k}{2}\right)} + \frac{2 \cdot 4^n}{\sqrt{\pi n}}.$$

We divide it by $(2n+1)t_n \sim 2 \cdot 4^n \pi^{-1/2} n^{-1/2}$ and obtain

Theorem 3. *The average register value of a node in a tree of size n is asymptotic to*

$$1 + \frac{1}{\sqrt{n}}H(\log_4 n)$$

where $H(x)$ is periodic with period 1 and the Fourier expansion

$$H(x) = \sum_{k \in \mathbb{Z}} h_k e^{2k\pi i x}$$

with

$$h_0 = \frac{-\sqrt{\pi}}{\log 2}$$

and

$$h_k = \frac{1}{\log 2}\frac{1}{\sqrt{\pi}}\Gamma\left(\frac{1+\chi_k}{2}\right)\zeta(\chi_k), \quad k \neq 0.$$

5. The average number of registers to evaluate r arithmetic expressions

Corresponding to the title we consider an ordered forest of r binary trees (r fixed) with altogether n internal nodes ("size n"). The register function extends trivially by taking the maximum of the register functions of the r trees.

To compute the average we consider the obvious generating functions (compare the introduction):

$$E(z) = \sum_{p \geq 1} B^r(z) - (B(z) - S_p(z))^r$$

$$= (1+u)^r \sum_{p \geq 1} 1 - \left(\frac{1-u^{2^p-1}}{1-u^{2^p}}\right)^r$$

$$= (1+u)^r \sum_{p \geq 1} 1 - \left(1 + \frac{(1-u^{-1})u^{2^p}}{1-u^{2^p}}\right)^r$$

$$= -(1+u)^r \sum_{\lambda=1}^{r} \binom{r}{\lambda} (1-u^{-1})^\lambda \sum_{p \geq 1} \frac{u^{2^p \lambda}}{(1-u^{2^p})^\lambda}.$$

Let us consider

$$E_\lambda(u) = \sum_{p \geq 1} \frac{u^{2^p \lambda}}{(1-u^{2^p})^\lambda};$$

$$\mathfrak{M} E_\lambda(e^{-t}) = \mathfrak{M} \sum_{p \geq 1} \frac{e^{-t\lambda 2^p}}{(1-e^{-t 2^p})^\lambda}$$

$$= \sum_{p \geq 1} 2^{-ps} \mathfrak{M} \frac{e^{-t\lambda}}{(1-e^{-t})^\lambda}$$

$$= \frac{1}{2^s - 1} \mathfrak{M} e^{-t\lambda} \sum_{m \geq 0} \binom{\lambda+m-1}{\lambda-1} e^{-mt}$$

$$= \frac{\Gamma(s)}{2^s - 1} \sum_{m \geq 1} \binom{m-1}{\lambda-1} m^{-s}$$

$$= \frac{\Gamma(s)}{2^s - 1} \frac{1}{(\lambda-1)!} \sum_{m \geq 1} \left(m^{\lambda-1} - \binom{\lambda}{2} m^{\lambda-2} + \ldots\right) m^{-s}$$

$$= \frac{\Gamma(s)}{2^s - 1} \frac{1}{(\lambda-1)!} \left[\zeta(s-\lambda+1) - \binom{\lambda}{2} \zeta(s-\lambda+2) + \ldots\right].$$

Using the Mellin inversion formula we find for $\lambda \geq 2$

$$E_\lambda(e^{-t}) \sim \frac{1}{2^\lambda - 1} t^{-\lambda} - \frac{1}{2^{\lambda-1} - 1} \frac{\lambda}{2} t^{-\lambda+1} + \dots .$$

For $\lambda = 1$ there is a double pole "in the second term"; this has also been computed previously:

$$E_1(e^{-t}) \sim \frac{1}{t} + \frac{1}{2} \log_2 t - \frac{1}{2} \log_2 2\pi + \frac{1}{4} + \frac{\gamma}{2 \log 2}$$

$$+ \frac{1}{\log 2} \sum_{k \neq 0} \Gamma(\chi_k) \zeta(\chi_k) t^{-\chi_k} .$$

Also

$$-(1+u)^r (1-u^{-1})^\lambda \sim (-1)^{\lambda+1} t^\lambda 2^r \left(1 - (r-\lambda) \frac{t}{2}\right).$$

Therefore

$$E(z) \sim \sum_{\lambda=1}^{r} \frac{1}{2^\lambda - 1} (-1)^{\lambda+1} 2^r \left(1 - (r-\lambda) \frac{t}{2}\right) \binom{r}{\lambda}$$

$$- \sum_{\lambda=2}^{r} \frac{1}{2^{\lambda-1} - 1} \frac{\lambda}{2} t 2^r (-1)^{\lambda+1} \binom{r}{\lambda}$$

$$+ rt 2^r \left[\frac{1}{2} \log_2 t - \frac{1}{2} \log_2 2\pi + \frac{1}{4} + \frac{\gamma}{2 \log 2} \right.$$

$$\left. + \frac{1}{\log 2} \sum_{k \neq 0} \Gamma(\chi_k) \zeta(\chi_k) t^{-\chi_k} \right].$$

The huge bracket is already well-known; from the two sums only the coefficient of t is of interest. It is

$$\sum_{\lambda=1}^{r} \frac{1}{2^\lambda - 1} (-1)^{\lambda+1} 2^r (-r+\lambda) \frac{1}{2} \binom{r}{\lambda}$$

$$- \sum_{\lambda=2}^{r} \frac{1}{2^{\lambda-1} - 1} \frac{\lambda}{2} 2^r (-1)^{\lambda+1} \binom{r}{\lambda}$$

$$= 2^{r-1} \sum_{\lambda=1}^{r-1} \frac{(-1)^\lambda}{2^\lambda - 1} \left[(r-\lambda) \binom{r}{\lambda} - (\lambda+1) \binom{r}{\lambda+1} \right] = 0.$$

Since

$$B^r(z) \sim (2-2\varepsilon)^r \sim 2^r - r2^r\varepsilon,$$

we see that the desired average is (with respect to the first two terms in the asymptotic expansion!) independent of r. Hence we have

Theorem 4. *The average number of registers to evaluate r arithmetic expressions of (altogether) size n is asymptotic to*

$$\log_4 n + D(\log_4 n), \quad n \to \infty$$

with the well-known function $D(x)$, being periodic with period 1.

6. How early is the register function reached

If we regard the labelling procedure which yields the register function at the root, it normally happens that this value appears already earlier. We now count the number of nodes above this first occurrence; this number is ≥ 0. In this section we are concerned with the average of this parameter.

Let $Q_p(z, w)$ be the generating function of trees with register function p where the coefficient of w^j refers to the value j of our desired parameter. Furthermore let $N_p(z) = \frac{\partial}{\partial w} Q_p(z, w)\Big|_{w=1}$ and $N(z) = \sum_{p \geq 0} N_p(z)$. Naturally, we are interested in

$$\frac{[z^n] N(z)}{[z^n] B(z)}.$$

If we regard the original recursion for the register function reg and translate it into an equation of generating functions we obtain:

$$Q_p = 2zwQ_p \sum_{j<p} R_j + zR_{p-1}^2, \quad p \geq 1, \quad Q_0 = 1$$

and therefore

$$N_p = 2z R_p(B - S_p) + 2z N_p(B - S_p)$$
$$= \frac{2z R_p(B - S_p)}{1 - 2z(B - S_p)}$$

$$= \frac{2(1+u)u^{2^p}(1-u^{2^p-1})}{(1-u^{2^p})(1+u^{2^p})^2}$$

$$= \frac{2(1+u)u^{2^p}}{(1-u^{2^{p+1}})^2} - \frac{2(1+u)^2}{u} \frac{u^{2^{p+1}}}{(1-u^{2^{p+1}})^2} + \frac{2(1+u)}{u} \frac{u^{3 \cdot 2^p}}{(1-u^{2^{p+1}})^2}.$$

Let us consider

$$M_a(u) = \sum_{p \geq 0} \frac{u^{a \cdot 2^p}}{(1-u^{2^{p+1}})^2},$$

then

$$N(z) = 2(1+u)M_1 - \frac{2(1+u)^2}{u} M_2 + \frac{2(1+u)}{u} M_3.$$

$$\mathfrak{M} M_a(e^{-t}) = \mathfrak{M} \sum_{p \geq 1} \frac{e^{-t\frac{a}{2} 2^p}}{(1-e^{-t 2^p})^2}$$

$$= \sum_{p \geq 1} 2^{-ps} \mathfrak{M} \frac{e^{-t\frac{a}{2}}}{(1-e^{-t})^2}$$

$$= \frac{1}{2^s - 1} \mathfrak{M} e^{-t\frac{a}{2}} \sum_{m \geq 0} (m+1) e^{-mt}$$

$$= \frac{\Gamma(s)}{2^s - 1} \sum_{m \geq 1} \left(m + \frac{a-2}{2}\right)^{-s} m.$$

Apart from the $\Gamma(s)/(2^s-1)$-factor we obtain for

$a=1$: $(2^{s-1}-1)\zeta(s-1) + \frac{1}{2}(2^s-1)\zeta(s)$

$a=2$: $\zeta(s-1)$

$a=3$: $(2^{s-1}-1)\zeta(s-1) - \frac{1}{2}(2^s-1)\zeta(s)$.

Applying the Mellin inversion formula again we find that $M_a(u)$ is asymptotic to

$$\sim \frac{1}{3}t^{-2} + \frac{1}{2}t^{-1} + \frac{\zeta(-1)}{2\log 2}\log t \quad (a=1)$$

$$\sim \frac{1}{3}t^{-2} - \frac{\zeta(-1)}{\log 2}\log t \qquad (a=2)$$

$$\sim \frac{1}{3}t^{-2} - \frac{1}{2}t^{-1} + \frac{\zeta(-1)}{2\log 2}\log t \qquad (a=3).$$

Hence

$$N(z) \sim \tfrac{3}{4}\zeta(-1)\log_2 t \sim -\tfrac{1}{16}\log_2 \varepsilon.$$

Theorem 5. *The average number of nodes "above" the node which first equals the register value is asymptotic to*

$$\frac{\sqrt{\pi}}{32\log 2}\sqrt{n}, \qquad n \to \infty.$$

7. Epilogue

From the explicit formulae for the generating functions mentioned in the introduction we find for instance

$$S_{p+1}(z) = (B(z)-1)^{2^p} R_p(z).$$

We will sketch a simple combinatorial argument for that: Take a tree with register function p and consider its unique largest subtree in the forest of critical nodes. This tree is a complete binary tree with 2^p leaves. If we replace each leaf by an arbitrary nonempty tree (counted by $B(z)-1$), we thereby obtain a tree with strictly larger register function. This mapping is injective; it is surjective as well. So we have a bijection and therefore the announced formula. (The inverse mapping could be described in a clumsy way by cutting down a tree with register function $>p$ in a certain sense of maximality, yielding the 2^p nonempty trees and a tree with register function p.)

We will now prove the explicit formulae starting from our just established equality. This is therefore a second easy derivation of the explicit formulae. (The first one is due to P. Kirschenhofer and H. Prodinger, see [12]). We have

$$S_{p+1} = u^{2^p} R_p \quad \text{and} \quad S_p = u^{2^{p-1}} R_{p-1};$$

taking differences we see:

$$R_p = u^{2^{p-1}} R_{p-1} - u^{2^p} R_p$$

$$= R_{p-1} \frac{u^{2^p-1}}{1+u^{2^p}}$$

$$= \frac{u^{1+2+4+\ldots+2^{p-1}}}{(1+u^2)(1+u^4)\ldots(1+u^{2^p})}$$

$$= \frac{u^{2^p-1}}{(1-u^{2^{p+1}})/(1-u^2)} = \frac{1-u^2}{u} \frac{u^{2^p}}{1-u^{2^{p+1}}}.$$

References

[1] P. Flajolet, Analyse d'algorithmes de manipulation d'arbres et de fichiers, Cahiers de BURO, **34-35** (1981), 1–209.

[2] P. Flajolet, J. -C. Raoult, J. Vuillemin, On the average number of registers required for evaluating arithmetic expressions, FOCS-Conference 1977, 196–205.

[3] P. Flajolet, J. -C. Raoult, J. Vuillemin, The number of registers required for evaluating arithmetic expressions, Theoretical Computer Science **9** (1979), 99–125.

[4] P. Flajolet, A. Odlyzko, The average height of binary trees and other simple trees, J. Comput. Syst. Sci. **25** (1982), 142–158.

[5] P. Flajolet, H. Prodinger, Register allocation for unary-binary trees, SIAM J. on Computing, **15** (1986), 629–640.

[6] P. Flajolet, M. Régnier, R. Sedgewick, Mellin transform techniques for the analysis of algorithms, in preparation.

[7] J. Francon, Sur le nombre de registres nécessaires à l'évaluation d'une expression arithmétique, RAIRO Inf. Theor. **18** (1984), 355–364.

[8] R. Kemp, The average number of registers to evaluate a binary tree optimally, Acta Informatica **11** (1979), 363–372.

[9] A. Meir, J. W. Moon, J. R. Pounder, On the order of random channel networks, SIAM J. Alg. Discr. Meth. **1** (1980), 25–33.

[10] J. W. Moon, On Horton's law for random channel networks, Annals of Discrete Mathematics **8** (1980), 117–121.

[11] H. Prodinger, *Abzählprobleme bei Bäumen*, Publ. de l'IRMA (Straßburg, G. Baron, P. Kirschenhofer, Ed.), (1983), 167–173.

[12] H. Prodinger, Die Bestimmung gewisser Parameter bei binären Bäumen mit Hilfe analytischer Methoden, Lecture Notes in Mathematics **1114** (1985), 118–133.

[13] H. Prodinger, *Quelques techniques analytiques pour l'etude asymptotique des parametres dans les arbres*, Proceedings Ecole de Printemps Ile de Ré, P. Flajolet, Ed.

[14] E. T. Whittaker, G. N. Watson, *A course of modern analysis*, Cambridge University Press, 1927.

BOUNDS FOR ALL-TERMINAL RELIABILITY IN PLANAR NETWORKS

Aparna RAMESH

University of Waterloo,
Waterloo, Ontario, Canada

Michael O. BALL*

University of Waterloo
Waterloo, Ontario, Canada

Charles J. COLBOURN

University of Waterloo
Waterloo, Ontario, Canada

A communication network can be modeled as a graph where the nodes of the graph represent the sites, and the edges represent the links between the sites. The edges of the graph operate with equal probability p. The all-terminal reliability of the network with n nodes and b edges can be written as $R(p) = \sum_{i=0}^{b} F_i p^{b-i} q^i$ where F_i is the number of connected subgraphs with $(b-i)$ edges and $q = (1-p)$. All known subgraph counting bounds for R use exact values for $F_0, \ldots, F_c$ and F_d where c is the cardinality of a minimum cut and $d = b - n + 1$. In this paper we derive upper and lower bounds for R for planar networks by using exact values of $F_{d-1}, F_{d-2}, F_{d-3}, F_{c+1}$ and F_{c+2} where approximations were used before. The effect of using these exact values instead of approximations on Kruskal-Katona bounds and Ball-Provan bounds is studied.

1. Introduction

Computing the reliability of a stochastic network has received significant attention in recent years. All known algorithms for computing network reliability exactly have running times which are exponential in the size of the network [B1, B5]. Virtually all network reliability problems of practical interest are known to be NP-hard [P1, V1]. A practical solution in such a case is to give efficiently

---

* On leave from *College of Business and Management, University of Maryland.*

computable upper and lower bounds on R. Many researchers have taken this approach (see, for example, [B2, B3, C1, C2, L2, V2]). Harms [H1] provides an excellent survey and practical comparisons of these bounds. Here we obtain improvements of the bounds for planar networks. The bounds we use here are the Kruskal-Katona bounds suggested by Van Slyke and Frank [V2], and the Ball-Provan bounds [B3]. These bounds apply to an important network problem called the "connectedness problem", or all-terminal reliability. The connectedness problem is computing the probability that all nodes of the network are able to communicate with each other.

A computer communication network is modeled as a *probabilistic graph* where the nodes of the graph represent the sites and the edges represent the links between the sites. The edges can be in one of the two states, *operative* or *failed*. With each edge is associated a positive real number which is the probability that the edge is *operative*. The all-terminal reliability of the network is the probability that all the nodes of the network are able to communicate with each other. We make two assumptions about the network.

(1) Edges fail randomly with equal probability $(1-p)$.
(2) Nodes of the network are perfectly reliable.

Under these assumptions the all-terminal reliability of the network with n nodes and b edges can be written as a polynomial in p, as $R = \sum_{i=0}^{b} F_i p^{b-i} q^i$ where

$$F_i = \begin{cases} \text{\# of connected subgraphs with } (b-i) \text{ edges OR} \\ \text{\# of subsets of } i \text{ edges whose complement contains a spanning tree.} \end{cases}$$

Although the reliability polynomial forms a concise representation, the actual calculation of R is quite difficult. In particular, Provan and Ball [P1] showed that computing R is NP-hard. Therefore it appears that computing R exactly is not practical for general graphs.

Even though computing R exactly for general graphs has been shown to be a hard problem, the problem is still open for *planar graphs*. Since there is no polynomial algorithm known to compute the exact reliability of planar graphs, it is useful to develop upper and lower bounds for R. Let c be the cardinality of a minimum cut in the network with n nodes. Since for any network with n nodes at least $n-1$ edges are needed to connect them, and it takes at least c edges to disconnect them, we have

$$F_d = t = \text{\# of spanning trees of the graph where } d = b-n+1$$
$$F_i = 0 \quad \text{for} \quad i > (b-n+1)$$

$$F_i = \binom{b}{c} \quad \text{for} \quad i < c.$$

$$F_c = \binom{b}{c} - n_c$$

where n_c is the number of minimum cardinality cuts. n_c can be computed in polynomial time [B2] and t can be computed in polynomial time [B4].

For planar networks, additional information is efficiently computable. For planar networks, $c \leq 5$; hence, a simple exhaustive technique provides a polynomial time algorithm for determining F_{c+k} for any fixed k. A recent theorem of Liu and Chow [L1] shows that F_{d-k} can be computed for any k via a number of determinant computations; the number of determinants evaluated is independent of n, but grows exponentially in k; nevertheless, for fixed k a polynomial time algorithm results.

The fact that additional efficiently computable coefficients are available for planar networks is well known; the purpose of this paper is to determine their effect on the existing bounds.

2. Kruskal-Katona bounds

Van Slyke and Frank [V2] use a theorem of Kruskal [K2] and Katona [K1] that applies to any *coherent* system, to obtain bounds on the network reliability polynomial. A network problem is *coherent* if when a network is in a failed state, no set of further component failures will return it to an operating state. In order to describe this theorem and subsequent bounds we first develop a number of combinatorial concepts. For any non-negative integer m, the *k-canonical representation* of m is given by:

$$m = \binom{m_k}{k} + \binom{m_{k-1}}{k-1} + \ldots + \binom{m_l}{l}$$

where $m_k > m_{k-1} > \ldots > m_l \geq l \geq 1$. The k-canonical representation always exists and is unique. For $k \geq l \geq 1$ and any $i \geq k$, the (i, k)th *lower pseudopower of* $(m_k, \ldots, m_l)$ is:

$$(m_k, \ldots, m_l)^{i/k} = \binom{m_k}{i} + \binom{m_{k-1}}{i-1} + \ldots + \binom{m_l}{i-k+l}.$$

We represent $(m_k, \ldots, m_l)^{i/k}$ as $m^{i/k}$. The theorem developed by Kruskal [K2] and Katona [K1] states that

a) $F_k^{i/k} \geq F_i$ when $i \geq k$

b) $F_k^{i/k} \leq F_i$ when $i \leq k$.

Thus, knowing F_c and using a) we get the upper bound on R as

$$R \leq \sum_{i=0}^{c-1} \binom{b}{i} p^{b-i} q^i + F_c p^{b-c} q^c + \sum_{i=c+1}^{d-1} F_c^{i/c} p^{b-i} q^i + F_d p^{b-d} q^d$$

and knowing F_{b-n+1} and using b) we get the lower bound on R as

$$R \geq \sum_{i=0}^{c-1} \binom{b}{i} p^{b-i} q^i + F_c p^{b-c} q^c + \sum_{i=c+1}^{d} F_d^{i/d} p^{b-i} q^i$$

where $d = b - n + 1$ and F_d is the number of spanning trees of the network.

3. Ball-Provan bounds

Ball and Provan [B2, B3] develop a set of bounds using results due to Stanley [S1]. Stanley's results apply to systems that are *shellable*. Complements of spanning connected subgraphs of a graph form a *matroid*, referred to as the co-graphic matroid. Provan and Billera have shown that all matroids are shellable [P2]. Therefore Stanley's result applies to them.

These bounds are formed in a manner which is analogous to the Kruskal-Katona bounds. A different form of reliability polynomial is used:

$$R = p^{n-1} \sum_{i=0}^{b} h_i q^i. \tag{1}$$

This conversion is performed by factoring out p^{n-1} and then recombining the remaining terms. The h_i terms can be obtained directly from the F_i terms using

$$h_i = \sum_{k=0}^{i} (-1)^{i-k} \binom{b-n-1-k}{i-k} F_k$$

for $i = 0, 1, \ldots, b-n+1$. The F_i's can be obtained from the h_i's using:

$$F_i = \sum_{j=0}^{i} \binom{d-j}{i-j} h_j$$

where $d = b - n + 1$.

Using the above relationships, the conditions given at the end of Section 1 become

$$h_i = \binom{n+i-2}{i} \quad \text{for } i < c$$

$$h_i = 0 \quad \text{for } i > d, \text{ where } d = b - n + 1.$$

$$h_c = \binom{n+c-2}{c} - n_c \quad \text{for } i = c,$$

$$\sum_{i=0}^{d} h_i = t.$$

The (i, k)th upper pseudopower of $(m_k, \ldots, m_l)$ is:

$$(m_k, \ldots, m_l)^{\langle i/k \rangle} = \binom{m_k - k + i}{i} + \binom{m_{k-1} - k + i}{i-1} + \ldots + \binom{m_l - k + i}{i - k + l}.$$

We represent $(m_k, \ldots, m_l)^{\langle i/k \rangle}$ as $m^{\langle i/k \rangle}$. Stanley [S1] defines an "h-vector" for general independence systems and gives several properties of it. Ball and Provan [B3] showed that the h-vector defined by (1) is identical to the h-vector defined by Stanley. This relationship showed the applicability of the following result, due to Stanley, to network reliability problems.

Theorem: [S1] *Let $(h_0, \ldots, h_d)$ be an integer vector. Then the following statements are equivalent:*

(i) $(h_0, \ldots, h_d)$ *is the h-vector for some rank d shellable independence system.*

(ii) $h_0 = 1$ *and* $0 \leq h_j \leq h_i^{\langle j/i \rangle}$ *for* $0 \leq i < j \leq d.$

Any vector that satisfies (ii) is called an *O-sequence*. We refer to (ii) as the *upper pseudopower constraint*. Before studying the resulting bounds, we define some notation. For any non-negative integers m, d and k the (k, d)-*factor* of m is the number x for which:

$$x - x^{\langle k/d \rangle} = m.$$

Let

$$a = t - \sum_{i=0}^{c} h_i$$

$$r_0 = \max \left(r : \sum_{i=c+1}^{r} h_c^{\langle i/c \rangle} \leq a \right)$$

$$m = (a_d - 1, \ldots, a_l - 1)$$

where $(a_d, \ldots, a_l)$ is the d-canonical vector for the (c, d)-factor of a. Then the Ball-Provan bounds are defined by

$$R \leq p^{n-1}\left(\sum_{i=0}^{c} h_i q^i + \sum_{i=c+1}^{r_0} h_c^{\langle i/c \rangle} q^i\right.$$

$$\left. + \left(a - \sum_{i=c+1}^{r_0} h_c^{\langle i/c \rangle}\right) q^{r_0+1}\right)$$

and

$$R \geq p^{n-1}\left(\sum_{i=0}^{c} h_i q^i + \sum_{i=c+1}^{d} m^{\langle i/d \rangle} q^i\right).$$

4. Using the additional coefficients

We now exploit the additional coefficients available for planar networks to tighten the Kruskal-Katona and Ball-Provan bounds. In all cases, the improvement in the bounds is evaluated using the graph in Figure 1 (p. 271). We compare old and new versions of the bounds with each other and with exact values. For the graph in Figure 1, the exact values of the reliability are obtained using results of Wald and Colbourn [W1].

4.1. Kruskal-Katona bounds

By using the exact values of the coefficients F_{c+1} and F_{c+2}, the Kruskal-Katona upper and lower bounds for R take the following form:

$$R \leq \sum_{i=0}^{c-1} \binom{b}{i} p^{b-i} q^i + \sum_{i=c}^{c+2} F_i p^{b-i} q^i$$

$$+ \sum_{i=c+3}^{d-1} F_{c+2}^{i/c+2} p^{b-i} q^i + F_d p^{b-d} q^d$$

and

$$R \geq \sum_{i=0}^{c-1} \binom{b}{i} p^{b-i} q^i + \sum_{i=c}^{c+2} F_i p^{b-i} q^i + \sum_{i=c+3}^{d} F_d^{i/d} p^{b-i} q^i.$$

The improvement obtained by using the above bounds for R is referred to as "Left" improvement.

By using the exact values of the coefficients F_{d-3}, F_{d-2} and F_{d-1}, the upper and lower bounds take the following form:

$$R \leqslant \sum_{i=0}^{c-1} \binom{b}{i} p^{b-i} q^i + F_c p^{b-c} q^c$$
$$+ \sum_{i=c+1}^{d-4} F_c^{i/c} p^{b-i} q^i + \sum_{i=d-3}^{d} F_i p^{b-i} q^i$$

and

$$R \geqslant \sum_{i=0}^{c-1} \binom{b}{i} p^{b-i} q^i + F_c p^{b-c} q^c$$
$$+ \sum_{i=c+1}^{d-4} F_{d-3}^{i/d-3} p^{b-i} q^i + \sum_{i=d-3}^{d} F_i p^{b-i} q^i.$$

We refer to the improvement obtained by the above as "Right" improvement.

Finally, by using exact values for F_{c+1}, F_{c+2}, F_{d-3}, F_{d-2} and F_{d-1} the upper and lower bounds become

$$R \leqslant \sum_{i=0}^{c-1} \binom{b}{i} p^{b-i} q^i + \sum_{i=c}^{c+2} F_i p^{b-i} q^i$$
$$+ \sum_{i=c+3}^{d-4} F_{c+2}^{i/c+2} p^{b-i} q^i + \sum_{i=d-3}^{d} F_i p^{b-i} q^i$$

and

$$R \geqslant \sum_{i=0}^{c-1} \binom{b}{i} p^{b-i} q^i + \sum_{i=c}^{c+2} F_i p^{b-i} q^i$$
$$+ \sum_{c+3}^{d-4} F_{d-3}^{i/d-3} p^{b-i} q^i + \sum_{i=d-3}^{d} F_i p^{b-i} q^i.$$

We refer to this improvement as "Both Sides Improvement".

Table 1 gives the lower and upper bounds for the reliability values computed both by the Kruskal-Katona original method and the Kruskal-Katona both sides improved method.

4.2. Ball-Provan bounds

Using the exact values of F_{c+1} and F_{c+2}, the values of h_{c+1} and h_{c+2} are easily computed. The Ball-Provan bounds with exact h_{c+1} and h_{c+2} values take the

Table 1.

Kruskal-Katona bounds

	COMPARISON OF RELIABILITY VALUES				
	Original		Both Sides Improved		ACTUAL RELIABILITY
p	lower bound	upper bound	lower bound	upper bound	
0.4	0.00175959	0.0490986	0.00618366	0.00861309	0.00675426
0.5	0.0111161	0.268058	0.0458488	0.0817167	0.0561276
0.6	0.0403344	0.612533	0.154581	0.331290	0.221177
0.7	0.111063	0.832196	0.320802	0.676763	0.511013
0.8	0.268535	0.927769	0.562679	0.881191	0.801013
0.9	0.632148	0.981000	0.896723	0.970781	0.963646
0.91	0.684243	0.984529	0.922665	0.976537	0.971852
0.92	0.737727	0.987712	0.944831	0.981696	0.978799
0.93	0.791297	0.990543	0.962895	0.986230	0.984572
0.94	0.843203	0.993016	0.976784	0.990117	0.989260
0.95	0.891259	0.995125	0.986721	0.993341	0.992955
0.96	0.932949	0.996864	0.993217	0.995898	0.995756
0.97	0.965723	0.998227	0.997013	0.997799	0.997761
0.98	0.987566	0.999208	0.998948	0.999076	0.999071
0.99	0.998038	0.999801	0.999779	0.999784	0.999784

form

$$R \leq p^{n-1}\left(\sum_{i=0}^{c+2} h_i q^i + \sum_{i=c+3}^{r_0} h_{c+2}^{\langle i/c+2\rangle} q^i \right.$$
$$\left. + (a - \sum_{i=c+3}^{r_0} h_{c+2}^{\langle i/c+2\rangle}) q^{r_0+1}\right)$$

and

$$R \geq p^{n-1}\left(\sum_{i=0}^{c+2} h_i q^i + \sum_{i=c+3}^{d} m^{\langle i/d\rangle} q^i\right)$$

where $m = (a_d - 1, \ldots, a_l - 1)$ and $(a_d, \ldots, a_l)$ is the d-canonical vector for the $(c+2, d)$-factor of $t - \sum_{i=0}^{c+2} h_i$ and t is the number of spanning trees of the graph.
We refer to the improvement achieved by these bounds as "Left" improvement.

Improving Ball-Provan bounds by knowing the F_i values from the right end of the polynomial is not straightforward. We describe a method that uses only the F_{d-1} coefficient. This method as well as the Ball-Provan bounds use the following Proposition.

Proposition: [B3] *Given two vectors of integers* $(\bar{h}_0, \bar{h}_1, \ldots, \bar{h}_d)$ *and* $(\underline{h}_0, \underline{h}_1, \ldots, \underline{h}_d)$, *if for all j*

$$\sum_{k=0}^{j} \bar{h}_k \geq \sum_{k=0}^{j} \underline{h}_k$$

then $\bar{g}(p) \geq \underline{g}(p)$ *for all* $0 < p < 1$, *where* $\bar{g}(p) = \sum_{k=0}^{d} \bar{h}_k q^k$ *and* $\underline{g}(p) = \sum_{k=0}^{d} \underline{h}_k q^k$.

Since $h_j \geq 0$, and $\sum_{i=0}^{d} h_i = F_d$, we can interpret assigning values to h_i's as placing F_d balls in $d+1$ buckets [B3]. Now the Proposition implies that to obtain an upper bound we would like to put as many balls as possible in the lowest numbered buckets and for a lower bound we would like to put as many balls as possible in the highest numbered buckets. In the original Ball-Provan method F_d balls are distributed in $d+1$ buckets satisfying only the upper pseudopower constraint. The following "Right" improved method incorporates the F_{d-1} constraint also, $F_{d-1} = (dh_0 + (d-1)h_1 + (d-2)h_2 + \ldots + h_{d-1})$.

The upper bound technique sets h_i for $i = c+1, \ldots$, according to the Ball-Provan upper bound. However, at each stage a check is made to see whether setting h_i to its upper bound would imply that the F_{d-1} value could not be achieved. That is, for all feasible assignments of the remaining h_i's

$$\sum_{j=0}^{d-1} \binom{d-j}{d-1-j} h_j > F_{d-1}.$$

If such an infeasibility is discovered the h_i value is reduced from the upper bound until the violation no longer exists. Given the values for $h_0, \ldots, h_i$ such a check can be made by computing $h_{i+1}, \ldots, h_d$ according to the Ball-Provan lower bound. This provides a valid check since the Ball-Provan lower bound simultaneously minimizes all F_k values (and, in particular, F_{d-1}). We define this lower feasibility check by:

lower feasibility check: Assuming $h_0, \ldots, h_i$ are fixed, set $h_{i+1}, \ldots, h_d$ according to the Ball-Provan lower bound. The distribution is feasible if

$$\sum_{j=0}^{d-1} \binom{d-j}{d-1-j} h_j \leq F_{d-1}.$$

The first h_i that does not satisfy the lower feasibility check is decremented until the check is satisfied. After that the remaining h_i's are set to the standard Ball-Provan upper bounds. The fact that this procedure gives a valid upper bound, subject to the constraint that the true F_{d-1} value is *less than or equal to* the given F_{d-1} value follows directly from the Proposition. We note that there certainly is room for improvement in this technique. In particular, the F_{d-1} value achieved by the upper bound could, in fact, be greater than the F_{d-1} given.

The lower bound is obtained in a similar manner, where the lower feasibility check is replaced by the following:

upper feasibility check: Assuming $h_0, \ldots, h_i$ are fixed, set $h_{i+1}, \ldots, h_d$ according to the Ball-Provan upper bound. The distribution is feasible if

$$\sum_{j=0}^{d-1} \binom{d-j}{d-1-j} h_j \geq F_{d-1}.$$

We now describe the bound generation algorithms.

Upper bound algorithm:

(i) Set $i = c+1$

(ii) Put the maximum number of balls in the ith bucket, i.e., set $h_i = h_{i-1}^{<i/i-1>}$

(iii) Execute the lower feasibility check. If satisfied set $i = i+1$ and goto step (ii). Otherwise goto step (iv)

(iv) Until the lower feasibility check is satisfied, set $h_i = h_i - 1$

(v) Assuming $h_0, \ldots, h_i$ are fixed, assign remaining balls according to Ball-Provan upper bound

Lower bound algorithm:

(i) Set $i = c+1$

(ii) Put the minimum number of balls into the ith bucket, i.e., assuming $h_0, \ldots, h_{i-1}$ are fixed, set h_i according to the Ball-Provan lower bound

(iii) Execute the upper feasibility check. If satisfied set $i = i+1$ and goto step (ii). Otherwise goto step (iv)

(iv) Until the upper feasibility check is satisfied set $h_i = h_i + 1$

(v) Assuming $h_0, \ldots, h_i$ are fixed assign remaining balls according to Ball-Provan lower bound

The upper and lower bounds obtained by the above method are referred to as "Right" Improved method. This method together with the additional coefficients

Table 2.

Ball-Provan bounds

p	COMPARISON OF RELIABILITY VALUES				ACTUAL RELIA-BILITY
	Original		Both Sides Improved		
	lower bound	upper bound	lower bound	upper bound	
0.4	0.00177344	0.012155	0.00206632	0.0108967	0.00675426
0.5	0.0114658	0.102968	0.0164341	0.0905016	0.0561276
0.6	0.0445923	0.375131	0.0821938	0.327927	0.221177
0.7	0.140279	0.723457	0.284918	0.645875	0.511013
0.8	0.379975	0.916913	0.649919	0.861356	0.801013
0.9	0.787771	0.980951	0.944290	0.968283	0.963646
0.91	0.827829	0.984509	0.958816	0.974814	0.971852
0.92	0.865241	0.987705	0.970561	0.980575	0.978799
0.93	0.899213	0.990541	0.979768	0.985555	0.984572
0.94	0.928976	0.993015	0.986736	0.989750	0.989260
0.95	0.953850	0.995125	0.991805	0.993168	0.992955
0.96	0.973324	0.996864	0.995330	0.995831	0.995756
0.97	0.987163	0.998227	0.997647	0.997780	0.997761
0.98	0.995555	0.999208	0.999054	0.999073	0.999071
0.99	0.999292	0.999801	0.999783	0.999784	0.999784

h_{c+1} and h_{c+2} gives the "Both Sides Improved" Ball-Provan method. Table 2 gives the reliability values computed by the original Ball-Provan method and Both Sides Improved Ball-Provan method.

5. Computational results

We have chosen just one test graph, in order to illustrate the points in detail. The test graph is a 17 node ladder which is shown in Figure 1.

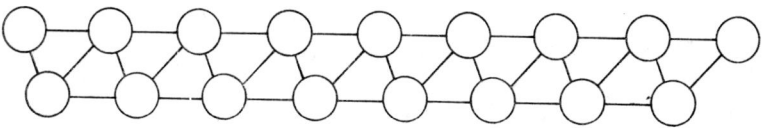

Figure 1. The ladder on 17 nodes.

All algorithms have been implemented in the language "C", on a VAX 11/780. Reliability values were computed using double precision arithmetic (sixty-four bits). Values are reported using enough digits of precision to demonstrate significant differences in the bounds being examined.

The most striking observation is that knowledge of additional coefficients causes a dramatic tightening of the bounds, as shown in Tables 1 and 2. In part, the magnitude of the improvement is a consequence of the small size of the test graph; nevertheless, worthwhile improvements are obtained even for much larger networks. It is no surprise that left improvements cause a great improvement when p is near 1, and right improvements are most useful when p is near 0.

Another important remark is that in the Kruskal-Katona improvements, left improvement only tightens approximation of the unknown coefficients in the upper bound, and right improvement only in the lower bound. In the Ball-Provan bounds, however, left improvement causes a tightening in both approximations. For example, consider F_6 for the seventeen node ladder. The Kruskal-Katona bounds yield

Bound	Lower	Upper
Original	173072	692055
Left improved	173072	664400
Right improved	370310	692055
Both improved	370310	664400

In the Ball-Provan case, the left improvement tightens from the original interval (241845, 692055) to the improved (508537, 652574). Here, both coefficient bounds are tightened.

Finally, it is important to remark that the current difficulty in performing a right improvement in the Ball-Provan bounds shows up strongly in the computational results. The right improved Kruskal-Katona approach typically outperforms the right improved Ball-Provan bounds.

6. Concluding remarks

As one would expect, the additional information available for planar graphs enables one to achieve significant improvements in the bounds on individual coefficients. This in turn provides a substantial tightening of the bounds. A current weakness of the approach here is that the information from knowing F_{d-1}, F_{d-2} and F_{d-3} is not fully exploited in the Ball-Provan bounds. We are currently examining techniques for further improvement using these coefficients.

Finally, it is worthwhile noting two problems. The complexity of all-terminal reliability for planar networks is unknown. A proof of #P-completeness or a polynomial time algorithm is surely needed here. Secondly, additional properties of the complexes associated with planar all-terminal reliability might well lead to improved bounds, as coherence and shellability have already done.

ACKNOWLEDGEMENTS

Research of the second author is supported by the U.S. Army Research Office. Research of the third author is supported by NSERC Canada under grant A0579.

References

[B1] M. O. Ball, "Computing network reliability", *Operations Research*, Vol. 27, 1979, pp. 823-838.

[B2] M. O. Ball and J. S. Provan, "Calculating bounds on reachability and connectedness in stochastic networks", *Networks*, Vol. 13, 1983, pp. 253-278.

[B3] M. O. Ball and J. S. Provan, "Bounds on the reliability polynomial for shellable independence systems", *SIAM J. Alg. Disc. Math.*, Vol. 3, 1982, pp. 166-181.

[B4] R. L. Brooks, C. A. B. Smith, A. H. Stone and W. T. Tutte, "The dissection of rectangles into squares", *Duke Mathematical Journal*, Vol. 7, 1940, pp. 312-340.

[B5] J. A. Buzacott, "A recursive algorithm for finding reliability measures related to the connection of nodes in a graph", *Networks*, Vol. 10, 1980, pp. 311-327.

[C1] C. J. Colbourn, "The reliability polynomial", *Ars Combinatoria*, to appear.

[C2] C. J. Colbourn and D. D. Harms, "Bounding all-terminal reliability in computer networks", CCNG Report E-123, University of Waterloo, 1985.

[H1] D. D. Harms, "An investigation into bounds on network reliability", M. Sc. Thesis, Department of Computational Science, University of Saskatchewan, 1984.

[K1] G. Katona, "A theorem of finite sets", Theory of Graphs (Proceedings of Tihany Colloquium, Sept. 1966), eds. P. Erdös and G. Katona, 1966, pp. 209-214.

[K2] J. B. Kruskal, "The number of simplicies in a complex", *Mathematical Optimization Techniques*, ed. R. Bellman, University of California Press, Berkeley, 1963, pp. 251-278.

[L1] C. I. Liu and Y. Chow, "Enumeration of connected spanning subgraphs of a planar graph", *Acta Mathematica Hungarica*, 41, 1983, pp. 27-36.

[L2] M. V. Lomonosov and V. P. Polesskii, "Lower bounds of network reliability", *Problems of Information Transmission*, Vol. 8, 1972, pp. 118-123.

[P1] J. S. Provan and M. O. Ball, "The complexity of counting cuts and of computing the probability that the graph is connected", *SIAM J. Computing*, Vol. 12, No. 4, 1983, pp. 747-788.

[P2] J. S. Provan and L. J. Billera, "Decompositions of simplicial complexes related to diameters of convex polyhedra", *Math. Oper. Res.*, 5, 1980, pp. 579-594.

[S1] R. P. Stanley, "Cohen-Macauley Complexes", in *Higher Combinatorics*, ed. M. Aigner, Reidel, Dordrecht, 1977, pp. 51-62.

[V1] L. G. Valiant, "The complexity of enumeration and reliability problems", *SIAM J. Computing*, Vol. 8, 1979, pp. 410-421.

[V2] R. Van Slyke and H. Frank, "Network reliability analysis: Part I", *Networks*, Vol. 1, 1972, pp. 279-290.

[W1] J. A. Wald and C. J. Colbourn, "Steiner trees in probabilistic networks", *Microelectronics and Reliability*, Vol. 23, No. 5, 1983, pp. 837-840.

INDUCED SUBGRAPHS IN A RANDOM GRAPH

Andrzej RUCIŃSKI

Adam Mickiewicz University,
60-769 Poznań, Poland,

and

St. John's University,
Staten Island, New York, N.Y. 10301, U.S.A.

For a class of sequences (F_k) of graphs with v_k vertices and e_k edges, $k=1, 2, \ldots$, we show that the size T_n of largest induced copy of a graph from $\{F_1, F_2, \ldots\}$ in a random graph $K(n, p)$ divided by $\log n$ tends to $2/A$, as $n \to \infty$, with probability 1, where $A = a \log 1/p + (1-a) \log 1/(1-p)$ and $a = \lim_{k \to \infty} e_k / \binom{v_k}{2}$.

If, in addition, F_k is an induced subgraph of F_{k+1}, $k=1, 2, \ldots$, then the size t_n of the smallest maximal induced copy of a graph from $F_1, F_2, \ldots$ in $K(n, p)$ divided by $\log n$ tends to $1/A$ in probability.

1. Introduction

We will be concerned with the problem of induced subgraphs of a random graph $K(n, p)$ on vertex set $\{1, \ldots, n\}$ whose edges are present independently, each with probability p, $0 < p < 1$, that does not depend on n.

Let $(\mathscr{F}_k)$ be a sequence of families of graphs on v_k vertices and with e_k edges, $k=1, 2, \ldots$, such that the sequence (v_k) is increasing to infinity. (If $\mathscr{F}_k = \{F_k\}$, we write (F_k).) Given a family of graphs, say $\mathscr{F}$, by an $\mathscr{F}$-graph we mean every graph that is isomorphic to a graph from $\mathscr{F}$. (If $\mathscr{F} = \{F\}$, we write F-graph.) Define $X_n(k)$ as a random variable counting all induced $\mathscr{F}_k$-graphs in $K(n, p)$ and denote by T_n the size of largest induced $\bigcup_{k=1}^{\infty} \mathscr{F}_k$-graph in $K(n, p)$, i.e.

$$T_n = \max \{v_k : X_n(k) > 0\}.$$

The main purpose of this paper is to establish an asymptotic behaviour of T_n as $n \to \infty$. It is done in Section 2, where we pay our attention mostly to the case

$\mathscr{F}_k = \{F_k\}$. The main result, Theorem 2, assumes a simple form for a wide class of sequences of graphs called here p-balanced. In Section 3 we establish p-balance of some sequences of graphs. In particular, since all sequences are $\frac{1}{2}$-balanced, we obtain an interesting consequence in graph enumeration (Corollary 1). Section 4 is a detailed elaboration of a sequence of r-partite complete graphs that fails to be p-balanced for some values of p. Finally, in Section 5 we investigate a random variable

$$t_n = \min\{v_k : Y_n(k) > 0\},$$

where $Y_n(k)$ stands for the number of maximal induced $\mathscr{F}_k$-graphs in $K(n, p)$ and "maximal" means here "not contained in any induced $\mathscr{F}_{k+1}$-graph in $K(n, p)$".

The behaviour of t_n, contrary to T_n, does not depend on the structure of graphs from $(\mathscr{F}_k)$ but only on (v_k) and (e_k). (The only obvious assumption is that each graph from $\mathscr{F}_k$ is an induced subgraph of a graph from $\mathscr{F}_{k+1}$.) This is one of the technical reasons why we get a weaker convergence for t_n than for T_n (compare Theorems 2 and 6 below). An interesting thing is that in all the cases covered by our results T_n is approximately twice t_n, which is also the case for complete subgraphs when $p = p(n) \to 0$ (see [13]).

Due to the general form of our theorems, as consequences we obtain some results from [1–3, 5–8, 10]. We are subject to the notation: $|G|$, $e(G)$ and $\text{aut}(G)$ for the number of vertices, edges, and automorphisms of a graph G, respectively. We write $H \subset G$, if H is an induced subgraph of G. By $P(A)$ we denote the probability of an event A, whereas EX and $\text{Var}(X)$ stand for expectation and variance of a random variable X. Everywhere log means the natural logarithm. The symbols $\lfloor x \rfloor$ and $\lceil x \rceil$ designate the largest integer not greater than x and the smallest integer not smaller than x, respectively.

2. Induced subgraphs of $K(n, p)$

The density $a(G)$ of a graph G is defined by

$$a(G) = \frac{e(G)}{\binom{|G|}{2}}.$$

Let $F_1, F_2, \ldots$ be a sequence of graphs with $|F_k| = v_k$ increasing to infinity as $k \to \infty$. For every $k = 1, 2, \ldots$ and for every $G \subset F_k$ we put

$$A(G) = \frac{|G|}{v_k}\left\{a(G)\log\frac{1}{p} + [1 - a(G)]\log\frac{1}{q}\right\},$$

where $q = 1-p$. Moreover, let

$$A_k = \max_{G \subset F_k} A(G), \quad k = 1, 2, \ldots$$

and

$$\bar{A}_{(F_k)} = \bar{A} = \overline{\lim_{k \to \infty}} A_k, \quad \underline{A}_{(F_k)} = \underline{A} = \underline{\lim_{k \to \infty}} A_k,$$

and $A_{(F_k)} = A = \lim_{k \to \infty} A_k$, if exists.

A sequence (G_k), $G_k \subset F_k$, $k = 1, 2, \ldots$ is called upper in (F_k) if $A(G_k) = A_k$, $k = 1, 2, \ldots$.

Recall that $X_n(k)$ is the number of induced F_k-graphs in $K(n, p)$.

Theorem 1. *Let* $m = m(n)$ *be such that* $\lim_{n \to \infty} m(n) = \infty$.

(a) *If* $\underline{\lim}_{n \to \infty} (v_m/\log n) > 2/\underline{A}$ *then*

$$P(X_n(m) > 0) = o(1).$$

(b) *If* $\overline{\lim}_{n \to \infty} (v_m/\log n) < 2/\bar{A}$ *then*

$$P(X_n(m) > 0) = 1 - o(1).$$

Proof. Let (G_k) be an upper sequence in (F_k). Denote $u_k = |G_k|$, $k = 1, 2, \ldots$. If there is a subsequence (u_{k_l}) such that $u_{k_l}/v_{k_l} = o(1)$ then $A(G_{k_l}) = o(1)$ and it contradicts the fact that (G_k) is upper, since $A(F_{k_l})$ does not converge to 0. Thus $v_k = O(u_k)$. Let $Z_n(k)$ count the induced G_k-graphs in $K(n, p)$. We have

$$P(X_n(m) > 0) \leq P(Z_n(m) > 0) \leq EZ_n(m)$$

$$= \binom{n}{u_m} \frac{u_m!}{\mathrm{aut}(G_m)} p^{e(G_m)} q^{\binom{u_m}{2} - e(G_m)}$$

$$\leq n^{u_m} p^{e(G_m)} q^{\binom{u_m}{2} - e(G_m)}$$

$$= \exp\left\{ u_m \left[\log n - \frac{v_m}{2} A(G_m) \right] + \frac{1}{2} v_m A(G_m) \right\}.$$

There are $\varepsilon > 0$ and $n_1 > 0$ such that

$$v_m > \log n \bigg/ \left(\frac{A}{2} - \varepsilon \right) \quad \text{for } n > n_1.$$

Also there is $n_2 > 0$ such that $A(G_m) > \underline{A} - \varepsilon$ for $n > n_2$. Hence for $n > \max\{n_1, n_2\}$,

$$P(X_n(m) > 0) \leq \exp\{-c' u_m \log n + c'' v_m\}$$
$$= o(n^{-t}), \quad t = 1, 2, \ldots, \tag{1}$$

where c' and c'' are appropriate positive constants.

To prove thesis (b) we make use of the inequality

$$P(X = 0) \leq \operatorname{Var}(X)/(EX)^2$$
$$= E[X(X-1)]/(EX)^2 + (EX)^{-1} - 1. \tag{2}$$

We assume $p \leq \frac{1}{2}$, since once our statement is proved in this case, it will follow for $p > \frac{1}{2}$ by considering the existence of induced complements of F_k-graphs in a random graph $K(n, q)$, $q = 1 - p$.

Let

$$a_i^{(k)} = \max_{\substack{G \subset F_k \\ |G| = i}} a(G)$$

and

$$A_i^{(k)} = \frac{i}{v_k}\left\{a_i^{(k)} \log \frac{1}{p} + [1 - a_i^{(k)}] \log \frac{1}{q}\right\}$$

$$k = 1, 2, \ldots, \quad i = 1, 2, \ldots, v_k.$$

Finally, let $e_k = e(F_k)$, $k = 1, 2, \ldots$. Then

$$EX_n(m) = \binom{n}{v_m} \frac{v_m!}{\operatorname{aut}(F_m)} p^{e_m} q^{\binom{v_m}{2} - e_m}$$

$$\geq v_m^{-v_m} \exp\left\{v_m\left[\log n - \frac{v_m}{2} A(F_m)\right]\right\}.$$

Since there are $\varepsilon > 0$ and $n_1 > 0$ such that $v_m < \log n/(\bar{A}/2 + \varepsilon)$ for $n > n_1$ and there is $n_2 > 0$ such that $A(F_m) \leq A_m < \bar{A} + \varepsilon$ for $n > n_2$, we arrive at

$$EX_n(m) \geq v_m^{-v_m} \exp\{v_m c' \log n\}, \quad n > \max(n_1, n_2),$$

where c' is a positive constant. Thus

$$(EX_n(m))^{-1} = o(n^{-t}), \quad t = 1, 2, \ldots. \tag{3}$$

Further, we have (Here is the only place we need the assumption $p \leq \tfrac{1}{2}$. However, the estimate (4) below remains true for all $0<p<1$.)

$$E\{X_n(m)[X_n(m)-1)]\}$$
$$\leq \binom{n}{v_m} \sum_{i=0}^{v_m} \binom{v_m}{i}\binom{n-v_m}{v_m-i} c_m^2$$
$$\times p^{2e_m} q^{2\binom{v_m}{2}-\binom{i}{2}-2e_m} (q/p)^{\binom{i}{2} a_i^{(m)}},$$

where c_m is defined below. Hence, since $\binom{n-v_m}{v_m} + v_m \binom{n-v_m}{v_m-1} < \binom{n}{v_m}$,

$$\frac{E\{X_n(m)[X_n(m)-1]\}}{(EX_n(m))^2}$$

$$\leq 1 + \sum_{i=2}^{v_m} \frac{\binom{v_m}{i}\binom{n-v_m}{v_m-i}}{\binom{n}{v_m}} q^{-\binom{i}{2}} (q/p)^{\binom{i}{2} a_i^{(m)}}$$

$$\leq 1 + \sum_{i=2}^{v_m} \frac{v_m^i n^{v_m-i}(1+o(1)) v_m!}{i!(v_m-i)! n^{v_m}} c^{-\binom{i}{2}}(q/p)^{\binom{i}{2} a_i^{(m)}}$$

$$\leq 1 + (1+o(1)) v_m^2 n^{-1} \sum_{i=2}^{v_m} \frac{1}{i!} \{v_m^2 n^{-1} q^{-i/2}(q/p)^{(i/2) a_i^{(m)}}\}^{i-1}$$

$$= 1 + (1+o(1)) v_m^2 n^{-1} \sum_{i=2}^{v_m} \frac{1}{i!} \{v_m^2 \exp[-\log n + v_m A_i^{(m)}/2]\}^{i-1}.$$

Obviously $A_i^{(m)} \leq A_m$, thus, for n sufficiently large, we can estimate the last expression by

$$1 + v_m^2 n^{-1} \sum_{i=2}^{\infty} \frac{1}{i!} \{v_m^2 \exp[-c' \log n]\}^{i-1}$$
$$= 1 + o(n^{-1-c''}),$$

where $0 < c'' < c'$. This together with (2) and (3) gives

$$P(X_n(m)=0) = o(n^{-1-c''}), \quad c'' > 0, \tag{4}$$

which completes the proof. □

We can restate Theorem 1 for sequences $(\mathscr{F}_k)$ of families of graphs on v_k vertices and with e_k edges as well. Put

$$A^* = \min_{(F_k),\, F_k \in \mathscr{F}_k} \bar{A}_{(F_k)}$$

and

$$A_* = \lim_{k \to \infty} \left\{ \frac{e_k}{\binom{v_k}{2}} \log \frac{1}{p} + \left[1 - \frac{e_k}{\binom{v_k}{2}}\right] \log \frac{1}{q} \right\}.$$

Let c_k be the number of $\mathscr{F}_k$-graphs one can form on given v_k vertices.

Theorem 1'. *Let $m = m(n)$ be such that $\lim_{n \to \infty} m(n) = \infty$.*

(a) *If for every $\varepsilon > 0$, $(c_m/v_m!)^{1/v_m} = o(n^\varepsilon)$ and*

$$\varliminf_{n \to \infty} (v_m/\log n) > 2/A_* \text{ then}$$

$$P(X_n(m) > 0) = o(1).$$

(b) *If $\varlimsup_{n \to \infty} (v_m/\log n) < 2/A^*$ then*

$$P(X_n(m) > 0) = 1 - o(1).$$

The proof follows the same line as that of Theorem 1 and therefore is omitted. It is based on inequalities $P(X_n(k) > 0) \leq EX_n(k)$ and $P(X_n(k) = 0) \leq P(U_n(k) = 0)$, where $U_n(k)$ counts the induced F_k-graphs in $K(n, p)$, $\bar{A}_{(F_k)} = A^*$. □

Note that Theorem 1' is weaker than Theorem 1 and it does not reduce to the latter one when $\mathscr{F}_k = \{F_k\}$. However, as we will see in Section 3, Theorem 1' will do in many applications. Let us only mention here that if $|\mathscr{F}_k| \leq n^c$ then one can strengthen it by replacing A^* with $\min_{(F_k),\, F_k \in \mathscr{F}_k} A_{(F_k)}$. Note also that Theorem 1 becomes sharp when $\bar{A} = A$ and then yields that $T_n/\log n$ converges in probability to $2/A$. (In fact, to prove this we need (1).)

Due to the strong form of estimates (1) and (4) we can prove even a better result following the approach of Grimmett and McDiarmid [5].

Theorem 2. *Let (F_k) be a sequence of graphs with $|F_k| = v_k$, $k = 1, 2, \ldots$, (v_k) increasing to infinity. If the limit A, introduced above, exists and $v_{k+1} - v_k = o(v_k)$*

then

$$T_n/\log n \to 2/A, \text{ as } n \to \infty,$$

with probability 1.

Comment. For the convergence with probability 1 to make sense we must embed $K(n, p)$ into a probabilistic space Ω of all infinite graphs ω on the set of natural numbers. Each such an ω corresponds to a sequence (ω_n) of finite graphs, where ω_n is induced in ω by $\{1, ..., n\}$. (For details see [5].)

Proof of Theorem 2. We first reduce the problem to the case $v_k = k$. Let us define F'_j as F_k plus $j - v_k$ isolated vertices, where $v_k \leq j < v_{k+1}, j = v_1, v_1 + 1, \ldots$. Given j, let $k(j)$ stand for the index k that satisfies the above double inequality. If (G'_j) is a subsequence of (F'_j) then

$$a(G'_j) = \frac{e_{k(j)}}{\binom{j}{2}} = \frac{e_{k(j)}}{\binom{v_{k(j)}}{2}(1+o(1))}$$

$$= a(G'_{k(j)})(1+o(1)), \quad j \to \infty.$$

Thus if $\lim_{j \to \infty} a(G'_j) = a^*$ then $\lim_{j \to \infty} a(F_{k(j)}) = a^*$. Using this fact one can easily deduce that $A_{(F'_j)} = A_{(F_k)}$.

Moreover, if $X'_n(j)$ is the number of induced F'_j-graphs in $K(n, p)$ and $T'_n = \max\{j : X'_n(j) > 0\}$ then, for every n, $T_n \leq T'_n < T_n + (v_{l+1} - v_l)$, where $v_l = T_n$. Hence, by the assumption $v_{k+1} - v_k = o(v_k)$, $0 \leq T'_n - T_n = o(T_n)$ for every $\omega \in \Omega$ and if $T'_n = O(\log n)$ with probability 1, then

$$\lim_{n \to \infty} T'_n/\log n = \lim_{n \to \infty} T_n/\log n$$

with probability 1. So, it is enough to prove Theorem 2 only in the case when $v_k = k, k = 1, 2, \ldots$.

Now we will show that

$$P(\overline{\lim_{n \to \infty}}\,(T_n/\log n) \leq 2/A) = 1. \tag{5}$$

Denote by $\mathscr{A}_N$ the event

$$\left\{\overline{\lim_{n\to\infty}}\,(T_n/\log n)<\left(2+\frac{1}{N}\right)\Big/A\right\}, \quad N=1,2,\ldots$$

and by $\mathscr{B}_{N,n}$ the event

$$\left\{T_n/\log n\geqslant\left(2+\frac{1}{N}\right)\Big/A\right\}, \quad N=1,2,\ldots,\ n=1,2,\ldots.$$

By (1) we get, as $n\to\infty$ and N is fixed,

$$P(\mathscr{B}_{N,n})\leqslant \sum_{l}{}^{*} P(X_n(l)>0)$$

$$\leqslant nP(X_n(m)>0)=o(n^{-2}),$$

where $\sum^*$ is taken over $I=\{l:(2+1/N)\log n/A\leqslant l\leqslant n\}$ and $m=m(n)$ is determined by $P(X_n(m)>0)=\max_{l\in I} P(X_n(l)>0)$. Thus $\sum_{n=1}^{\infty}P(B_{N,n})<\infty$ and by the Borel-Cantelli lemma $P(\mathscr{A}_N)=1$, $N=1,2,\ldots$. Next, since $\mathscr{A}_N\supset\mathscr{A}_{N+1}$, $N=1,2,\ldots$, we arrive at

$$P(\overline{\lim_{n\to\infty}}\,(T_n/\log n)\leqslant 2/A)=P(\bigcap_{N=1}^{\infty}\mathscr{A}_N)=1.$$

Finally, we will prove that

$$P(\varliminf_{n\to\infty}(T_n/\log n)\geqslant 2/A)=1. \qquad (6)$$

Denote by $\mathscr{C}_N$ the event

$$\left\{\varliminf_{n\to\infty}(T_n/\log n)>\left(2-\frac{1}{N}\right)\Big/A\right\}, \quad N=1,2,\ldots$$

and by $\mathscr{D}_{N,n}$ the event

$$\left\{T_n/\log n\leqslant\left(2-\frac{1}{N}\right)\Big/A\right\}, \quad N=1,2,\ldots,\ n=1,2,\ldots.$$

By (4) we get, as $n \to \infty$ and N is fixed

$$P(\mathcal{D}_{N,n}) \leq P\left(X_n\left(\left[\left(2-\frac{1}{2N}\right)\log n/A\right]\right)=0\right)$$
$$= o(n^{-1-\varepsilon}), \quad \varepsilon > 0.$$

Thus $\sum_{n=1}^{\infty} P(\mathcal{D}_{N,n}) < \infty$ and $P(\mathcal{C}_N) = 1$, $N = 1, 2, \ldots$ which leads to

$$P(\lim_{n \to \infty}(T_n/\log n) \geq 2/A) = P(\bigcap_{N=1}^{\infty} \mathcal{C}_N) = 1.$$

Facts (5) and (6) imply our thesis. □

Theorem 2'. *Let $(\mathcal{F}_k)$ be a sequence of graphs on v_k vertices, $v_{k+1} - v_k = o(v_k)$, and with e_k edges, $k = 1, 2, \ldots$. If*

(a) *for every sequence $m = m(n)$ satisfying $2 \log n/A_* \leq v_m \leq n$, and for every $\varepsilon > 0$*

$$(c_m/v_m!)^{1/v_m} = o(n^\varepsilon), \text{ and}$$

(b) $A^* = A_*$,

where c_k, A^*, and A_* are defined as above, then

$$T_n/\log n \to 2/A^*, \text{ as } n \to \infty,$$

with probability 1. □

3. p-balanced sequences of graphs

It may be a hard problem sometimes to determine the parameters $\bar{A}$, $\underline{A}$, A, A^*, and A_* involved in the results from the previous section. Fortunately, in many cases the sequence (F_k) is either upper itself or at least "asymptotically upper" in the sense that

$$A = a \log \frac{1}{p} + (1-a) \log \frac{1}{q}, \text{ where}$$

$$a = \lim_{k \to \infty} a(F_k) \text{ exists}.$$

(7)

We call a sequence (F_k) of graphs p-balanced, $0 < p < 1$, if it satisfies (7), where $a(.)$ and A are defined in Section 2. Note that if, for given $(\mathcal{F}_k)$, all sequences (F_k), $F_k \in \mathcal{F}_k$, $k = 1, 2, \ldots$ are p-balanced then $A^* = A_* = A$.

Let us start with the observation that every sequence of graphs (F_k) is $\frac{1}{2}$-balanced and that $A = \log 2$ for $p = \frac{1}{2}$. This simple fact, when applied to Theorem 1, has consequences in graphs enumeration, since in the probabilistic space $K(n, \frac{1}{2})$ all graphs on n labeled vertices have the same probability $2^{-\binom{n}{2}}$.

Corollary 1. *Let (F_k) be a sequence of graphs on v_k vertices, v_k increasing to infinity as $k \to \infty$, and let $m = m(n)$ be such that $\lim_{n \to \infty} m(n) = \infty$.*

(a) *If $\overline{\lim}_{n \to \infty} (v_m/\log n) < 2/\log 2$ then almost all graphs on $\{1, \ldots, n\}$ contain at least one induced subgraph isomorphic to F_m.*

(b) *If $\underline{\lim}_{n \to \infty} (v_m/\log n) > 2/\log 2$ then almost all graphs on $\{1, \ldots, n\}$ contain no induced subgraph isomorphic to F_m.*

This is a weaker version of a result of Bollobás and Thomason (see [15, p. 319]).

Next, consider the case when $0 < p \leq \frac{1}{2}$ and (F_k) consists of strongly balanced graphs. A graph G is strongly balanced if for every $H \subset G$, $|H| > 1$, $e(H)/(|H|-1) \leq e(G)/(|G|-1)$ (see [14]).

In this case

$$A(F_k) - A(G) \geq \left(a(F_k) - \frac{|G|}{v_k} a(G) \right) \left(\log \frac{1}{p} - \log \frac{1}{q} \right)$$

$$\geq \frac{2}{v_k} \left(\frac{e_k}{v_k - 1} - \frac{e(G)}{|G|-1} \right) \left(\log \frac{1}{p} - \log \frac{1}{q} \right) \geq 0$$

for all $G \subset F_k$. Thus (F_k) is upper and so p-balanced if a exists.

A graph G is said to be balanced (see [4]) if for every $H \subset G$, $e(H)/|H| \leq e(G)/|G|$. A strongly balanced graph is balanced but the inverse is not true in general. However, although a sequence of balanced graphs (F_k) may not be upper, it is still p-balanced assuming a exists. Indeed, let (G_k) be an upper sequence in (F_k). Then

$$0 \geq A(F_k) - A_k \geq \left[a(F_k) - \frac{|G_k|}{v_k} a(G_k) \right] \left(\log \frac{1}{p} - \log \frac{1}{q} \right)$$

$$\geq \frac{2}{v_k - 1} \left[\frac{e_k}{v_k} - \frac{e(G_k)}{|G_k|} - \frac{(v_k - |G_k|) e(G_k)}{v_k(|G_k|-1)|G_k|} \right] \left(\log \frac{1}{p} - \log \frac{1}{q} \right)$$

$$\geq -\frac{\log \frac{1}{p} - \log \frac{1}{q}}{v_k - 1} \to 0 \quad \text{as } k \to \infty.$$

Thus $\lim_{k\to\infty} (A(F_k) - A_k) = 0$ and if a exists then $\lim_{k\to\infty} A(F_k) = A$. By similar argument one gets that every sequence (F_k) of graphs with balanced complements is p-balanced for $p \geq \frac{1}{2}$.

Two other classes of p-balanced sequences are those with density tending either to 0 or to 1. In the first case, for any sequence (G_k) that is upper in (F_k), $a(G_k) \to 0$ as $k \to \infty$, since

$$a(G_k)|G_k|(|G_k|-1)/v_k(v_k-1) \leq a(F_k)$$

and $v_k = O(|G_k|)$. Hence

$$\log \frac{1}{q} = \lim_{k\to\infty} A(F_k)$$

$$\leq \lim_{k\to\infty} A_k \leq \log \frac{1}{q} = A.$$

The case when $a = 1$ follows by symmetry with $A = \log \frac{1}{p}$.

Let us summarize the above observations.

Theorem 3. *A sequence of graphs (F_k) is p-balanced if a exists and at least one of the following holds:*

(a) $p = \frac{1}{2}$,

(b) $p \leq \frac{1}{2}$ and F_k are balanced,

(c) $p \geq \frac{1}{2}$ and the complements of F_k are balanced,

(d) $a = 0$ or $a = 1$. □

Let us conclude this section with the examples of r-trees and r-regular graphs.

The most convenient definition of r-trees is the recursive one: a complete graph on r vertices is the smallest r-tree. Given an r-tree, a new one is obtained by joining a new vertex to any r mutually adjacent vertices. We claim that every r-tree is strongly balanced. To see this, label the vertices of an r-tree T with numbers $1, 2, \ldots, |T|$ in the order they appear in any of the constructions of T that follows the recursive definition. Then put arrows on the edges of T always in the direction of the smaller label. In each subgraph $\vec{H}$ of the obtained directed graph the number of edges equals the sum of all out-degrees in $\vec{H}$ that cannot exceed $\binom{r}{2} + (|\vec{H}| - r)r$. Hence, for each $H \subset T$, $e(H) \leq \binom{r}{2} + (|H| - r)r$. The function

$f(x) = \left[\binom{r}{2} + (x-r)r\right]/(r-1)$ is increasing and so, $e(H)/(|H|-1) \leq e(T)/(|T|-1)$. Since 1-tree is a tree and $|T|$-tree is a complete graph, all trees and complete graphs are strongly balanced. Thus the next corollary covers the already known results about trees [3, 7, 8] and complete graphs [2, 5].

Corollary 2. *Let $\mathcal{F}_k$ be a family of r-trees on k vertices, $r = r(k)$, $k = 1, 2, \ldots$ and let T_n be the size of the largest induced r-tree in $K(n, p)$ that belongs to $\bigcup_{k=1}^{\infty} \mathcal{F}_k$.*

(a) *If for every $\varepsilon > 0$, $r = o(k^{\varepsilon})$, then*

$$T_n/\log n \to 2/\log\frac{1}{q}, \quad 0 < p < 1.$$

(b) *If $r/k \to c$, $\frac{1}{2} \leq c < 1$, then*

$$T_n/\log n \to 2\left[c(2-c)\log\frac{1}{p} + (1-c)^2\log\frac{1}{q}\right],$$

$$0 < p \leq \frac{1}{2}.$$

(c) *If $r/k \to 1$, then*

$$T_n/\log n \to 2/\log\frac{1}{p}, \quad 0 < p < 1,$$

and the convergences are with probability 1.

Proof. Every r-tree on k vertices has $\binom{r}{2} + (k-r)r$ edges and so, $a = 2c - c^2$, where $c = \lim_{k \to \infty} \frac{r}{k}$. The number of all r-trees on given k vertices is $\binom{k}{r}[r(k-r) + 1]^{k-r-2}$ (see, for example, [9]). Thus $(c_k/k!)^{1/k} = O(r^{1-2r/k})$ and the restrictions on r in part (a) and on c in part (b) allow us to use Theorem 2. Note that an alternative version of Corollary 2 might be obtained by replacing the two restrictions with a suitable one on c_k. □

Not every regular graph is strongly balanced (for instance, when disconnected) but all regular graphs are balanced. Indeed, if H is a subgraph of a regular graph then $e(H)/|H| \leq (\frac{1}{2}r|H|)/|H| = \frac{1}{2}r$. Since the complement of a regular graph is regular as well we are in position to get p-balance of a sequence of regular graphs for all values of p and a by combining cases (b) and (c) of Theorem 3. The corollary below generalizes the result about cycles one can find in [6].

Corollary 3. *Let $\mathscr{F}_k$ be a family of r-regular graphs on v_k vertices, $r = r(k)$, $k = 1, 2, \ldots$, and let T_n be the size of the largest induced subgraph of $K(n,p)$ that belongs to $\bigcup_{k=1}^{\infty} \mathscr{F}_k$. If the assumption (a) of Theorem 2' holds, $v_{k+1} - v_k = o(v_k)$ and $r/v_k \to c$ as $k \to \infty$, $0 \leqslant c \leqslant 1$, then*

$$T_n / \log n \to 2 \bigg/ \left[c \log \frac{1}{p} + (1-c) \log \frac{1}{q} \right]$$

as $n \to \infty$

with probability 1, $0 < p < 1$. □

The family of all r-regular graphs on k vertices satisfies the assumptions of Corollary 3 only when $r = 2$, since otherwise c_k is too large (for an appropriate formula see [12]). We cannot apply Corollary 3 for k-cubes either, since $v_k = 2^k$ and $v_{k+1} - v_k = v_k$.

4. r-partite complete graphs

This section is entirely devoted to investigations of the sequence $K(u_1, \ldots, u_r)$ of r-partite complete graphs with $u_i = u_i(k)$ vertices of color i, $i = 1, \ldots, r$, $u_1 + \ldots + u_r = k$,

$$\lim_{k \to \infty} \frac{u_i(k)}{k} = c_i,$$

$i = 1, \ldots, r$, $1 \geqslant c_1 \geqslant c_2 \geqslant \ldots \geqslant c_r \geqslant 0$, $\sum_{i=1}^{r} c_i = 1$.

This choice is motivated by the fact that r-partite complete graphs have well defined structure and the sequence of them is not always p-balanced. However, we are still able to find the value of A, the parameter that is crucial for Theorem 2. Let

$$c = \min_{1 \leqslant s \leqslant r-1} \frac{(1 - \sum_{i=1}^{s} c_i) \sum_{i=1}^{s} c_i}{\sum_{i=1}^{s} c_i^2 - \sum_{i=1}^{s} c_i \sum_{i=1}^{r} c_i^2} \quad \text{for } c_1 > c_r,$$

and

$$A_s = (\sum_{i=1}^{s} c_i - \sum_{i=1}^{s} c_i^2 / \sum_{i=1}^{s} c_i) \log \frac{1}{p}$$
$$+ \sum_{i=1}^{s} c_i^2 / \sum_{i=1}^{s} c_i \log \frac{1}{q},$$
$$s = 1, \ldots, r.$$

Theorem 4. *The sequence $(K(u_1, \ldots, u_r))$ is p-balanced if and only if*

$$p^{c+1} + p \leqslant 1 \quad \text{or} \quad c_1 = c_r. \tag{8}$$

Moreover, there are numbers

$$\frac{\sqrt{5}-1}{2} \leqslant p_{r-1} \leqslant p_{r-2} \leqslant \ldots \leqslant p_1 \leqslant 1$$

such that $A = A_s$ for $p_s \leqslant p \leqslant p_{s-1}$, $s = 1, \ldots, r$, where $p_r = 0$, $p_0 = 1$, p_{r-1} is the unique solution of (8) and $p_1 < 1$, if $c_1 > c_r$.

Proof. We will prove first that $K(u_1, \ldots, u_r)$ is balanced. In fact, let $G \subset K(u_1, \ldots, u_r)$ have v_i vertices of color i, $i = 1, \ldots, r$. Then

$$\frac{e(G)}{|G|} = \frac{\sum\sum_{i<j} v_i v_j}{\sum_i v_i} \leqslant \frac{\sum\sum_{i<j} u_i u_j}{k},$$

since for

$$f(x_1, \ldots, x_r) = \frac{\sum\sum_{i<j} x_i x_j}{\sum_i x_i}$$

and for each $l = 1, \ldots, r$, $\partial f / \partial x_l > 0$. Obviously, $a(K(u_1, \ldots, u_r)) \to 1 - \sum_{i=1}^{r} c_i^2$ and by Theorem 3, $(K(u_1, \ldots, u_r))$ is p-balanced for $p \leqslant \tfrac{1}{2}$. Now assume $p \geqslant \tfrac{1}{2}$ and denote $\bar{v} = \sum_{i=1}^{r} v_i$, $\bar{\bar{v}} = \sum_{i=1}^{r} \binom{v_i}{2}$. Then

$$A(G) = k^{-1} \bar{v} \left\{ \bar{\bar{v}} \binom{\bar{v}}{2}^{-1} \log \frac{1}{q} + \left[1 - \bar{\bar{v}} \binom{\bar{v}}{2}^{-1}\right] \log \frac{1}{p} \right\}$$

$$= k^{-1}\log\frac{1}{p}[2\bar{v}(\bar{v}-1)^{-1}(x-1)+\bar{v}]$$

$$= k^{-1}\log\frac{1}{p}g(v_1,\ldots,v_r),$$

where $x=\log\frac{1}{q}/\log\frac{1}{p}\geqslant 1$. For each $l=1,\ldots,r$

$$\frac{\partial g}{\partial v_l}=\frac{(2v_l-1)(\bar{v}-1)-2\bar{v}}{(\bar{v}-1)^2}(x-1)+1$$

$$=\frac{xv_l^2+2xv_l(\bar{v}-v_l-1)-(\sum_{i\neq l}v_i^2-1)(x-1)+(\bar{v}-v_l-1)^2}{(\bar{v}-1)^2}.$$

Thus for every fixed $v_1,\ldots,v_{l-1},v_{l+1},\ldots,v_r$ there is a number v_l' such that g as a function of v_l is decreasing for $v_l\leqslant v_l'$ and increasing for $v_l\geqslant v_l'$. Hence the maximum may be achieved only at $(v_1,\ldots,v_r)$ with each $v_i=0$ or u_i. Notice also that if $(v_1(k),\ldots,v_r(k))$ gives the maximum for $A(G)$, $G\subset K(u_1(k),\ldots,u_r(k))$, then $v_i(k)\leqslant v_i(k+1)$, $i=1,\ldots,r$. The two facts together imply that for every $0<p<1$ there is an upper sequence (G_k) in $(K(u_1,\ldots,u_r))$, and $k_0>0$ such that for each $i=1,\ldots,r$, $V(G_k)$ either contains the set of all $u_i(k)$ vertices of color i or is disjoint from it, $k=k_0, k_0+1,\ldots$. This means that A exists and is equal to

$$A_I=\log\frac{1}{p}[(\sum_{i\in I}c_i^2/\sum_{i\in I}c_i)y+\sum_{i\in I}c_i]$$

for some $I\subset\{1,\ldots,r\}$, where $y=\log\frac{1}{q}/\log\frac{1}{p}-1$. If $p\leqslant(\sqrt{5}-1)/2$ then $x\leqslant 2$ and $\partial g/\partial v_l\geqslant 0$ unless there is $s\neq l$ such that $v_i=0$ for $i\neq s$. Thus, for $p\leqslant(\sqrt{5}-1)/2$, $A=A_I$, where $I=\{1,\ldots,r\}$ or $I=\{s\}$. But in the latter case

$$A_{\{1,\ldots,r\}}=\sum_{i=1}^r c_i^2\log\frac{1}{q}+(1-\sum_{i=1}^r c_i^2)\log\frac{1}{p}$$

$$\geqslant\frac{1}{2}(1+\sum_{i=1}^r c_i^2)\log\frac{1}{q}\geqslant\frac{1}{2}(1+c_1^2)\log\frac{1}{q}$$

$$\geqslant c_1\log\frac{1}{q}=A_{\{1\}}\geqslant A_{\{s\}}.$$

By checking a few inequalities one can eliminate many sets I from the list of possible solutions of $A = A_I$. Let $f_I(y) = a_I y + b_I$, where $a_I = \sum_{i \in I} c_i^2/b_I$, $b_I = \sum_{i \in I} c_i$ and $y = y(p) = \log \frac{1}{q} / \log \frac{1}{p} - 1$, and let I and J be two subsets of $\{1, \ldots, r\}$. If

$$|I| = |J|, \quad I - J = \{s\}, \quad J - I = \{t\}, \quad s \leq t \quad \text{and} \tag{9}$$
$$f_J(y) \geq f_I(y)$$

then $f_J(y) \leq f_{I \cap J}(y)$. Indeed, denoting $c_I = a_I b_I$, the first inequality is equivalent to

$$y \geq \frac{(c_s - c_t) b_I b_J}{(c_{I \cap J} + c_t^2)(b_{I \cap J} + c_s) - (c_{I \cap J} + c_s^2)(b_{I \cap J} + c_t)} = y_1$$

and the second one to

$$y \geq \frac{b_{I \cap J} b_J}{c_{I \cap J} - b_{I \cap J} c_t} = y_2.$$

Finally, $y_1 \geq y_2$. Hence for any p and for any $I \neq \{1, \ldots, |I|\}$ there is $J = \{1, \ldots, |J|\}$ such that $f_I \leq f_J$. The last is true by repeated applications of (9) and then of the obvious fact that if $\max \{i : i \in I\} \leq \min \{i : i \in J\}$ then $a_I \leq a_J$, $b_I \leq b_J$, and so $f_I \leq f_J$. To complete the proof note that if $I = J - \min \{i : i \in J\}$ then $a_I \geq a_J$ with equality only when all c_i, $i \in J$, are equal. In turn, if $a_I > a_J$ then there is p' such that for all $p > p'$, $f_I > f_J$. □

The above theorem does not tell us in how many nonempty intervals the numbers $p_{r-1} \leq \ldots \leq p_1 \leq 1$ partition the interval $(0, 1)$. The actual evolution of the size of the largest induced r-partite complete graph in $K(n, p)$ that belongs to $\{K(u_1(k), \ldots, u_r(k)) : k = 1, 2, \ldots\}$ depends on particular values of $c_1, \ldots, c_r$. All we know is that A is a continuous function of p satisfying the thesis of Theorem 4.

At the end of this section consider the cases $r = 3$ and $r = 2$. For $r = 3$, $A = A_I$ only for $I = \{1\}, \{1, 2\}, \{1, 2, 3\}$. If $c_1 = c_2 = c_3 = 1/3$ then $A = A_{\{1, 2, 3\}}$ for all p. If $c_1 > c_3$ then $A = A_{\{1, 2, 3\}}$ for $p \leq p_2$ and $A = A_{\{1\}}$ for $p_1 \leq p < 1$. Whether or not $A = A_{\{1, 2\}}$ for some p, i.e. whether or not $p_2 < p_1$ depends on c_1, c_2, c_3. If, for instance $c_1 = \frac{1}{2}$, $c_2 = c_3 = \frac{1}{4}$, then

$$A = \begin{cases} \frac{3}{8} \log \frac{1}{q} + \frac{5}{8} \log \frac{1}{p} & \text{if } 1 - p - p^5 \geq 0 \\ \frac{1}{2} \log \frac{1}{q} & \text{if } 1 - p - p^5 \leq 0. \end{cases}$$

If $c_1=\frac{1}{2}$, $c_2=\frac{1}{3}$, $c_3=\frac{1}{6}$, then

$$A = \begin{cases} \dfrac{1}{18}\log\dfrac{1}{q} + \dfrac{11}{18}\log\dfrac{1}{p} & \text{if } 1-p-p^{19/4} \geq 0 \\[2mm] \dfrac{13}{30}\log\dfrac{1}{q} + \dfrac{11}{15}\log\dfrac{1}{p} & \text{if } 1-p-p^{19/4} \leq 0 \\ & \text{but } 1-p-p^6 \geq 0 \\[2mm] \dfrac{1}{2}\log\dfrac{1}{q} & \text{if } 1-p-p^6 \leq 0. \end{cases}$$

Finally, in the bipartite case ($r=2$) with $c_1=c$, $c_2=1-c$,

$$A = \begin{cases} c(1-c)\log\dfrac{1}{p} + [c^2+(1-c)^2]\log\dfrac{1}{q} & \\ & \text{if } p \leq p_1 \\[2mm] c\log\dfrac{1}{q} & \text{otherwise} \end{cases}$$

where p_1 is the unique solution of $1-p-p^{2c/(2c-1)}=0$.

5. Maximal induced subgraphs of a random graph

Let $F_1 \subset F_2 \subset \ldots$ and $v_1 < v_2 < \ldots$. An induced F_k-graph in $K(n,p)$ is called maximal (in respect to (F_k)) if it is not a subgraph of any induced F_{k+1}-graph in $K(n,p)$. In this section we briefly investigate the smallest size t_n of a maximal induced subgraph of $K(n,p)$ that belongs to $\{F_1, F_2, \ldots\}$.

For every k, let x_k count in how many ways one can extend F_k to an F_{k+1}-graph F so that $F_k \subset F$ when $v_{k+1}-v_k$ additional vertices are given. For example, if F_k is a path on k vertices then $x_k=2$. Further, let

$$b_k = \frac{2(e_{k+1}-e_k)}{(v_{k+1}-v_k)(v_{k+1}+v_k-1)} \quad \text{and}$$

$$b = \lim_{k \to \infty} b_k.$$

By the Stolz criterion, if b exists then $a = \lim_{k \to \infty} a(F_k)$ exists too and $a=b$, but the inverse is not true in general. Also, let

$$B_k = b_k \log\frac{1}{p} + (1-b_k)\log\frac{1}{q},$$

$$\overline{B} = \overline{\lim_{k\to\infty}} B_k, \quad \underline{B} = \underline{\lim_{k\to\infty}} B_k, \quad B = \lim_{k\to\infty} B_k.$$

Finally, $Y_n(k)$ stands for the number of maximal induced F_k-graphs in $K(n, p)$.

Theorem 5. *Let $m = m(n) \to \infty$ as $n \to \infty$. If $x_m = o(n^\varepsilon)$ for every $\varepsilon > 0$ and*

$$1/\underline{B} < \underline{\lim_{n\to\infty}} (v_m/\log n)$$

$$\leqslant \overline{\lim_{n\to\infty}} (v_m/\log n) < 2/\overline{A}$$

then

$$P(Y_n(m) > 0) = 1 - o(1).$$

For the proof we will need a method introduced in [11] and stated below as a lemma.

Lemma. *Let $f = f(n) \to \infty$ as $n \to \infty$ and let $X_i(n)$ and $Y_i(n)$, $i = 1, \ldots, f$, $n = 1, 2, \ldots$ be zero-one random variables satisfying $Y_i(n) \leqslant X_i(n)$. Denote $X(n) = \sum_{i=1}^{f} X_i(n)$ and $Y(n) = \sum_{i=1}^{f} Y_i(n)$. If*

(a) *for every n, $P(Y_i(n) = 1/X_i(n) = 1)$ does not depend on i,*

(b) $P(Y_1(n) = 1/X_1(n) = 1) = 1 - o(1)$, *and*

(c) $\mathrm{Var}\, X(n) = o((EX(n))^2)$

then $P(Y(n) > 0) = 1 - o(1)$. □

Proof of Theorem 5. Let $F_m^{(1)}, \ldots, F_m^{(f)}$ be all F_m-graphs in K_n, the complete graph on $\{1, \ldots, n\}$,

$$X_i(n) = \begin{cases} 1 & \text{if } F_m^{(i)} \subset K(n, p) \\ 0 & \text{otherwise,} \end{cases}$$

and

$$Y_i(n) = \begin{cases} 1 & \text{if } F_m^{(i)} \subset K(n,p) \text{ and} \\ & F_m^{(i)} \text{ is maximal,} \\ 0 & \text{otherwise,} \end{cases}$$

$i=1,\ldots,f, n=1,2,\ldots$. Obviously, $X_n(m) = \sum_{i=1}^{f} X_i(n)$ and $Y_n(m) = \sum_{i=1}^{f} Y_i(n)$. Assumption (a) of the lemma is trivially fulfilled, whereas assumption (c) was checked when proving Theorem 1 (see Section 2).

Now we will show that also (b) holds. Let $K^*(n,p)$ be a random graph obtained from $K(n,p)$ by adding all the edges of $F_m^{(1)}$ and removing all the edges of the complement of $F_m^{(1)}$. Thus $F_m^{(1)} \subset K^*(n,p)$. Denote by $Z_n(m)$ the number of induced F_{m+1}-graphs in $K^*(n,p)$ that contain $F_m^{(1)}$. Then

$$P(Y_1(n)=1/X_1(n)=1) = P(Z_n(m)=0).$$

But

$$P(Z_n(m)>0) \leqslant EZ_n(m)$$
$$\leqslant n^{v_{m+1}-v_m} x_m p^{e_{m+1}-e_m} q^{\binom{v_{m+1}}{2}-\binom{v_m}{2}-(e_{m+1}-e_m)}$$
$$= x_m \exp\{\tfrac{1}{2}(v_{m+1}-v_m)(v_{m+1}+v_m-1)[2\log n/(v_{m+1}+v_m-1)-B_m]\}$$
$$\leqslant x_m \exp\{-c\log n\},$$

for n large enough, where c is an appropriate constant. □

Unfortunately, we are able to complete Theorem 5 with the statement that $\varlimsup_{n\to\infty} \dfrac{v_m}{\log n} < \dfrac{1}{B}$ implies $P(Y_n(m)>0) = o(1)$ only when $v_{k+1}-v_k=1$. This leads to the main result of this section.

Theorem 6. *If $v_k=k$, $k=1,2,\ldots$, b exists, $2B > \bar{A}$ and $x_{c\log n} = o(n^\varepsilon)$ for all $c>0$ and $\varepsilon > 0$ then*

$$t_n/\log n \to 1/B = 1/A \text{ in probability,}$$

where $t_n = \min\{k : Y_n(k) > 0\}$.

Proof. If $\varlimsup_{n\to\infty} \frac{m(n)}{\log n} < \frac{1}{B}$ then

$$P(Y_n(m) > 0) \leq EY_n(m)$$

$$= \binom{n}{m} \frac{m!}{\operatorname{aut}(F_m)} p^{e_m} q^{\binom{m}{2} - e_m}$$

$$\times (1 - x_m p^{e_{m+1} - e_m} q^{m - (e_{m+1} - e_m)})^{n-m}$$

$$\leq n^m \exp\{-x_m (n-m) p^{e_{m+1} - e_m} q^{m - (e_{m+1} - e_m)}\}$$

$$\leq n^m \exp\{-x_m e^{\log n/m - B} + mx_m\}$$

$$\leq n^m \exp\{-n^c\}, \quad c > 0. \tag{10}$$

For every $\varepsilon > 0$,

$$P(|t_n/\log n - 1/B| > \varepsilon)$$

$$\leq P(t_n > (1/B + \varepsilon)\log n) + P(t_n < (1/B - \varepsilon)\log n)$$

$$\leq P(Y_n(\lfloor(1/B + \varepsilon)\log n\rfloor) = 0) + \sum_{i=1}^{M} P(Y_n(i) > 0) = o(1)$$

by Theorem 5 and (10) with $m = m(n)$ determined by $P(Y_n(m) > 0) = \max_{1 \leq i \leq M} P(Y_n(i) > 0)$, where $M = \lfloor(1/B - \varepsilon)\log n\rfloor$. □

As a consequence of Theorems 2 and 6 for all $0 < p < 1$ and for all p-balanced sequences of graphs $F_1 \subset F_2 \subset \ldots$ such that

(i) $v_{k+1} - v_k = 1$, $k = 1, 2, \ldots$,

(ii) $\lim_{k \to \infty} \frac{e_{k+1} - e_k}{v_k}$ exists,

(iii) $x_{c \log n} = o(n^\varepsilon)$ for all $c, \varepsilon > 0$,

we have that almost surely the largest induced $\{F_1, F_2, \ldots\}$-graph in $K(n, p)$ is approximately twice as large as the smallest maximal induced $\{F_1, F_2, \ldots\}$-graph in $K(n, p)$.

Instead of trying to generalize Theorem 6 for sequences of families of graphs and to the case $v_{k+1} - v_k > 1$, we illustrate the two possible generalizations with examples.

Example 1. Let $(\mathscr{F}_k)$ be the sequence of families of all r-trees on k vertices, r-fixed. Then $b = \lim_{k \to \infty} \frac{r}{k} = 0$ and $x_k = (k-1-r)r+1$. Thus, for all $0 < p < 1$, Theorem 6 holds with $B = \log 1/q$. This covers some results from [3, 7, 8].

Example 2. Let $F_k = K(k, \ldots, k)$, the r-partite complete graph with k vertices of color i, $i = 1, \ldots, r$, r fixed. Because $v_k = rk$, we cannot apply Theorem 6 directly. However, by Theorem 5, for every $\varepsilon > 0$,

$$P(t_n > (1/B+\varepsilon)\log n)$$
$$\leq P(Y_n(\lfloor (1/B+\varepsilon)\log n \rfloor) = 0) = o(1)$$

where

$$t_n = \min\{rk : Y_n(k) > 0\},$$

$$B = \left(1 - \frac{1}{r}\right)\log\frac{1}{p} + \frac{1}{r}\log\frac{1}{q}.$$

(Notice that $x_k = r!$, $k = 1, 2, \ldots$.) Now, consider the sequence $K(1, \ldots, 1)$, $K(2, 1, \ldots, 1)$, $K(2, 2, 1, \ldots, 1)$, $\ldots$, $K(2, 2, \ldots, 2)$, $K(3, 2, \ldots, 2)$, $\ldots$, i.e. $(K(u_1, \ldots, u_r))$ with $u_i = u_i(k) = \lceil (k-i+1)/r \rceil$, $i = 1, \ldots, r$, $k = 1, 2, \ldots$. Since $u_i(k)/k \to 1/r$, $i = 1, \ldots, r$, the new sequence gives the same parameter B. Obviously, we can apply Theorem 6 to the new sequence, thus obtaining $t'_n/\log n \to 1/B$ in probability, where t'_n is the size of smallest maximal induced subgraph of $K(n, p)$ that is an element of $(K(u_1, \ldots, u_r))$. But the two random variables are related by

$$t'_n - r < t_n, \tag{11}$$

so, $t_n/\log n \to 1/B$ in probability as well. One can find more about the case $r = 2$ in [10].

Comment. The above method of embedding a sequence (F_k) with $v_{k+1} - v_k > 1$ into a sequence (F'_k) with $v'_{k+1} - v'_k = 1$ does not work in general as it did for T_n when proving Theorem 2. The point is that although (11) remains true as long as

$$F_k = F'_{v_k} \subset F'_{v_k+1} \subset \ldots \subset F'_{v_{k+1}-1} \quad F'_{v_{k+1}} = F_{k+1}, \quad k = 1, 2, \ldots$$

the limit (b) may not exist for (F'_k).

ACKNOWLEDGEMENTS

The author wishes to express his thanks to M. Karoński and Z. Palka for inspiration and discussions, as well as to T. Łuczak and W. Fernandez de la Vega for valuable remarks.

References

[1] A. Blass, F. Harary, Properties of almost all graphs and complexes, *J. Graph Theory* 3 (1979), 225–241.

[2] B. Bollobás, P. Erdös, Cliques in random graphs, *Math. Proc. Camb. Phil. Soc.* 80 (1976), 419–427.

[3] P. Erdös, Z. Palka, Trees in random graphs, *Discrete Math.* 46 (1983), 145–150. Addendum to "Trees...", *Discrete Math.* 48 (1984), 331.

[4] P. Erdös, A. Rényi, On the evolution of random graphs, *Publ. Math. Inst. Hung. Acad. Sci.* 5 (1960), 17–61.

[5] G. R. Grimmett, C. J. H. McDiarmid, On colouring random graphs, *Math. Proc. Camb. Phil. Soc.* 77 (1975), 313–324.

[6] M. Karoński, *Balanced Subgraphs of Large Random Graphs*, Adam Mickiewicz University Press, Poznań, 1984.

[7] M. Karoński, Z. Palka, On the size of a maximal induced tree in a random graph, *Mathematica Slovaca* 30 (1980), 151–155, Addendum and erratum to the paper "On the size...", *Mathematica Slovaca* 31 (1981), 107–108.

[8] A. Marchetti-Spaccamela, M. Protasi, The largest tree in a random graph, *Theor. Comp. Sci.* 23 (1983), 273–286.

[9] J. W. Moon, The number of labelled k-trees, *J. Combinatorial Theory* 4 (1960), 206–207.

[10] Z. Palka, Bipartite complete induced subgraphs of a random graph, *Annals of Discrete Mathematics*, 28 (1985), 209–219.

[11] Z. Palka, A. Ruciński, J. Spencer, On a method for random graphs, *Discrete Math.* 61 (1986), 253–258.

[12] E. M. Palmer, *Graphical Evolution. An Introduction to the Theory of Random Graphs*, John Wiley and Sons, 1985.

[13] A. Ruciński, in preparation.

[14] A. Ruciński, A. Vince, Strongly balanced graphs and random graphs, *J. Graph Theory*, 10 (1986), 251–264.

[15] B. Bollobás, *Random Graphs*, Academic Press, 1985.

ON A NONUNIFORM RANDOM RECURSIVE TREE

Jerzy SZYMAŃSKI

Technical University of Poznań,
Poznań, Poland

1. Introduction

A *tree* is a connected graph which has no cycles (see [1] or [5] for definitions not given here). A tree R with n vertices labelled $1, 2, \ldots, n$ is a *recursive tree* if for each k such that $2 \leqslant k \leqslant n$ the labels of vertices in the unique path from the 1st vertex to the kth vertex of a tree form an increasing subsequence of $\{1, 2, \ldots, n\}$. Such a tree can also be defined as a result of successively joining the ith vertex to one of the first $i-1$ vertices for $i = 2, 3, \ldots, n$.

A *random recursive tree* with n vertices (n-RRT) is a tree picked at random from the family of all recursive trees with n vertices. We do not assume that all possible choices of a tree are equiprobable. Hence, we shall consider the probability space $(\mathcal{R}_n, P)$, where $\mathcal{R}_n$ consists of all $(n-1)!$ recursive trees with n vertices.

In this paper we deal with a random recursive tree such that the probability of joining a new vertex to the vertex i depends only on the degree of vertex i. Let us denote by $p_{nd} = 1 - q_{nd}$ the probability that $(n+1)$st vertex is joined to a given vertex of degree d in an n-RRT. We have to assume that for each tree $R \in \mathcal{R}_n$

$$\sum_{i=1}^{n} p_{n, D_{ni}}(R) = 1, \qquad (1)$$

where $D_{ni}(R)$ denotes the degree of the ith vertex in an recursive tree R. This model of a random recursive tree was first introduced in an unpublished paper of Dondajewski and Szymański [3].

In this paper we shall consider two special cases of the random recursive tree. If $p_{nd} = 1/n$ is independent of d then we obtain a *uniform random recursive tree*. In this case $P(R) = 1/(n-1)!$ for each tree $R \in \mathcal{R}_n$. In the literature of random recursive trees only the uniform case was investigated.

Let us observe that for $p_{nd}=d/(2n-2)$ relation (1) holds. In this case the probability that a new vertex is joined to vertex i is proportional to the degree of i. We shall call this case the model with the attraction of vertices proportional to their degrees.

Let $\alpha_i(R)$ for $R \in \mathcal{R}_n$ denote the recursive tree obtained from tree R by joining the $(n+1)$st vertex to the vertex i of R; moreover, let $\beta(R)$ denote the recursive tree obtained from R by deleting the nth vertex and its incident edge. It is easy to see that in both models if R is an n-RRT then $\beta(R)$ is an $(n-1)$-RRT and if i is chosen randomly then $\alpha_i(R)$ is an $(n+1)$-RRT.

In the next section we deal with the number of vertices of a given degree in the general model. In the last section we consider the model with $p_{nd}=d/(2n-2)$ and compare the results with the results for the uniform case.

2. Vertex degrees in the general case

Let X_{nk} denote the number of vertices of degree k in a recursive tree with n vertices. Moreover, let $Z_{nk}=X_{n1}+X_{n2}+\ldots+X_{nk}$. Of course, X_{nk} and Z_{nk} are random variables.

Let D'_{ni} denote the degree of the unique vertex $j<i$ such that vertices i and j are joined by an edge.

Theorem 1. *For* $n \geqslant 3$

$$E(X_{n1})=\sum_{i=0}^{n-3} Q_{n-1}(i)+Q_{n-1}(n-3), \qquad (2)$$

where

$$Q_n(k)=\prod_{i=n-k+1}^{n} q_{i1} \quad \text{for} \quad 1\leqslant k \leqslant n-1, \quad Q_n(0)=1.$$

Proof. From the construction of recursive tree one can deduce that

$$P(X_{n+1,1}=k)$$
$$=P(D'_{n+1,n+1}\neq 2 \mid X_{n1}=k-1)P(X_{n1}=k-1)$$
$$+P(D'_{n+1,n+1}=2 \mid X_{n1}=k)P(X_{n1}=k).$$

It is easy to see that $P(D'_{n+1,n+1}=2 \mid X_{n1}=k)=kp_{n1}$ and $P(D'_{n+1,n+1}\neq 2 \mid X_{n1}=k-1)=1-(k-1)p_{n1}$, so

$$P(X_{n+1,1}=k)=(1-(k-1)p_{n1})P(X_{n1}=k-1)$$
$$+kp_{n1}P(X_{n1}=k).$$

Using this recurrence relation and the definition of the expectation we get for $n \geq 3$

$$E(X_{n+1,1})$$
$$= \sum_k k P(X_{n1}=k-1) - p_{n1} \sum_k k(k-1) P(X_{n1}=k-1)$$
$$+ p_{n1} \sum_k k^2 P(X_{n1}=k)$$
$$= q_{n1} E(X_{n1}) + 1.$$

Relation (2) is the solution to the above recurrence relation with boundary condition $E(X_{3,1}) = 2$. □

Let us now deal with the expected number of vertices of degree > 1. The following theorem gives us recurrence relations for this expectation.

Theorem 2. *For $n \geq 2$ and $2 \leq k \leq n$*

$$E(X_{n+1,k}) = q_{nk} E(X_{nk}) + p_{n,k-1} E(X_{n,k-1}) \tag{3}$$

and

$$E(Z_{n+1,k}) = q_{nk} E(Z_{nk}) + p_{nk} E(Z_{n,k-1}) + 1. \tag{4}$$

Proof. We only prove relation (3) because the proof of (4) is similar. From the definition of the expectation

$$E(X_{n+1,k}) = \sum_{R \in \mathcal{R}_{n+1}} X_{n+1,k}(R) P(R).$$

For fixed tree $R \in \mathcal{R}_n$ there exist n trees of the form $\alpha_i(R) \in \mathcal{R}_{n+1}$ ($i=1, 2, \ldots, n$). Notice that $P(\alpha_i(R)) = P(R) p_{n, D_{ni}(R)}$, where $D_{ni}(R)$ denotes the degree of vertex i, so

$$E(X_{n+1,k})$$
$$= \sum_{R \in \mathcal{R}_n} P(R) \sum_{i=1}^{n} X_{n+1,k}(\alpha_i(R)) p_{n, D_{ni}(R)}. \tag{5}$$

There are three cases to consider:

— if $D_{ni}(R) = k$ then $X_{n+1,k}(\alpha_i(R)) = X_{nk}(R) - 1$,

- if $D_{ni}(R) = k-1$ then $X_{n+1,\,k}(\alpha_i(R)) = X_{nk}(R) + 1$,
- if $D_{ni}(R) \neq k$ and $D_{ni}(R) \neq k-1$ then
$$X_{n+1,\,k}(\alpha_i(R)) = X_{nk}(R).$$

Using these relations we find that

$$\sum_{i=1}^{n} X_{n+1,\,k}(\alpha_i(R)) p_{n,\,D_{ni}(R)}$$

$$= (X_{nk}(R) - 1) p_{nk} X_{nk}(R)$$

$$+ (X_{nk}(R) + 1) p_{n,\,k-1} X_{n,\,k-1}(R)$$

$$+ X_{nk}(R) \sum_{\substack{r=1 \\ r \neq k \\ r \neq k-1}}^{n-1} p_{nr} X_{nr}(R)$$

$$= q_{nk} X_{nk}(R) + p_{n,\,k-1} X_{n,\,k-1}(R)$$

since $\sum_{r=1}^{n-1} X_{nr}(R) p_{nr} = 1$. If we substitute this in equation (5), we get (3). □

Let us notice that Theorems 1 and 2 allow us to find values of $E(X_{nk})$ and $E(Z_{nk})$ for arbitrary n and k.

3. Model with the attraction of vertices proportional to their degrees

Let us now deal with the case $p_{nd} = d/(2n-2)$. It is possible to find the probability that we obtain a given tree $R \in \mathscr{R}_n$.

Theorem 3. *In the model with $p_{nd} = d/(2n-2)$ if $n \geq 2$ then for any given tree $R \in \mathscr{R}_n$*

$$P(R) = \frac{\prod_{i=1}^{n} (D_{ni}(R) - 1)!}{2^{n-2}(n-2)!}.$$

Proof. Immediate, by induction. □

Figure 1 shows recursive trees with 4 vertices and their probabilities.

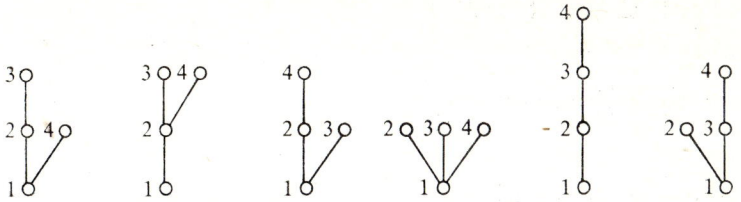

Figure 1. Recursive trees with 4 vertices and their probabilities in the model with $p_{nd} = d/(2n-2)$.

Let $u_n = 4^n / \binom{2n}{n} = \prod_{i=1}^{n} \frac{2i}{2i-1}$. It is possible to show that if $n \to \infty$ then

$$u_n = \sqrt{\pi n} \{1 + \tfrac{1}{8}n^{-1} + \tfrac{1}{128}n^{-2} + O(n^{-3})\}$$

and

$$u_n^{-1} = (\pi n)^{-1/2} \{1 - \tfrac{1}{8}n^{-1} + \tfrac{1}{64}n^{-2} + O(n^{-3})\}.$$

Using the numbers u_n we first find $E(X_{n1})$.

Theorem 4. *If $p_{nd} = d/(2n-2)$ then*

$$E(X_{n1}) = \tfrac{2}{3}(n+1) + \tfrac{4}{3}u_{n-2}^{-1}. \qquad (6)$$

Furthermore, if $n \to \infty$ then

$$E(X_{n+2, 1}) = \tfrac{2}{3}(n+1) + \tfrac{4}{3}(\pi n)^{-1/2}\{1 - \tfrac{1}{8}n^{-1} + \tfrac{1}{64}n^{-2} + O(n^{-3})\}.$$

Proof. If $p_{nd} = d/(2n-2)$ then

$$Q_n(k) = \prod_{i=n-k}^{n-1} \frac{2i-1}{2i} = u_{n-k-1}/u_{n-1}.$$

Using formula (2) we get

$$E(X_{n1}) = u_{n-2}^{-1} \{ \sum_{j=1}^{n-2} u_j + 2 \}.$$

It is possible to show that for $n \geq 1$

$$\sum_{i=1}^{n} u_i = \tfrac{2}{3}\{(n+1)u_n - 1\}.$$

and therefore (6) holds. □

Now we will deal with degrees greater than 1.

Theorem 5. *If $p_{nd} = d/(2n-2)$ and $d = o(n)$ for $n \to \infty$ then*

$$E(X_{nd}) \sim c_d(n-1), \qquad (7)$$

and

$$E(Z_{nd}) \sim s_d(n-1), \qquad (8)$$

where $c_d = \dfrac{4}{d(d+1)(d+2)}$ *and* $s_d = c_1 + c_2 + \ldots + c_d = \dfrac{d(d+3)}{(d+1)(d+2)}$.

Proof. Let us define $\varepsilon(n, d)$ by

$$E(Z_{nd}) = s_d(n-1) + \varepsilon(n, d).$$

Using (4), one can show that $\varepsilon(n, d)$ satisfies the recurrence relation

$$\varepsilon(n+1, d) = q_{nd}\varepsilon(n, d) + p_{nd}\varepsilon(n, d-1)$$

with the boundary condition $\varepsilon(n, 1) = \tfrac{4}{3}u_{n-2}^{-1}$. Let

$$M_n = \max_{d}\{\varepsilon(n, d)\}$$

and

$$m_n = \min_{d}\{\varepsilon(n, d)\}.$$

It is easy to show that M_n is a decreasing sequence, hence $M_n \leq M_2 = 4/3$, and $m_n > 0$ and $m_n \to 0$. This completes the proof of (8). Relation (7) is a consequence of the relation $E(X_{nd}) = E(Z_{nd}) - E(Z_{n, d-1})$.

It is possible to get a stronger bound for an error term, namely for $n \to \infty$

$$E(X_{nd}) = c_d(n-1) + O(n^{-\frac{1}{2}})$$

and for $d = o(n^{1/2})$

$$E(Z_{nd}) = s_d(n-1) + O(dn^{-\frac{1}{2}}). \qquad \Box$$

Let us mention that in the uniform case if $n \to \infty$ and $d = o(n)$ then $E(X_{nd}) \sim n/2^d$ and $E(Z_{nd}) \sim n(1 - 2^{-d+1})$ (see [2], [4], [6] and [7]).

Let $P(U_n = k)$ be the probability that a randomly chosen vertex from an n-RRT has a degree k. Of course, $P(U_n = k) = \frac{1}{n} E(X_{nk})$, hence there exists a limiting random variable U such that $P(U = k) = \lim_{n \to \infty} P(U_n = k) = c_k$. This random variable is defined correctly because $\sum_{k=1}^{\infty} c_k = 1$. It is easy to show the following result.

Theorem 6. *If U is a random variable such that $P(U=k) = c_k$ for $k = 1, 2, \ldots$, then $E(U) = 2$ and $\mathrm{Var}(U) = \infty$.* $\Box$

It is curious that in the uniform case all moments of U exist. In fact, $P(U=k) = 2^{-k}$, and $E_n(U) = 2n!$, where $E_n(U)$ denotes the kth factorial moment of the random variable U. Hence $E(U) = \mathrm{Var}(U) = 2$.

Let us now consider the degree of a fixed vertex.

Theorem 7. *If $p_{nd} = d/(2n-2)$ then for $2 \leqslant i \leqslant n$*

$$E(D_{ni}) = \frac{n-1}{i-1} \frac{u_{i-1}}{u_{n-1}} \qquad (9)$$

and

$$E(D_{ni}^2) = \frac{n-1}{i-1} + \frac{n-1}{2(i-1)} u_{i-1} \sum_{j=i}^{n-1} \frac{1}{ju_{j-1}}. \qquad (10)$$

Proof. The way of adding a new vertex to a recursive tree implies that

$$P(D_{n+1,i} = k) = q_{nk} P(D_{ni} = k) + p_{n,k-1} P(D_{ni} = k-1).$$

Using this relation and the definition of expectation, after elementary calculations one can get

$$E(D_{n+1,i}) = \frac{2n-1}{2i-1} E(D_{ni}).$$

If we solve this equation with the boundary condition $E(D_{ii})=1$, we get (9). Similarly, we can show that

$$E(D^2_{n+1,i}) = \frac{n}{n-1} E(D^2_{ni}) + E(D_{ni})/(2n-2).$$

Using standard methods to solve this equation with the boundary condition $E(D^2_{ii})=1$ we get (10). □

Let us mention that Theorem 7 does not take into account the degree of the root of a tree, but it is easy to see that the distributions of the degrees of vertices 1 and 2 are identical, hence $E(D^k_{n,1}) = E(D^k_{n,2})$.

Theorem 7 allows us to find the asymptotic behaviour of $E(D_{ni})$ and $\mathrm{Var}(D_{ni})$.

Corollary 7.1. *If $n \to \infty$ then*

$$E(D_{ni}) \sim \begin{cases} a_i n^{\frac{1}{2}} & \text{for fixed } i, \\ (n/i)^{\frac{1}{2}} & \text{for } i \to \infty \end{cases}$$

and

$$\mathrm{Var}(D_{ni}) \sim \begin{cases} b_i n & \text{for fixed } i, \\ n/i & \text{for } i \to \infty, i = o(n), \\ 1/c - 1/\sqrt{c} & \text{for } i = cn, 0 < c < 1, \end{cases}$$

where

$$a_i = \pi^{-\frac{1}{2}} u_{i-1}/(i-1) \text{ and}$$

$$b_i = \frac{1}{i-1} + \frac{u_{i-1}}{2i-2} \sum_{j=i}^{\infty} \frac{1}{j u_{j-1}} - \frac{u^2_{i-1}}{\pi(i-1)^2}. \quad \Box$$

Let us now compare this result with the uniform case, where for fixed i, $E(D_{ni}) \sim \mathrm{Var}(D_{ni}) \sim \ln(n)$, and for $i \to \infty$, $i = o(n)$ $E(D_{ni}) \sim \mathrm{Var}(D_{ni}) \sim \ln(n/i)$.

In the model $p_{nd} = d/(2n-2)$ we have that $E(D_{ni}) \leqslant E(D_{n1}) = 2(n-1)/u_{n-1}$

$\sim 2(n/\pi)^{1/2}$. This implies that the maximal degree Δ_n in an n-RRT satisfies

$$E(\Delta_n) \geqslant E(D_{n1}) \sim cn^{\frac{1}{2}},$$

where $c = 1.2838\ldots$. In the uniform case the maximal degree is less and $E(\Delta_n) \sim c' \ln(n)$, where $1 \leqslant c' \leqslant 1/\ln 2$.

Let G_{ni} denote the number of vertices in the subtree of a recursive tree rooted at vertex i. The following theorem gives two moments of the random variable G_{ni}.

Theorem 8. *If $p_{nd} = d/(2n-2)$ then for $i \leqslant n$*

$$E(G_{ni}) = \frac{n-1}{2i-2} + \frac{1}{2} \tag{11}$$

and

$$\mathrm{Var}(G_{ni}) = \frac{(n-1)(n-i)(2i-3)}{4i(i-1)^2}.$$

Proof. The way of adding a new vertex to a recursive tree implies that

$$E(G_{n+1,i}) = \sum_{R \in \mathcal{R}_n} P(R) \sum_{j=1}^{n} G_{n+1,i}(\alpha_j(R)) \frac{D_{nj}(R)}{2n-2}.$$

Since

$$G_{n+1,i}(\alpha_j(R)) = \begin{cases} G_{ni}(R) + 1 & \text{for } j \text{ from } i\text{th subtree}, \\ G_{ni}(R) & \text{otherwise}. \end{cases}$$

Therefore

$$E(G_{n+1,i}) = \sum_{R \in \mathcal{R}_n} P(R) \left\{ G_{ni}(R) + \sum' \frac{D_{nj}(R)}{2n-2} \right\},$$

where Σ' denotes the sum over all vertices j that belong to the ith subtree. It is easy to see that

$$\sum' \frac{D_{nj}(R)}{2n-2} = \frac{2G_{ni}(R) - 1}{2n-2},$$

so

$$E(G_{n+1,i}) = \frac{n}{n-1} E(G_{ni}) - \frac{1}{2n-2}.$$

Solving this recurrence equation with the boundary condition $E(G_{ii})=1$ we get (11). Similarly we can find a recurrence relation for the kth factorial moment, namely

$$E_k(G_{n+1,i}) = \sum_{R\in\mathcal{R}_n} P(R)\left\{(G_{ni}(R))_k + ((G_{ni}(R)+1)_k - (G_{ni}(R))_k) {\sum}' \frac{D_{nj}(R)}{2n-2}\right\}$$

$$= \frac{n+k-1}{n-1} E_k(G_{ni}) - \frac{k(2k-3)}{2n-2} E_{k-1}(G_{ni})$$

which we can solve for each k with the boundary condition $E_k(G_{i+k-2,i})=0$

$$E_k(G_{ni}) = \langle n-1\rangle_k \frac{k(2k-3)}{2} \sum_{j=i+k-2}^{n-1} E_{k-1}(G_{ji})/\langle j-1\rangle_{k+1},$$

where $\langle x\rangle_k = x(x+1)\ldots(x+k-1)$. If we put $k=2$, after some calculations we get the required formula for $\mathrm{Var}(G_{ni})$. □

In the uniform case the value of $E(G_{ni})=n/i$ as well as that of $\mathrm{Var}(G_{ni})$, is approximately twice the respective values of the nonuniform case.

References

[1] C. Berge, *Graphs and Hypergraphs*, (North-Holland, 1973).

[2] M. Dondajewski and J. Szymański, On the distribution of vertex-degrees in a strata of a random recursive tree, *Bull. Acad. Polon. Sci., Ser. Sci. Math.* Vol. XXX, No. 5-6 (1982) 205-209.

[3] M. Dondajewski and J. Szymański, On a generalized random recursive tree, (unpublished).

[4] J. L. Gastwirth, A probability model of a pyramid scheme, *Amer. Statist.* 31 (1977) 79-82.

[5] F. Harary and E. M. Palmer, *Graphical Enumeration*, (Academic Press, New York, 1973).

[6] A. Meir and J. W. Moon, Path edge-covering constants for certain families of trees, *Util. Math.* 14 (1978) 313-333.

[7] H. S. Na and A. Rapoport, Distribution of nodes of a tree by degree, *Math. Biosci.* 6 (1970) 313-329.

PSEUDO-RANDOM GRAPHS

Andrew THOMASON

University of Exeter,
Exeter, England

We call a graph G (p, α)-*jumbled* if, for every induced subgraph H of G, $\left| e(H) - p \binom{|H|}{2} \right| \leq \alpha |H|$ holds; here p and α are real numbers with $0 < p < 1 \leq \alpha$, and $e(H)$ is the number of edges in H. We show that a (p, α)-jumbled graph behaves in many ways like a random graph with edge probability p, and some aspects of this similarity are examined.

Since its introduction by Erdös and Rényi, the theory of random graphs has been greatly developed (for a modern treatment of the subject, see Bollobás [2]), and many properties of a random graph have been studied in detail. Random graphs have proved of great interest for various reasons, of which we mention just two: they provide the best known extremal graphs for several extremal problems, such as subcontractions [23], Zarankiewicz's problem [19] and Ramsey's theorem [13], and they offer examples of graphs with certain properties, giving us say expanders [11], graphs of small diameter [9] and parallel sorting algorithms [7]. In all these cases it would be useful to have a criterion by which to decide whether a *specific* graph behaves like a random graph, that is, has the property (of almost all graphs) that interests us. Such a criterion might also be used to describe the class of extremal graphs in the problems mentioned, which may perhaps give useful information about the problems themselves.

The purpose of this paper is to offer such a criterion and to explore some of its consequences. As a result we are able to extend and simplify many earlier results. For instance, results obtained by Bollobás and Thomason [6] using the Riemann hypothesis for algebraic curves over finite fields, by Alon and Milman [1] using eigenvalue methods, and by Gurevich and Shelah [18] using random graph techniques, can be obtained in more general settings by elementary arguments. In the latter two cases, these are described elsewhere in more detail [25], [26].

A graph G is said to be (p, α)-*jumbled* if p, α are real numbers satisfying $0 < p < 1 \leq \alpha$ and if every induced subgraph H of G satisfies

$$\left| e(H) - p\binom{|H|}{2} \right| \leq \alpha |H|.$$

Here $e(H)$ is the number of edges in H, following [2]. Equivalently, if $d(H)$ is the average degree inside H we may say

$$|d(H) - p(|H| - 1)| \leq 2\alpha$$

holds for every induced subgraph H. We think of a (p, α)-jumbled graph as behaving somewhat like a random graph where each edge is chosen with probability p. Of course, it is possible to suggest other definitions for a "pseudo-random" graph but it seems they often reduce to this one. Note that if G is (p, α)-jumbled then every induced subgraph is (p, α)-jumbled and the complement of G is $(1-p, \alpha)$-jumbled. Observe too that the clique number of G is at most $1 + 2\alpha(1-p)^{-1}$ and the independence number is at most $1 + 2\alpha p^{-1}$.

Naturally every graph of order n is $(p, n/2)$-jumbled, so the definition begins to be interesting when α is small compared to n. Although we require only $\alpha \geq 1$, a theorem of Erdös and Spencer [16] shows that α is at least of order $(pn)^{1/2}$. (Their proof is stated only for $p = \frac{1}{2}$ but the extension to other values of p with $pn \to \infty$ and $(1-p)n \to \infty$ is straightforward.) Our results will be stated for all values of p and α but they are best understood by thinking of α of order $(pn)^{1/2}$, and $p > n^{-1/3}$; the latter since many results require $\delta(G) \sim pn$ and $p\delta > \alpha$.

In this paper we show ways in which jumbled graphs behave like random graphs, and illustrate ways in which specific graphs may be shown to have random behaviour. The paper falls into three parts. First we give examples of (p, α)-jumbled graphs with small α; these are typical of the "explicit random graphs" we have in mind. Then we describe two ways of testing whether a graph is jumbled. The first is more or less a degree condition, and can be applied to a specific graph very easily. The second is a global condition. It tells us that if a graph G is *not* (p, α)-jumbled it contains a subgraph H with $\left| e(H) - p\binom{|H|}{2} \right| > \alpha |H|$ and $|H|$ large. To this extent it gives information about the extremal graphs in the problems cited earlier. Finally, we derive several properties of jumbled graphs. These properties are chosen mainly because they have been studied frequently in the random graph case and/or because they illustrate the techniques used in dealing with jumbled graphs.

Examples

Here are some examples of jumbled graphs. Some of them exhibit various kinds of pathological behaviour and most will be used later to illustrate particular points and to test the strength of theorems.

(1) Let $G \in \mathscr{G}(n, p)$, that is, the edges of G are chosen at random with probability p. Then G is almost surely $(p, 2(pn)^{1/2})$-jumbled (provided $pn \to \infty$ and $(1-p)n \to \infty$. The constant 2 here is generous).

(2) Choose a graph in $\mathscr{G}(n, p)$, select a subset X of the vertices, with $|X| = \lfloor(pn)^{1/2}\rfloor$, and join each pair of vertices in X. Then G is almost surely $(p, 3(pn)^{1/2})$-jumbled.

(3) As in (2), but with $|X| = \lfloor(pn)^{3/4}\rfloor$. Then G is almost surely $(p, (pn)^{3/4})$-jumbled.

(4) Let $V(G) = \{x_1, \ldots, x_{n/2}, y_1, \ldots, y_{n/2}\}$, where n is an even integer. For each pair $i < j$, join x_j to exactly one of x_i and y_i, chosen at random. Do the same for y_j. The graphs spanned by the x_*'s and the y_*'s are therefore randomly chosen elements of $\mathscr{G}(n/2, \tfrac{1}{2})$. G itself is $(\tfrac{1}{2}, n^{1/2})$-jumbled.

(5) Let $V(G)$ be as in (4). For each pair $i < j$, insert at random one of the four paths $x_i x_j y_i$, $x_i y_j y_i$, $x_j x_i y_j$, $x_j y_i y_j$. Add also the edges $x_i y_i$. Then G is $(\tfrac{1}{2}, n^{1/2})$-jumbled.

(6) Let $k \geq 2$ be an integer, let $n = 2kl + 1$ be a prime power, and let F_n be the field of order n. Construct a graph G of order n by setting $V(G) = \mathsf{F}_n$ and joining x to y if $x - y$ is a kth power in F_n. If $k = 2$ the graph we get, often called the Paley graph of order n, is $(\tfrac{1}{2}, n^{1/2})$-jumbled. If $k > 2$ the graph is $(k^{-1}, 2n^{3/4})$-jumbled. The justification for these assertions will be provided later by Theorem 1.1. (The appropriate values of μ required by the conditions of Theorem 1.1 can be taken as $\mu = \tfrac{1}{4}$ for $k = 2$ and $\mu = n^{1/2}$ for $k > 2$. These values can be verified using the elementary theory of characters over finite fields.)

(7) As in (6), but join x to y if $x + y$ is a kth power. Again we get a $(k^{-1}, 2n^{3/4})$-jumbled graph, or $(\tfrac{1}{2}, 2n^{1/2})$-jumbled if $k = 2$. This construction is more natural than that of (6), except when $k = 2$, and has an obvious generalisation to hypergraphs. Since we do not need -1 to be a kth power this time, it is enough if $n = kl + 1$.

(8) Let q be a prime power and let $V(G)$ be the elements of a vector space of dimension two over F_q; so G has $n = q^2$ vertices. Partition the $q + 1$ lines of the space into two sets P and N, where $|P| = k = p(q+1)$, $1 \leq k \leq q$, and $|N| = q + 1 - k$. Join x to y if $x - y$ is parallel to a line in P. Then G is $(p, n^{3/4})$-jumbled. This example (when $p = \tfrac{1}{2}$) is due to Delsarte and Goethals and to Turyn (see [22]). G is in

fact strongly regular with parameters $(k(q-1), (k-1)(k-2)+q-2, k(k-1))$. If $p = \frac{1}{2}$ we can choose the set P so as to obtain the Paley graph of order n.

(9) Let q be a prime and let $V(G)$ be the set F_q. Let t be an integer, $1 < t < q$. Join x to y if the fractional part of $(x-y)^2/q$ is at most t/q. By Theorem XIII. 16 of [2] and Theorem 1.1 below, this graph is $(p, n^{3/4} \log n)$-jumbled where $p = t/q$ and $n = q$. An infinite analogue of this graph was shown to have analogous properties by Pinch [21].

(10) Let q be a prime power and let $V(G)$ be the vertices of the projective geometry of dimension k over F_n. Then G has $n = (q^{k+1}-1)(q-1)^{-1}$ elements. Join $x = x_0 : \ldots : x_k$ to $y = y_0 : \ldots : y_k$ if $x_0 y_0 + \ldots + x_k y_k = 0$. Then (again by Theorem 1.1) G is $(1/q, 2(n/q)^{1/2})$-jumbled. This graph is sometimes called the Erdös-Rényi graph in the case $k = 2$.

(11) The graph of example (10) may be viewed, when $q = 2$, as formed by taking as vertices the non-empty subsets of a set of order $k+1$, two vertices being joined if their intersection has even order. Two subgraphs of this graph are particularly interesting. For the first, let G be the graph whose vertices are the non-empty *even* subsets of a set of order $k+1$, where k is even. Join two vertices if their intersection is also even. Then G has order $n = 2^k - 1$ and is $(\frac{1}{2}, 2n^{1/2})$-jumbled. The reason for choosing k even is just that the graph is fractionally easier to analyse if we do. In fact in this case G is strongly regular with parameters $((n-3)/2, (n-11)/4, (n-3)/4)$.

(12) For the second subgraph, let G be the graph whose vertices are the *odd* subsets of a set of order $k+1$, two vertices being adjacent if their intersection is even. Then G has order $n = 2^k$ and is $(\frac{1}{2}, 2n^{1/2})$-jumbled.

(13) Let H be a (p, α)-jumbled graph of order m, and let $k \geq 1$ be an integer. For each vertex $x \in H$, let $x_1, \ldots, x_k$ be vertices of the graph G. Join x_i to y_j in G whenever $xy \in H$ and $1 \leq i, j \leq k$. Then G has order km and is $(p, k\alpha + k)$-jumbled.

There are other specific examples of jumbled graphs with less dense edge sets. For instance, in the graph of example (10) we may require more equations to be satisfied before we join two vertices. We do not concentrate on these examples since, as we mentioned earlier, our results are most effective if p is not too small.

§1. Conditions implying a graph is jumbled

We begin by considering two ways of checking whether a given graph is (p, α)-jumbled. The first, Theorem 1.1, is a local approach, the second, Theorem 1.4, is a global approach. We shall make use of the notation $B(x)$, where x is a non-negative integer, to signify any real number y of absolute value at most x. Thus $y = B(x)$ means $|y| \leq x$, and $0 \leq z \leq x$ implies $B(z) = B(x)$. In this sense the notation behaves like Landau's $O(x)$ notation. Using this, we may rewrite the definition of a (p, α)-jumbled graph G in the form

$$e(H) = p\binom{|H|}{2} + B(\alpha |H|)$$

for all induced $H \subset G$.

Theorem 1.1. *Let n be an integer, and let $0 < p < 1$ and $\mu \geq 0$ be real numbers. Let G be a graph of order n with minimum degree pn in which no two vertices have more than $p^2 n + \mu$ common neighbours. Then G is (p, α)-jumbled, where $2\alpha = (pn + (n-1)\mu)^{\frac{1}{2}} + p$.*

Proof. Let H be a subgraph of G of order $k \leq n$, and let the average degree in H be d. Let $a_1, \ldots, a_k$ be the degree sequence of H, and let $b_1, \ldots, b_{n-k}$ be the numbers of edges between H and each of the $n - k$ vertices of $G - H$. Then

$$\sum_{i=1}^{k} a_i = kd$$

and

$$\sum_{i=1}^{n-k} b_i \geq \sum_{i=1}^{k} (pn - a_i) = k(pn - d).$$

Moreover, since no two vertices have more than $p^2 n + \mu$ common neighbours, we have

$$\sum_{i=1}^{k} \binom{a_i}{2} + \sum_{i=1}^{n-k} \binom{b_i}{2} \leq \binom{k}{2}(p^2 n + \mu),$$

so

$$k\binom{d}{2} + (n-k)\binom{k(pn-d)/(n-k)}{2} \leq \binom{k}{2}(p^2 n + \mu).$$

Rearranging gives

$$(d - pk)^2 \leq \frac{n-k}{n}[(k-1)\mu + np(1-p)],$$

which is somewhat stronger than the result claimed. □

Obviously, many of our examples (such as (6)–(12)) of concrete random graphs are shown by this theorem to be $(p, c(pn)^{1/2})$-jumbled, for appropriate p and constant c. For random graphs themselves though this theorem isn't so effective, since in this case μ is of order $n^{1/2}$.

It is curious to note that the minimum degree condition in Theorem 1.1 can be dropped if we require every pair of vertices to have $p^2n(1+o(1))$ common neighbours (here we imagine $n\to\infty$). For let v be a vertex of G with degree d. Then each vertex in $G-v$ has $p^2n(1+o(1))$ neighbours in $\Gamma(v)$, so the number of paths of length 2 in $G-v$ with both ends in $\Gamma(v)$ is $(n-1)\binom{p^2n}{2}(1+o(1))$. But this number is $\binom{d}{2}p^2n(1+o(1))$, and we have the minimum degree condition back again.

However, if we weaken the conditions of the theorem to require only every *edge* to be in at most $p^2n+\mu$ triangles, the conclusion fails to hold. Suppose for instance we are given a vertex-transitive pk-regular graph F of order k in which every edge is in p^2k triangles. (We will construct such a graph shortly.) Now for each $v \in F$ take a set $X(v)$ of order m, where m is some integer, and form a graph G with vertex set $\bigcup_{v\in F} X(v)$ with $xy \in G$ if $x \in X(u)$, $y \in X(v)$ and $uv \in F$. Then G has order $n=mk$, is vertex-transitive and pn-regular, and every edge of G is in precisely p^2n triangles. But G is not (p, α)-jumbled for any small value of α since its independence number is at least n/k.

An example of a graph F of the type described can be constructed as follows in the case $p=1/6$ and $k=36$. First take a K_4 with vertex set $\{v_1, v_2, v_3, v_4\}$ and edge colour it with colours 0, 1 and 2. Now construct four 9-cycles C^i, $1 \leq i \leq 4$, with vertex sets $\{a_j^i; 1\leq j \leq 9\}$ where a_j^i is joined to $a_{j\pm1}^i$. Construct H from the union of these four cycles by joining a_t^i to a_t^j if $t \equiv c \pmod 3$, where c is the colour of v_iv_j. Then H has order 36, is 3-regular, vertex transitive and has girth 7. Form F from H by setting $V(F)=V(H)$ and joining u to v in F if $d_H(u,v)=2$. Then F is 6-regular, vertex-transitive and every edge is in exactly one triangle.

In contrast to Theorem 1.1 we can show a graph is jumbled if we have only large scale information about it, namely, when we are given the number of edges in subgraphs of some large fixed order.

Lemma 1.2. *Let p, η, m, n be positive real numbers with $0<p$, $\eta<1$, such that ηn is an integer with $2 \leq \eta n \leq n-2$. Let G be a graph of order n in which for every induced subgraph H of order ηn,*

$$\left| e(H) - p\binom{\eta n}{2} \right| \leq m$$

holds. Then

$$\left|e(H)-p\binom{k}{2}\right|\leqslant 80\ mn^{-2}(1-\eta)^{-2}$$

for every induced subgraph H of order k.

Proof. Let H be a subgraph of order $k\geqslant\eta n$. If we count the number of edges in each of the $\binom{k}{l}$ subgraphs L of H of order $l=\eta n$, we get

$$e(H)=\binom{k-2}{l-2}^{-1}\sum_{L\subset H}e(L)=\binom{k-2}{l-2}^{-1}\sum_{L\subset H}\left(p\binom{l}{2}+B(m)\right)$$

$$=p\binom{k}{2}+\frac{k(k-1)}{l(l-1)}B(m)=p\binom{k}{2}+B\left(\frac{2k^2m}{l^2}\right)$$

$$=p\binom{k}{2}+B\left(\frac{80m}{\eta^2(1-\eta)^2}\right), \quad \text{as claimed}.$$

Now suppose H is a subgraph of order $k\leqslant\min\{(1-\eta)n,\eta n\}$. Let F be a subgraph of $G-H$ of order ηn, and let L be a subgraph of H of order l, where $1\leqslant l\leqslant k$. Then by the above,

$$e(H\cup F)=p\binom{k+\eta n}{2}+B\left(\frac{2(k+\eta n)^2}{(\eta n)^2}m\right) \qquad (a)$$

and

$$e(L\cup F)=p\binom{l+\eta n}{2}+B\left(\frac{2(l+\eta n)^2}{(\eta n)^2}m\right). \qquad (b)$$

Holding l fixed and summing over all $\binom{k}{l}$ subgraphs L, we have

$$\sum_L e(L\cup F)=\sum_L\{e(L)+e(L,F)+e(F)\}$$

$$=\binom{k-2}{l-2}e(H)+\binom{k-1}{l-1}e(H,F)+\binom{k}{l}e(F),$$

and dividing by $\binom{k}{l}$ and using (b) gives

$$\frac{l(l-1)}{k(k-1)}e(H)+\frac{l}{k}e(H,F)+e(F)$$
$$=p\binom{l+\eta n}{2}+B\left(\frac{2(l+\eta n)^2}{(\eta n)^2}m\right).$$

By means of (a) and the equation $e(F)=p\binom{\eta n}{2}+B(m)$ we can solve for $e(H)$ to find

$$e(H)=p\binom{k}{2}+B\left[\frac{k-1}{k-l}2m\left(\frac{(k+\eta n)^2}{(\eta n)^2}+\frac{k}{l}\frac{(l+\eta n)^2}{(\eta n)^2}+\frac{k}{2l}+\frac{1}{2}\right)\right].$$

Writing $l=\lambda k$, and since $k \leqslant \eta n$,

$$e(H)=p\binom{k}{2}+B\left(\frac{2m}{1-\lambda}\left(4+\frac{(1+\lambda)^2}{\lambda}+\frac{1}{2\lambda}+\frac{1}{2}\right)\right).$$

Choosing l so that $\tfrac{1}{3} \leqslant \lambda \leqslant \tfrac{1}{2}$, which we can if $k \geqslant 2$, we have

$$e(H)=p\binom{k}{2}+B(40m)=p\binom{k}{2}+B\left(\frac{80m}{\eta^2(1-\eta)^2}\right).$$

Finally, suppose $(1-\eta)n \leqslant k \leqslant \eta n$ (this of course happens only if $\eta \geqslant \tfrac{1}{2}$). Using the result of the previous paragraph we may sum the number of edges in all subgraphs L of order $l=(1-\eta)n$ to get

$$\binom{k-2}{l-2}e(H)=\binom{k}{l}\left[p\binom{l}{2}+B(40m)\right]$$

so

$$e(H)=p\binom{k}{2}+\frac{k(k-1)}{l(l-1)}B(40m)$$
$$=p\binom{k}{2}+B\left(\frac{80m}{\eta^2(1-\eta)^2}\right). \quad \square$$

Obviously the constant 80 in this lemma could be reduced. However if η is small the argument in the first part of the proof shows that $O(m\eta^{-2})$ is the right bound for the error, and if η is large, an examination of the graph $2K_{n/2}$ shows that $O(m(1-\eta)^{-2})$ is the right order (take $k=n/2$, say). Hence the bound $O(m\eta^{-2}(1-\eta)^{-2})$ cannot in general be improved.

An example of a direct application of Lemma 1.2 is as follows. Suppose we have a sequence $G_1, G_2, \ldots$ where G_n is a graph of order n such that every induced subgraph H of G_n of order $n/2$ satisfies

$$\left| e(H) - p \binom{n/2}{2} \right| \leq n^{3/2}.$$

For instance, G_n may be the graph of example (3). Then by Lemma 1.2 we can say G_n is $(p, 13n^{3/4})$-jumbled, since

$$13n^{3/4}k \geq \min\left\{ 320n^{3/2}, \binom{k}{2} \right\}$$

for all k. Moreover the graphs in this example show that G_n is not (p, n^β)-jumbled for any $\beta < 3/4$, so this result is best possible. However, in this example the subgraphs with large error are fairly localised, and most subgraphs have smaller error. In fact we shall now show (Theorem 1.4) that if $\omega(n) \to \infty$ is a function of n then the given conditions on G_n imply that G_n contains a subgraph G_n^* of order $n(1+o(1))$ which is $(p, \omega n^{1/2})$-jumbled, so that the errors in G_n^* are much smaller.

Before proving Theorem 1.4 we need an analogue of Lemma 1.2 for multipartite subgraphs.

Lemma 1.3. *Let $r \geq 2$ be an integer and let p, η, m, n and G be as in the statement of Lemma 1.2. Let H be an induced r-partite subgraph of G whose vertex classes have orders $k_1, k_2, \ldots, k_r$. Then*

$$\left| e(H) - p \sum_{1 \leq i < j \leq r} k_i k_j \right| \leq 360 m \eta^{-2} (1-\eta)^{-2}.$$

Proof. Choose an integer l with $r/3 \leq l \leq r/2$. There are $\binom{r}{l}$ partitions of the vertex classes into two groups, one with l classes and the other with $r-l$ classes. Each edge of H joins classes in different groups for $2\binom{r-2}{l-1}$ of these partitions. Consider some fixed partition, and let H_1 be the subgraph formed by the classes in the first group. Put $h=|H_1|$ and $H_2=H-H_1$. By Lemma 1.2

$$e(H_1)=p\binom{h}{2}+B(2A)$$

and

$$e(H_2)=p\binom{k-h}{2}+B(2A)$$

where $k=|H|$ and $A=40\,mn^2(1-\eta)^2$. Hence $e(H_1, H_2)=ph(k-h)+B(4A)$. Summing over all partitions gives

$$2\binom{r-2}{l-1}e(H)=2\binom{r-2}{l-1}p\sum_{i<j}k_ik_j+\binom{r}{l}B(4A)$$

so

$$e(H)=p\sum_{i<j}k_ik_j+\frac{r(r-1)}{2(r-l)l}B(4A)$$

$$=p\sum_{i<j}k_ik_j+B(9A). \qquad \square$$

Theorem 1.4. Let $p, \eta, \alpha, n, \omega$ be positive real numbers with $0<p, \eta<1$ such that ηn is an integer with $2\leq \eta n\leq n-2$. Let G be a graph of order n in which for every induced subgraph H of order ηn,

$$\left|e(H)-p\binom{\eta n}{2}\right|\leq \eta n\alpha$$

holds. Then G contains a subgraph G^* of order at least

$$\left(1-\frac{880}{\eta(1-\eta)^2\omega}\right)n$$

which is $(p, \omega\alpha)$-jumbled.

Proof. We first construct G_0 by repeatedly removing "dense" subgraphs H_1, H_2, ..., H_r such that $e(H_i)-p\binom{k_i}{2}>k_i\omega\alpha$, where $|H_i|=k_i$ and $H_j\subset G-\bigcup_{i<j}H_i$. Stop when it is no longer possible to choose another H_*, and let $G_0=G-\bigcup_{i=1}^{r}H_i$.

Let $H = \bigcup_{i=1}^{r} H_i$ and $k = |H| = \sum_{i=1}^{r} k_i$. By Lemma 1.2 $e(H) \leq p \binom{k}{2} + 2A$, where $A = 40\eta^{-1}(1-\eta)^{-2} n\alpha$, and by Lemma 1.3

$$e(H) - \sum_{i=1}^{r} e(H_i) \geq p \sum k_i k_j - 9A,$$

so

$$\sum_{i=1}^{r} e(H_i) \leq \sum_{i=1}^{r} p \binom{k_i}{2} + 11A,$$

giving $\sum_{i=1}^{r} k_i \omega \alpha \leq 11A$ and $k \leq 11A/\omega\alpha$. Now construct G^* by removing from G_0 'sparse' subgraphs $F_1, \ldots, F_s$ such that $e(F_i) - p\binom{f_i}{2} < -f_i \omega \alpha$, where $f_i = |F_i|$. By a similar argument, $|G_0 - G^*| < 11A/\omega\alpha$. Thus $|G - G^*| < 22A/\omega\alpha$ as asserted. $\square$

There are a couple of ways in which one might wish to weaken the conditions of Theorem 1.4 but they fail to give the desired conclusion. The first is to require $\left| e(H) - p\binom{\eta n}{2} \right| \leq \eta n \alpha$ holds not for all H of order ηn but for almost all. However, this is a requirement satisfied by a complete bipartite graph $K_{n/2, n/2}$ (assuming $p = \frac{1}{2}$) and so it cannot be strong enough. The second way to weaken the conditions is to ask only that the root mean square value of $\left| e(H) - p\binom{\eta n}{2} \right|$ be small. This too is inadequate, since the r.m.s. value depends only on the number of pairs of edges and this in turn depends only on the degree sequence. Indeed if G has order n and has $E = p\binom{n}{2}$ edges, with degree sequence $(p(n-1) + \varepsilon_i)_{i=1}^{n}$, then looking at all subgraphs H of order k one finds

$$\binom{n}{k}^{-1} \sum_{H} \left(e(H) - p\binom{k}{2} \right)^2$$

$$= \frac{1}{2} p(n-k) \binom{k}{2} \binom{n-2}{2}^{-1} \left\{ (1-p)(n-k-1) + \frac{k-2}{E} \sum_{i=1}^{n} \varepsilon_i^2 \right\}.$$

Once again this condition fails to discriminate against bipartite graphs.

Finally, suppose we take a typical jumbled graph, say a $(\frac{1}{2}, n^{1/2})$-jumbled graph G of order n. One might ask, in the spirit of Theorem 1.4, whether G contains a subgraph G^* of order almost n in which the errors in subgraphs of order k are small, provided k is small (say $k<n^{\frac{1}{2}}$; of course in G we have no information about such small subgraphs). The answer is no. Consider say the graph of example (6) when $k=2$ and n is a square. The $\frac{1}{2}\binom{n}{2}$ edges are covered by several complete subgraphs of order $\sqrt{n}$, each edge being in the same number of these. Thus if $|G^*|=(1-\varepsilon)n$, one of these complete subgraphs meets G^* in at least $(1-2\varepsilon)\sqrt{n}$ vertices, for otherwise a short calculation shows $G-G^*$ would have more than $\binom{\varepsilon n}{2}$ edges. We can therefore say no more about small subgraphs in G^* than we can in G.

§2. Properties of jumbled graphs

We now examine consequences of the definition of jumbled graphs. These graphs have many properties which are well known to hold for random graphs. The following two lemmas will be fundamental in our study.

Lemma 2.1. *Let G be a (p, α)-jumbled graph of order n, and let $0<\varepsilon<1$. Then at least $(1-\varepsilon)n$ of the vertex degrees of G lie in the range $p(n-1)\pm 10\alpha\varepsilon^{-1}$.*

Proof. Let S be a subgraph of order s, and let the sum of the degrees (in G) of the vertices of S be sd. Then $sd=2e(S)+e(S, G-S)$. But

$$e(S)+e(S, G-S)=e(G)-e(G-S)$$
$$=p\binom{n}{2}-p\binom{n-s}{2}+B(\alpha n)+B(\alpha(n-s))$$

and

$$e(S)=p\binom{s}{2}+B(\alpha s),$$

so

$$sd=2p\binom{s}{2}+ps(n-s)+B(\alpha n)+B(\alpha(n-s))+B(\alpha s)$$
$$=ps(n-1)+B(2\alpha n).$$

Thus $d=p(n-1)+B(2\alpha n s^{-1})$. Taking S to be the $\lfloor\varepsilon n/2\rfloor$ vertices of smallest degree, we see that the average of these degrees is at least $p(n-1)-10\alpha\varepsilon^{-1}$ (since the lemma is vacuous if $\varepsilon n<10$, and otherwise $\lfloor\varepsilon n/2\rfloor\geq 2\varepsilon n/5$). The proof is completed by then taking S to be the $\lfloor\varepsilon n/2\rfloor$ vertices of largest degree. □

Lemma 2.2. *Let G be a (p, α)-jumbled graph of order n, and let $0<\varepsilon<1$. Let H be an induced subgraph of G of order k. Then at least $n-\varepsilon k$ of the vertices of G have between $pk-21\alpha\varepsilon^{-1}$ and $pk+21\alpha\varepsilon^{-1}$ neighbours in H.*

Proof. By Lemma 2.1 at least $(1-\varepsilon/2)k$ of the vertices of H have degree $p(k-1)+B(20\alpha\varepsilon^{-1})=pk+B(21\alpha\varepsilon^{-1})$ in H. If $G-H\neq\emptyset$, let S be a non-empty subgraph of $G-H$ with $|S|=s$. Define d by $sd=e(S, H)$. Then

$$sd = e(S\cup H)-e(S)-e(H)$$

$$=p\binom{s+k}{2}-p\binom{s}{2}-p\binom{k}{2}+B(\alpha(s+k))+B(\alpha s)+B(\alpha k)$$

$$=psk+B(2\alpha(s+k)),$$

so

$$d=pk+B(2\alpha(1+ks^{-1})).$$

Choose S to be the $\lfloor \varepsilon k/4 \rfloor$ vertices of $G-H$ which send the least number of edges to H. Since we may as well assume $k\varepsilon \geq 21$ we see $\lfloor \varepsilon k/4 \rfloor \geq \varepsilon k/4 - 1 \geq \varepsilon k/5$ so $B(2\alpha(1+ks^{-1}))=B(2\alpha(1+5\varepsilon^{-1}))=B(12\alpha\varepsilon^{-1})$. Thus all but $\varepsilon k/4$ vertices of $G-H$ have at least $pk-12\alpha\varepsilon^{-1}$ neighbours in H, and likewise all but $\varepsilon k/4$ vertices have at most $pk+12\alpha\varepsilon^{-1}$ neighbours. □

For a set U of vertices of a graph G, denote by $N(U)$ the set of vertices of $G-U$ which are joined to every vertex in U, and denote by $\overline{N(U)}$ those vertices of $G-U$ joined to none of the vertices of U. If U_1 and U_2 are disjoint sets of vertices, we denote $|N(U_1)\cap \overline{N(U_2)}|$ by $v(U_1, U_2)$.

Under the conditions of Theorem 1.1 it is possible to show easily that almost all k-tuples of vertices have about $p^k n$ common neighbours. This is done by enumerating the k-tuples, letting d_i be the number of neighbours of the ith k-tuple, and estimating $\sum d_i$ and $\sum \binom{d_i}{2}$. This works for k up to $\log_b n$, where $b=1/p$. The following theorem does the same for any jumbled graph, though the argument works only as far as $k=\frac{1}{2}\log_b n$.

Theorem 2.3. *Let G be a (p, α)-jumbled graph of order n, let $k, l \geq 0$ be integers and let $0<\varepsilon<1$. Then*

$$|v(U_1, U_2)-p^k q^l n|<21(k+l)^2 \alpha \varepsilon^{-1}$$

for at least $(1-\varepsilon)\binom{n}{k}\binom{n-k}{l}$ choices of sets U_1 and U_2, where $|U_1|=k$, $|U_2|=l$ and $q=1-p$.

Proof. Let $\delta = \varepsilon(k+l)^{-1}$. First choose $u_1^1 \in G$ with degree $pn + B(21\alpha\delta^{-1})$; by Lemma 2.2 there are at least $(1-\delta)n$ choices for u_1^1. Then choose $u_1^2 \in G - U_1$ with $p^2n + 2B(21\alpha\delta^{-1})$ neighbours in common with u_1^1. By Lemma 2.2 there are at least $(1-\delta)(n-1)$ choices for u_1^2. Repeating the procedure in $G - \{u_1^1, u_1^2\}$ shows there are at least $(1-\delta)^k n(n-1)...(n-k+1)$ choices of sequences u_1^1, $u_1^2, ..., u_1^k$ with $p^k n + kB(21\alpha\delta^{-1})$ common neighbours. Let $U_1 = \{u_1^1, ..., u_1^k\}$. Likewise we may choose a sequence $u_2^1, ..., u_2^l$ such that if $U_2 = \{u_2^1, ..., u_2^l\}$ then $v(U_1, U_2) = p^k q^l n + (k+l)B(21\alpha\delta^{-1})$, and this may be done in at least $(1-\delta)^l(n-k)...(n-k-l+1)$ ways. Since U_1 and U_2 may arise from $k!l!$ different choices of sequences, we get at least $(1-\delta)^{k+l}\binom{n}{k}\binom{n-k}{l} \geq (1-\varepsilon)\binom{n}{k}\binom{n-k}{l}$ different choices of U_1 and U_2 such that

$$|v(U_1, U_2) - p^k q^l n| < 21(k+l)^2 \alpha \varepsilon^{-1}. \quad \square$$

Note that even under the conditions of Theorem 1.1 we cannot show $|N(U)| \sim p^k n$ for all k-subsets U, even for $k=3$. In the graph of example (9), for instance, $|N(U)| \sim 2^{-d} n$ where d is the dimension of the subspace spanned by U; d may take any value between $\lceil \log_2 |U| \rceil$ and $|U|$. However, for the Paley graphs (example (6)) Bollobás and Thomason [6] have shown $|N(U)| \sim p^k n$ for all $k \ll \frac{1}{2} \log_2 n$.

Given a set $U \subset V$, let $\Gamma(U)$ denote the set of vertices of $G - U$ joined to at least one vertex of U. Then $\Gamma(U) = G - U - \overline{N(U)}$. A graph is called an *expander* graph if, loosely speaking, $\Gamma(U)$ is as large as possible for every set U. Under our definition of a jumbled graph we can speak usefully of $\Gamma(U)$ for *all* U only if $|U| \gg \alpha$. However, under the conditions of Theorem 1.1 we get information about $\Gamma(U)$ for smaller U, as is discussed in [25].

The diameter

The next few graph properties we shall look at, such as diameter and connectivity, are only interesting when the graph has no isolated vertices or vertices of low degree. For this reason we shall consider (p, α)-jumbled graphs of order n and minimum degree pn. To apply these results to jumbled graphs in general, note by Lemma 2.1 a jumbled graph has most vertex degrees near pn and so contains a large subgraph of large minimum degree, to which the following theorems will apply.

It is easy to see that the diameter of a graph in $\mathcal{G}(n, p)$ is almost surely at most 2 if $p^2n - 2\log n \to \infty$. For a jumbled graph we can do almost as well.

Theorem 2.4. *Let G be a (p, α)-jumbled graph. Let $u, w \in G$ be vertices with degree at least d. If $pd > 4\alpha$ there is a $u-w$ path of length at most 3 in G. In particular, if $\delta(G) > 4\alpha p^{-1}$ then G has diameter at most 3.*

Proof. Choose $U \subset \Gamma(u)$, $W \subset \Gamma(w)$ with $|U| = |W| = d$. If $U \cap W \neq \emptyset$, we are home; otherwise the number of $U-W$ edges is

$$p|U||W| + B(|U|\alpha) + B(|W|\alpha) + B((|U| + |W|)\alpha)$$
$$\geq pd^2 - 4d\alpha > 0,$$

so there is a $u-w$ path of length at most 3. $\square$

Connectivity

For random graphs it is well known that the connectivity is almost surely the same as the minimum degree (see Bollobás and Thomason [8] for a proof that this holds over the entire range $0 \leq p \leq 1$). For jumbled graphs we have the following.

Theorem 2.5. *Let G be a (p, α)-jumbled graph of order n. Then*

$$\kappa(G) > \delta(G) - 4\alpha p^{-1} + 1.$$

Proof. Let S be a vertex cut of G. As in the proof of Theorem 2.4, if $k > 4\alpha p^{-1}$ there is at least one edge between any two subgraphs of order k, so a smallest component of $G - S$ has order at most $4\alpha p^{-1}$. But such a component together with S contains at least $\delta(G) + 1$ vertices. $\square$

Note that the expanding properties referred to after Theorem 2.3 of a jumbled graph G satisfying the conditions of Theorem 1.1 allow us to rule out very small components of $G - S$ and so to show $\kappa(G) = \delta(G)$ for such graphs.

Hamilton cycles

A graph in $\mathcal{G}(n, p)$ is almost surely hamiltonian if $np - \log n - \log \log n \to \infty$. Likewise if p is not too small a jumbled graph is hamiltonian, and indeed has many hamilton cycles. We will find all these cycles by means of the following lemma.

Lemma 2.6. Let G be a graph, and let P be a path in G of length $l \geq 0$. (A path of length 0 is just a vertex.) If G has no independent set of order $\kappa(G) - l + 1$ then G has a hamilton cycle containing P.

Proof. A theorem of Chvátal and Erdös [12] says if G has no independent set of order $\kappa(G)$ then G has a hamilton cycle between any two specified vertices. Applying this to the graph G', obtained from G by removing the middle $l-1$ vertices of P (or removing P if $l=0$) gives the result claimed. □

Theorem 2.7. Let G be a (p, α)-jumbled graph, and let P be a path in G of length $l \geq 0$. If

$$\delta(G) \geq 6\alpha p^{-1} + l,$$

G has a hamilton cycle containing P.

Proof. The largest independent set in G has order at most $2\alpha p^{-1} + 1$, which by Theorem 2.5 is no larger than $\kappa(G) - l$. Apply Lemma 2.6. □

Corollary 2.8. Let G be a (p, α)-jumbled graph of order n with minimum degree at least pn. If

$$(p - k/n)^2 n \geq 6(\alpha + 2k),$$

where k is a nonnegative integer, then G has a set of $k+1$ edge-disjoint hamilton cycles.

Proof. Theorem 2.7 gives the case $k=0$, and shows G has a hamilton cycle. Removing the edges of this gives a graph G' of order n which is $(p - 2/n, \alpha + 2)$-jumbled with minimum degree $pn - 2$. We again apply Theorem 2.7 to find a hamilton cycle, and repeating k times gives the result claimed. □

Calkin [10] asked whether there is an exponentially large number of hamilton cycles in the Paley graph (example (6)). By Corollary 2.8 we can find a set of $n/100$ edge-disjoint cycles and Corollary 2.2 of [23] then assures us of at least $(n/100)^2$ hamilton cycles. But we can do much better.

Corollary 2.9. Let G be a (p, α)-jumbled graph of order n, with minimum degree $pn \geq m = \lceil 6\alpha p^{-1} \rceil$. Then G has at least $\frac{1}{2}(pn)!/m!$ hamilton cycles.

Proof. Choose a vertex x. There are at least $pn(pn-1)\ldots(pn-l+1)$ paths of length l beginning at x. If $l=pn-m$, Theorem 2.7 shows each of these paths is in a hamilton cycle, and each cycle contains at most two such paths. □

Applying this to the Paley graphs gives us almost $(n/2)!$ hamilton cycles.

Induced subgraphs

We now examine small induced subgraphs of a jumbled graph. There are various reasons for this. For instance, Erdös [14] has conjectured the minimum number of monochromatic K_k's in an edge 2-colouring of K_n is $\binom{n}{k} 2^{1-\binom{k}{2}}(1+o(1))$, and it is known that this bound is attained if the colouring is random. Theorem 2.10 shows it is enough for the colouring to be jumbled; this was shown by Giraud [17] in the case $k=4$ for graphs satisfying the conditions of Theorem 1.1. Theorem 2.10 also shows that a $(\frac{1}{2}, \alpha)$-jumbled graph of order n contains an induced copy of *every* graph of order r if $n \geqslant cr^4 2^{2r}$ and α is of order $n^{\frac{1}{2}}$. Such graphs were called *r-full* by Bollobás and Thomason [6], who in answer to a question of Rosenfeld showed that several of our examples of jumbled graphs are r-full.

The restriction to $p \leqslant \frac{1}{2}$ in the following theorem is for convenience only, and for graphs with $p > \frac{1}{2}$ the theorem can be applied to the complement.

Theorem 2.10. *Let F be a graph of order $r \geqslant 3$ with m edges, and let z be the order of its automorphism group. Let G be a (p, α)-jumbled graph of order n, where $p \leqslant \frac{1}{2}$. Suppose ε satisfies $0 < \varepsilon < 1$ and $\varepsilon^2 p^r n \geqslant 42 \alpha r^2$. Then the number of induced subgraphs of G which are isomorphic to F lies between*

$$(1-\varepsilon)^r p^m q^{\binom{r}{2}-m} z^{-1} n^r$$

and

$$(1+\varepsilon)^r p^m q^{\binom{r}{2}-m} z^{-1} n^r,$$

where $q = 1-p$.

Proof. Define $\delta > 0$ by

$$(1+\delta)^{(r+1)/2} = 1+\varepsilon.$$

Then

$$1-\varepsilon < (1-\delta)^{(r+1)/2},$$

and, since $\varepsilon<1$,

$$(1-\delta)^{r-1}>(2-2^{2/(r+1)})^{r-1}>\tfrac{1}{4}$$

holds. Moreover $\delta>\varepsilon/r$. We conclude therefore

$$\delta^2(1-\delta)^{r-1}p^{r-1}n\geqslant \varepsilon^2p'n/4r^2p\geqslant 21\alpha,$$

and it is the outer inequality we will use.

Let $V(F)=\{w_1,\ldots,w_r\}$. We estimate in how many ways we may choose a sequence $x_1,\ldots,x_r$ of vertices of G spanning an induced subgraph isomorphic to F (so x_i corresponds to w_i etc.).

To begin with, we have n choices for x_1, and of these all but at most $21\alpha/\delta p$ are 'normal', that is, have degrees in the range $pn(1+B(\delta))$; this comes from Lemma 2.2 using the value $21\alpha/\delta pn$ for the ε there. Suppose now we have chosen $x_1,\ldots,x_j$, $j<r$, and let X_i^j, $j<i\leqslant r$, be the set of vertices of $G-\{x_1,\ldots,x_j\}$ such that $x\in X_i^j$ is joined to x_k, $k\leqslant j$, if and only if $w_iw_k\in F$. In other words, when we later choose x_i, we will have to choose it from X_i^j. Note that the X_i^j need not be distinct. We now have exactly $|X_{j+1}^j|$ choices for x_{j+1}. Lemma 2.2 once again shows that, for each $i>j+1$, all but at most $21\alpha/\delta p$ choices have $p|X_i^j|(1+B(\delta))$ neighbours in X_i^j; for these choices we will have

$$|X_i^{j+1}|=p|X_i^j|(1+B(\delta))$$

or

$$|X_i^{j+1}|=q|X_i^j|(1+B(\delta))$$

according as $w_{j+1}w_i$ is or is not an edge of F. We call x_{j+1} normal if it has $p|X_i^j|(1+B(\delta))$ neighbours in X_i^j for each i, so there are at most $(r-j-1)21\alpha/\delta p$ abnormal choices for x_{j+1}.

For $2\leqslant i\leqslant r$, let $f(i)$ be the number of neighbours of w_i among $\{w_1, w_2,\ldots,w_{i-1}\}$. If $x_1,\ldots,x_j$ are all normal choices, we have

$$|X_{j+1}^j|\geqslant (1-\delta)^j p^f q^{j-f} n,$$

where $f=f(j+1)$, so there are at least

$$(1-\delta)^j p^f q^{j-f} n - (r-j-1)21\alpha/\delta p$$
$$\geqslant (1-\delta)^{j+1} p^f q^{j-f} n$$

normal choices for x_{j+1}; this inequality holds since (for $j+1<r$)

$$\delta(1-\delta)^j p^f q^{j-f} n \frac{\delta p}{(r-j-1)21\alpha}$$

$$\geq \frac{\delta^2(1-\delta)^{r-1} p^{j+1} n}{21(r-j-1)\alpha}$$

$$\geq \frac{\delta^2(1-\delta)^{r-1} p^{r-1} n}{21\alpha} \geq 1,$$

and if $j+1=r$ all choices are normal. Hence there are at least

$$(1-\delta)^{\binom{r+1}{2}} p^m q^{\binom{r}{2}-m} n^r$$

normal choices for the sequence $x_1, \ldots, x_r$. Likewise we have

$$|X_{j+1}^j| \leq (1+\delta)^j p^f q^{j-f} n,$$

so we obtain at once that there are at most

$$N = (1+\delta)^{\binom{r}{2}} p^m q^{\binom{r}{2}-m} n^r$$

normal choices for $x_1, \ldots, x_r$. It can be seen that each edge $w_j w_i$ of F contributes a factor $(1+\delta)p$ to N, since

$$|X_i^j| \leq (1+\delta) p |X_i^{j-1}|,$$

and likewise each non-edge contributes a factor $(1+\delta)q$. However, if x_j is an abnormal choice we may say only $|X_i^j| \leq |X_i^{j-1}|$, and so the factors $(1+\delta)p$ or $(1+\delta)q$ are lost for any edge or non-edge incident with a vertex chosen abnormally. So the number of choices of $x_1, \ldots, x_r$ where some fixed subsequence $x_{i_1}, \ldots, x_{i_k}$ is chosen abnormally is at most

$$N\left[(1+\delta)^{(r-k)k+\binom{k}{2}} p^t q^{(r-k)k+\binom{k}{2}-t}\right]^{-1} n^{-k} \left(\frac{21\alpha r}{\delta p}\right)^k,$$

where t is the number of edges of F incident with at least one of $w_{i_1}, \ldots, w_{i_k}$.

Hence the number of choices of $x_1, \ldots, x_r$ with k abnormal choices is at most

$$\binom{r}{k}[(1+\delta)p]^{-(r-k)k-\binom{k}{2}}n^{-k}\left(\frac{21\alpha r}{\delta p}\right)^k N$$

$$\leq \binom{r}{k}\left[\frac{21\alpha r}{\delta pn[(1+\delta)p]^{(r-1)/2}}\right]^k N$$

$$\leq \binom{r}{k}\delta^k\left[\frac{21\alpha r^3}{\varepsilon^2 np^{(r+1)/2}}\right]^k N \leq \binom{r}{k}\delta^k N,$$

so the total number of choices for $x_1, \ldots, x_r$ does not exceed $(1+\delta)^r N$.

We have now shown that the number of ways to construct the sequence $x_1, \ldots, x_r$ lies between

$$(1-\delta)^{\binom{r+1}{2}} p^m q^{\binom{r}{2}-m} n^r$$

and

$$(1+\delta)^{\binom{r+1}{2}} p^m q^{\binom{r}{2}-m} n^r.$$

The proof is completed by noting that

$$(1+\delta)^{\binom{r+1}{2}} = (1+\varepsilon)^r,$$

$$(1-\delta)^{\binom{r+1}{2}} \geq (1-\varepsilon)^r,$$

and that each subgraph isomorphic to F corresponds to exactly z sequences $x_1, \ldots, x_r$. □

This theorem shows that $(\frac{1}{2}, \alpha)$-jumbled graphs with α of order $n^{\frac{1}{2}}$ are r-full for r up to about $\frac{1}{2}\log_2 n$. It is natural to ask whether the estimates of Theorem 2.10 could be extended to larger values of r by using a slightly less blunt method. Alas, this is not so. Let us estimate the number of K_r's in the graph of example (11) with $k=2s$ and $r=s+t$. Let W be a subset of the vertex set. Regarding the vertex set as elements of a vector space, let U be the subspace generated by W. Then the set of vertices joined to every element of W is precisely $U^\perp$, the subspace orthogonal to U under the dot product. Now dim $U^\perp = k - \dim U$ and if W spans a complete subgraph then $U = U^\perp$ so dim $U \leq s$. If we construct K_r's by first choosing s independent mutually orthogonal vectors and then choosing t more in the subspace spanned by the first s, we have at least

$$\frac{1}{r!}(2^{k-1}-1)(2^{k-2}-2)(2^{k-3}-4)\ldots(2^{s+1}-2^{s-1})(2^s-s-1)\ldots(2^s-r)$$

$$\sim \frac{\pi}{r!} n^s 2^{-\binom{s}{2}} n^t 2^{-st} = \frac{\pi}{r!} n^r 2^{-\binom{r}{2}+\binom{t}{2}}$$

K_r's, where $\pi = \prod_{i>0}(1-4^{-i})$, which is $2^{\binom{t}{2}}\pi$ times the expected number suggested by Theorem 2.10.

Cliques and the chromatic number

Theorem 2.10 shows that the clique number of a jumbled graph must be at least $\frac{1}{2}\log_b n$ (if α is of order $(pn)^{\frac{1}{2}}$). The graph of example (12) shows the clique number need not exceed $\log_b n$, since a clique corresponds to a set of linearly independent vectors. This contrasts with the case of a random graph, where the clique number is known very precisely (see Matula [20] and Bollobás and Erdös [5]), and is around $2\log_b n$. More striking is the possibility of large cliques. We mentioned that a clique has order at most $2\alpha(1-p)^{-1}+1$. Several of our examples have such large cliques. In the Paley graph (example (6)) if n is a square, the elements in the subfield of order $n^{\frac{1}{2}}$ form a clique.

The graph of example (11), looked at in the previous section, is particularly interesting from the point of view of maximal cliques, since every maximal clique is the same size. Indeed if W is the set of vertices of a clique then W is contained in a subspace of dimension $k/2$ which spans a clique. Thus the greedy algorithm always produces a clique of the maximum order. The Paley graph has fewer large cliques, and many small maximal ones. In the case $n=p^2$ and $p \equiv 3 \pmod{4}$, every edge is contained in a clique of order $n^{\frac{1}{2}}$. Another large clique is given by the vertices z^i, $1 \leq i \leq (p+1)/2$, together with 0; here $z = g^{2(p-1)}$ and g is a primitive root for the prime p. To see that these vertices form a clique, it is enough to show that $1-z^i$ is a square. Putting $z^i = y$, we see

$$(1-y)^{(p^2-1)/2} = (1-y)^{(p-1)(p+1)/2}$$

$$= \left[\frac{(1-y)^p}{(1-y)}\right]^{(p+1)/2}$$

$$= \left[\frac{1-y^p}{1-y}\right]^{(p+1)/2} = \left[\frac{1-y^{-1}}{1-y}\right]^{(p+1)/2}$$

$$=(-y)^{(p+1)/2}=1,$$

the fourth equality holding since $y^{p+1}=1$.

Bounds on the chromatic number follow at once from bounds on the clique (or independence) number. Thus for a $(\frac{1}{2}, n^{\frac{1}{2}})$-jumbled graph of order n, the chromatic number is at most $2n/\log_2 n$, need not be less than $n/\log_2 n$, and must be at least $\frac{1}{4}n^{\frac{1}{2}}$. The greedy algorithm may use as many as $n/2$ colours, as for instance on the graph of example (4) if the vertices are ordered $x_1, y_1, \ldots, x_{n/2}, y_{n/2}$.

Contractions and subdivisions

Given a graph G, we define the *contraction-clique number* $ccl(G)$ to be the largest integer k such that G has a subcontraction to K_k. Bollobás, Catlin and Erdös [4] showed that for $G \in \mathcal{G}(n, p)$, p constant,

$$ccl(G) = n(\log(1-p)/\log n)^{\frac{1}{2}}(1+o(1))$$

almost surely, and so proved that almost all graphs satisfy Hadwiger's conjecture. Thomason [24] proved that graphs of order n with $p\binom{n}{2}$ edges satisfy

$$ccl(G) > pn(\log_2(pn))^{-\frac{1}{2}}/6 \, ;$$

in fact, using the technique of [24] and Lemma 2.3 one can show that, for (p, α)-jumbled graphs with p constant and $\alpha = O(n^{1-\varepsilon})$,

$$ccl(G) \geq n(\log(1-p)/\log n)^{\frac{1}{2}}(1+o(1)).$$

Note that the graph of example (5) shows we may have $ccl(G) \geq pn$.

Finally, we consider the *topological clique number* of G, denoted $tcl(G)$, which is the largest k for which G contains a subdivision of K_k. By estimating $tcl(G)$ for $G \in \mathcal{G}(n, p)$ Erdös and Fajtlowicz [15] showed that almost every graph is a counterexample to Hajós' conjecture. Bollobás and Catlin [3] showed

$$tcl(G) = 2(n/(1-p))^{\frac{1}{2}}(1+o(1))$$

for almost every $G \in \mathcal{G}(n, p)$, p constant. It is interesting that we are able to give good bounds on $tcl(G)$ for a jumbled graph.

Lemma 2.11. *Let G be a (p, α)-jumbled graph of order n. Then*

$$tcl(G) < 2(\alpha + n^{\frac{1}{2}})(1-p)^{-1}.$$

Proof. Let W be the set of branch vertices of a subdivision of K_k contained in G. Then $|W| = k$. There are at most $p\binom{k}{2} + \alpha k$ edges of G in W, so G contains at least $(1-p)\binom{k}{2} - \alpha k$ disjoint paths (of length at least 2) joining the vertices of W. Thus

$$n - k \geqslant (1-p)\binom{k}{2} - \alpha k,$$

or

$$k\left[k - \frac{2\alpha}{1-p} + \frac{1+p}{1-p}\right] - \frac{2n}{1-p} \leqslant 0.$$

But if

$$k \geqslant 2(\alpha + n^{\frac{1}{2}})(1-p)^{-1}$$

this inequality fails to hold. □

The lower bound we give could be sharpened with more work or if the conditions of Theorem 1.1 were assumed. But still the upper and lower bounds we give are of the same order if p is constant and α is of order $n^{1/2}$.

Theorem 2.12. *Let G be a (p, α)-jumbled graph of order n, and suppose $\varepsilon > 0$ satisfies $\varepsilon p^2 n > 40\alpha$. Then*

$$tcl(G) \geqslant \lfloor (1-\varepsilon)(pn)^{\frac{1}{2}} \rfloor.$$

Proof. We may assume $\varepsilon < 1$; let $k = \lfloor (1-\varepsilon)(pn)^{1/2} \rfloor$. By Lemma 2.1 there are $n/3 \geqslant k$ vertices of degree at least $pn - 20\alpha$. Let W be a set of such vertices with $|W| = k$. We will construct one by one a set of $\binom{k}{2}$ paths of length at most 3 joining the vertices of W such that W is thereby the set of branch vertices of a subdivided K_k. To do this, let $u, w \in W$ and let G^* be the graph obtained from G by removing $W - \{u, w\}$ along with the other vertices of the paths constructed so far. Then we have removed at most

$$k - 2 + 2\left[\binom{k}{2} - 1\right] < k^2$$

vertices, so in G^* u and w have degree at least d, where

$$d \geq pn - 20\alpha - k^2 > \varepsilon pn/2 > 4\alpha p^{-1}.$$

By Theorem 2.4 there is a $u-w$ path of length at most 3 in G^*, and this is the path we require. □

References

[1] N. Alon and V. D. Milman, Eigenvalues, expanders and superconcentrators, in Proc. 25th Annual Symp. on Foundations of Computer Science, Florida. pp. 320 - 322.

[2] B. Bollobás, *Random Graphs*, Academic Press, 1985.

[3] B. Bollobás and P. Catlin, Topological cliques of random graphs, J. Combinatorial Theory, Ser. B, 30 (1981) 224–227.

[4] B. Bollobás, P. Catlin and P. Erdös, Hadwiger's conjecture is true for almost every graph, Europ. J. Combinatorics 1 (1980) 195–199.

[5] B. Bollobás and P. Erdös, Cliques in random graphs, Math. Proc. Cambridge Phil. Soc. 80 (1976) 419–427.

[6] B. Bollobás and A. Thomason, Graphs which contain all small graphs, Europ. J. Combinatorics 2 (1981) 13–15.

[7] B. Bollobás and A. Thomason, Parallel sorting, Discrete Appl. Math. 6 (1983) 1–11.

[8] B. Bollobás and A. Thomason, Random graphs of small order, Annals of Discrete Math. 28 (1985) 47–97.

[9] B. Bollobás and W. F. de la Vega, The diameter of random regular graphs, Combinatorica 2 (1982) 125–134.

[10] N. Calkin, personal communication.

[11] F. R. K. Chung, On concentrators, superconcentrators, generalisers and non-blocking networks, Bell System Tech. J. 58 (1978) 1765–1777.

[12] V. Chvátal and P. Erdös, A note on hamiltonian circuits, Discrete Math. 2 (1972) 111–113.

[13] P. Erdös, Some remarks on the theory of graphs, Bull. Amer. Math. Soc. 53 (1947) 292–294.

[14] P. Erdös, On the number of complete subgraphs contained in certain graphs, Publ. Math. Inst. Hungar. Acad. Sci. VII, Ser. A, 3 (1962) 459–464.

[15] P. Erdös and S. Fajtlowicz, On the conjecture of Hajós, Combinatorica 1 (1981) 141–143.

[16] P. Erdös and J. Spencer, Imbalances in k-colorations, Networks 1 (1972) 379–385.

[17] G. Giraud, Une minoration du nombre de quadrangles unicolores et son application à la majoration des nombres Ramsey binaires-bicolores, C. R. Acad. Sci. Paris, Sér. A, 276 (1973) 1173–1175.

[18] Y. Gurevich and S. Shelah, Expected computation time for hamiltonian path problem and clique problem (to appear).

[19] P. Kövári, V. T. Sós and P. Turán, On a problem of K. Zarankiewicz, Colloq. Math. 3 (1954) 50–57.

[20] D. Matula, On the complete subgraphs of a random graph, *Combinatory Mathematics and its Applications* (Chapel Hill, N.C. 1970) 356–369.

[21] R. G. E. Pinch, Sequences well distributed in the square, Math. Proc. Cambridge Phil. Soc. 99 (1986) 19–22.

[22] J. J. Seidel, A survey of two-graphs, Colloquio Internazionale sulle Teorie Combinatorie, Atti dei Convegni Lincei 17, Accad. Naz. Lincei, Roma I (1976) 481–511.

[23] A. Thomason, Hamiltonian cycles and uniquely edge colourable graphs, Annals of Discrete Math. 3 (1978) 259–268.

[24] A. Thomason, An extremal function for contractions of graphs, Math. Proc. Cambridge Phil. Soc. 95 (1984) 261–265.

[25] A. Thomason, Dense expanders (to appear).

[26] A. Thomason, A simple linear expected time hamiltonian path algorithm (to appear).

ON THE INDEPENDENCE NUMBER OF RANDOM SUBGRAPHS OF THE n-CUBE

Karl WEBER

Ingenieurhochschule für Seefahrt,
2530 Warnemünde, German Democratic Republic

We show that for $pn \to \infty$ the independence number of random induced (spanning) subgraphs of the n-cube is $p2^{n-1}(1+o(1))$ and $2^{n-1}(1+o(1))$, respectively, with probability tending to 1 as $n \to \infty$.

The n-cube Q_n is the graph consisting of the 2^n vertices $(a_1, a_2, \ldots, a_n)$, $a_i \in \{0, 1\}$, and the $n2^{n-1}$ edges between vertices differing in exactly one coordinate. A spanning subgraph g of Q_n has the same vertex set as Q_n. An induced subgraph f of Q_n with the vertex set $A \subseteq Q_n$ contains exactly those edges of Q_n that join two vertices in A. (Note that by Q_n or f are not only denoted the graphs but also their vertex sets, g stands for the edge set of g too.) Choosing the edges of g (the vertices of f) at random, independently of each other with the same probability p, we arrive at a random spanning (induced) subgraph whose probabilities are defined as $\text{Prob}(g) = p^{|g|} q^{n2^{n-1}-|g|}$ and $\text{Prob}(f) = p^{|f|} q^{2^n-|f|}$, respectively, where $q = 1-p$. We say g (or f) has a given property almost surely (a.s.) if the probability that g (or f) has this property tends to 1 as $n \to \infty$.

For example, for fixed $p > 1/2$ ($p < 1/2$) f is a.s. (not) connected (cf. [4]). If $p = 1/2$ then f has a.s. only isolated vertices outside the largest component. More precisely, the number of isolated vertices is asymptotically Poisson distributed with mean $\lambda = 1/2$, that is particularly $\text{Prob}(f \text{ is connected}) \sim \text{Prob}(f \text{ has no isolated vertices}) \sim 1/\sqrt{e}$ ([3, 7]).

Analogous results can be derived for random spanning subgraphs. One has to replace only $\lambda = 1/2$ by $\lambda = 1$ and so $1/\sqrt{e}$ by $1/e$ ([1, 2, 5, 8]).[1]

Let (X_n), $n = 1, 2, \ldots$, be a sequence of non-negative integer-valued random variables. Then by Markov's inequality

$$\text{Prob}(X_n < \varphi_n EX_n) \to 1 \tag{1}$$

---

[1] Recently M. E. Dyer, A. M. Frieze and L. R. Foulds combined these two models of random subgraphs of the n-cube into one (see pp. 17–40).

as $n \to \infty$, where EX_n is the expectation of X_n and $\varphi_n \to \infty$ arbitrarily slowly as $n \to \infty$. In particular, (1) implies that

$$\text{Prob}(X_n = 0) \to 1 \text{ if } EX_n \to 0. \tag{2}$$

(All limits, asymptotics, etc., are understood as $n \to \infty$.)

If there are sequences $\varepsilon_n \to 0$ and α_n such that $\text{Prob}(|X_n - \alpha_n| < \varepsilon_n \alpha_n) \to 1$ we will say "X_n is a.s. asymptotically α_n" and use the short notation "$X_n \sim \alpha_n$ a.s.".

As usual, a set of pairwise non-adjacent vertices of a graph G is called an independent (vertex) set. The maximum cardinality of an independent set of G is said to be the independence number $\beta_0(G)$. Analogously the edge independence number $\beta_1(G)$ is the maximum number of pairwise non-adjacent edges of G. (An independent set of edges is also called a matching.) Recall that by a well-known theorem of König,

$$\beta_0(G) + \beta_1(G) = |V| \tag{3}$$

for every bipartite graph $G = (V, E)$.

Theorem 1. *If $pn \to \infty$ then*

$$\beta_0(f) \sim p2^{n-1} \text{ a.s.}$$

Proof. The number $|f|$ of vertices of f is binomially distributed with parameters p and 2^n so that $|f| \sim p2^n$ and consequently $\beta_1(f) \leq p2^{n-1}(1 + o(1))$ a.s. We shall show that $\beta_1(f) \sim p2^{n-1}$ a.s. if $pn \to \infty$. Then the theorem follows by (3).

In order to construct a matching with asymptotically $p2^{n-1}$ edges the following algorithm of Saposhenko (see also [6]) is used: Let $x_1, x_2, \ldots, x_n$ denote the coordinates of the n-cube. At the first step we take all edges of f joining vertices which differ in (exactly) the first coordinate x_1 (shortly called "edges in direction x_1"), then we take all edges of f in direction x_2 without common vertices with already taken edges and so on up to the direction x_n. Define the conditional probabilities

$$\delta_i(a) = \text{Prob}(a \in Q_n \text{ is not incident with any chosen edge}$$
$$\text{after step } i \mid a \in f),$$
$$i = 1, 2, \ldots, n.$$

Then the expectation for the number Z_n of vertices of f being not incident with any chosen edge after all n steps of the described algorithm is

$$EZ_n = 2^n \text{ Prob}(\boldsymbol{a} \in f \text{ and } \boldsymbol{a} \text{ is not incident with any chosen edge})$$
$$= \delta_n p 2^n$$

(obviously $\delta_i(\boldsymbol{a})$ is independent of $\boldsymbol{a}$ so that we can omit the argument). Thus by (1) Z_n is a.s. not greater than $\varphi_n \delta_n p 2^n$, where $\varphi_n \to \infty$ arbitrarily slowly.

Next we derive an upper bound for δ_n. The recursion

$$\delta_i = \delta_{i-1}(1 - p + p(1 - \delta_{i-1}))$$
$$= \delta_{i-1} - p\delta_{i-1}^2, \quad 1 \leq i \leq n-1; \quad \delta_0 = 1,$$

is obvious since the events that $\boldsymbol{a}$ and $\boldsymbol{b}$ are not incident with any chosen edge after step $i-1$ are independent, provided $\boldsymbol{a}$ and $\boldsymbol{b}$ differ only at the ith position. Considering that $(\delta_i - \delta_{i-1})/p\delta_{i-1}^2 = -1$ we have

$$n = -\sum_{i=1}^{n} \frac{\delta_i - \delta_{i-1}}{p\delta_{i-1}^2}$$

$$\leq \int_{\delta_n}^{\delta_0} \frac{dx}{px^2} = -\frac{1}{p} + \frac{1}{p\delta_n}$$

and so $\delta_n \leq \dfrac{1}{pn+1}$. Hence assuming $pn \to \infty$ we may choose $\varphi_n \to \infty$ in such a way that $\varphi_n \delta_n \to 0$, i.e. $Z_n = o(p2^n)$. Thus after application of all n steps of the above algorithm a.s. there remain $o(p2^n)$ vertices of f only which are not incident with any chosen edge. Consequently we constructed a matching with $p2^{n-1}(1 - o(1))$ edges. □

It does hold an analogous result for random spanning subgraphs but the proof is quite different from the preceding one.

Theorem 2. *If $pn \to \infty$ then*

$$\beta_0(g) \sim 2^{n-1} \text{ a.s.}$$

Proof. The lower bound $\beta_0(g) \geq 2^{n-1}$ is evident for all g. Let us show that $\beta_0(g) \leq 2^{n-1}(1 + o(1))$ a.s. To do that put $m = 2^{n-1}(1 + \varepsilon)$, $\varepsilon > 0$, and let $Y_m(g)$

denote the number of independent sets of g with cardinality m. Then for the expectation we have $EY_m = \sum q^{e(A)}$, where the sum is taken over all vertex sets $A \subseteq Q_n$ with $|A| = m$ and $e(A)$ denotes the number of edges of Q_n joining vertices within A. Elementary calculations show that assuming $|A| \geq 2^{n-1}$ we have $e(A) \geq n(|A| - 2^{n-1})$ and thus

$$\log_2 EY_m \leq 2^n - \varepsilon n 2^{n-1} \log_2(1/q).$$

Hence for a fixed probability p we obtain $\log_2 EY_m \to -\infty$ already for $\varepsilon = c/n$, where c is a sufficiently large constant. If $p \to 0$ then putting $p = \varphi/n$ and considering that $\log_2(1/q) \geq p \log_2 e$ we obtain $\log_2 EY_m \to -\infty$ for $1/\varphi = o(\varepsilon)$. That means, for every sequence $\varphi = \varphi_n \to \infty$ we find an $\varepsilon = \varepsilon_n \to 0$ such that $EY_m \to 0$ for $m = 2^{n-1} \times (1 + \varepsilon)$, i.e. by (2) there is a.s. no independent set of this cardinality. □

Conjecture. If $p > 1/2$ fixed then a.s.

$$\beta_0(g) = 2^{n-1}$$

and

$$\beta_0(f) = |f|/2$$

provided that $|f|$ is even.

Recall that by (3) the equalities for the independence number given in the conjecture are equivalent to the existence of a 1-factor. The conjecture is sharp in the sense that for $p = 1/2$ there exist isolated vertices with positive probability (cf. the results on connectedness mentioned above).

Finally note that for p with $pn \to 0$ asymptotically all vertices both of f and g are isolated so that for these probabilities $\beta_0(f) \sim p2^n$ and $\beta_0(g) \sim 2^n$ a.s. We do not know β_0 if p has the order of magnitude $1/n$.

Open problem. Determine β_0 for $p = c/n$.

References

[1] Yu. D. Burtin. On the probability of connectedness of random subgraphs of the n-cube. Problemy Pered. Inform. 13 (1977) 2, 90–95

[2] P. Erdös, J. Spencer, Evolution of the n-cube. Comput. Math. Appl. 5 (1979) 33–39

[3] A. A. Saposhenko. Geometric structure of almost all Boolean functions. Problemy Kibernet. 30 (1975) 227–261 (in Russian)

[4] E. Toman. Geometric structure of random Boolean functions. Problemy Kibernet. 35 (1979) 111–132 (in Russian)

[5] —. On the probability of connectedness of random subgraphs of the n-cube. Math. Slovaca 30 (1980) 3, 251–265 (in Russian)

[6] K. Weber. The length of random Boolean functions. Elektron. Informationsverarb. u. Kybernet. EIK 18 (1982) 12, 659–668

[7] —. Subcubes of random Boolean functions. EIK 19 (1983) 7/8, 365–374

[8] —. Subcube coverings of random spanning subgraphs of the n-cube. Math. Nachr. 120 (1985) 327–345

DIRECTED SITE PERCOLATION AND DUAL FILLING MODELS*

John C. WIERMAN

Department of Mathematical Sciences, The Johns Hopkins University, Baltimore, Maryland 21218, U.S.A.

1. Introduction

Since the origins of mathematical percolation theory in the 1950's dual percolation models have played an important role in the theory of two-dimensional undirected percolation models. For a bond percolation model on a planar graph G, the dual percolation model is defined on G^*, the Whitney dual graph of G. Sykes and Essam identified a class of (non-planar) two-dimensional graphs for which the site percolation model has a dual percolation model defined on a matching graph. In each case, circuits in one model correspond to barriers which prevent infinite clusters in its dual model, and this property leads to the conclusion that the critical probabilities of the two models sum to one. For more information on undirected models, see the monograph by Kesten (1982).

A duality theory for directed percolation has just recently begun to develop. Dhar, Barma, and Phani (1981) introduced the "diode-resistor" model. The dual model is not a standard percolation model, but does exhibit a threshold or critical probability. The fact that the critical probabilities of the two models sum to one was used to improve the lower bound for the square lattice directed bond percolation critical probability.

Wierman (1985) formulated Dhar, Barma, and Phani's concept of duality in the setting of directed site percolation models on planar graphs, referring to the dual as a "filling" model. The formulation applies to directed planar bond models

---

* AMS 1980 Subject Classification: 60K35

Key Words and Phrases: *directed percolation, critical probability, filling model, dual graph, matching graph.*

Research supported by the National Science Foundation under Grant No. MCS-8303238.

also, since any directed planar bond model may be transformed into an equivalent directed planar site model. Restrictions on the directedness of the edges in the percolation model are needed to establish the existence of a dual filling model. Also, two different dual filling models may exist for a given directed percolation model.

This paper extends the results of Wierman (1985) to a broader class of directed percolation models, including models on certain non-planar graphs and models with some undirected (two-way) edges, such as the "stiff" model discussed in Smythe and Wierman (1978), p. 167.

A description of the directed percolation models is provided in section 2, with restrictions on the class of graphs which allow the definition of a filling model. The formulation of the filling model and the statement of the principal result are in section 3. The proof itself is given in section 4.

2. Directed percolation models

Directed graph

A directed graph D is an ordered pair $(V(D), E(D))$ such that $E(D)$ is a subset of ordered pairs of $V(D)$. Elements of $V(D)$ and $E(D)$ are called *vertices* and *edges* respectively. The ordered pair $(a, b) \in E(D)$ is called the *edge directed from a to b*. For $a, b \in V(D)$, both (a, b) and (b, a) may be elements of $E(D)$, in which case we say the edge between a and b is undirected. The edge (a, b) is called an *exit* from a and an *entrance* to b. For a directed graph D, we let $G \equiv G(D)$ denote the undirected graph with $V(G) = V(D)$ and $E(G) = \{\{a, b\} : (a, b) \in E(D) \text{ or } (b, a) \in E(D)\}$.

A *path* in the directed graph D is an alternating sequence $v_0, e_1, v_1, e_2, \ldots, e_n, v_n$ such that $e_i = (v_{i-1}, v_i) \in E(D)$ for all $i = 1, 2, \ldots, n$. A *doubly-infinite* path in D is doubly-infinite alternating sequence $\ldots, v_{-1}, e_0, v_0, e_1, v_1, \ldots$ such that every finite segment of the sequence is a path in D.

Periodic graph

A directed graph D imbedded in $\mathbb{R}^d$ is *translation invariant* if its edge and vertex sets are invariant under translation by each of a set of d linearly independent vectors $\{\xi_1, \ldots, \xi_d\}$ in $\mathbb{R}^d$. For a translation invariant directed graph D, $V(D)$ may be partitioned into disjoint sets $V_i = V_i(D)$ such that for each i, $v \in V_i$ if and only if $v + \sum_{j=1}^{d} k_j \xi_j \in v_i \; \forall \; k_j \in \mathbb{Z}, 1 \leq j \leq d$. Each V_i is called a periodic class of vertices.

We say D is *periodically-connected* if for every pair of periodic classes X and Y, there exist $x, x' \in X$, and $y, y' \in Y$ such that there exists a directed path from x to y and a directed path from y' to x'.

A directed graph D is *periodic* in $\mathbf{R}^d$ if the following hold:

(i) D is translation-invariant.

(ii) D is periodically-connected.

(iii) Every vertex D is an endpoint of at most z edges ($z < \infty$).

(iv) Each compact subset of $\mathbf{R}^d$ contains finitely many vertices of D.

A periodic directed graph has finitely many periodic classes of vertices.

Directed percolation model

For a directed graph D and $0 \leq p \leq 1$, let $\{X_v : v \in V(D)\}$ be a set of independent random variables with $P(X_v=1) = p = 1 - P(X_v=0)$. If $X_v=0$ ($X_v=1$), the vertex v is called *open* (*closed*, respectively). An edge is open if and only if both its endpoint vertices are open. One may consider the graph D to represent a medium, in which fluid is permitted to flow through the network of open vertices and edges. The terms fluid and medium may be broadly interpreted, to include phenomena such as petroleum in porous rock, electrons on an atomic lattice, or disease in a population. The fluid may flow through a directed edge only in the specified direction, perhaps due to the influence of gravity or a magnetic field or time.

An important characteristic of a percolation model on an infinite graph is the existence of a threshold value of p, the critical probability, above which the fluid penetrates the medium. In applications, the critical probability is often interpreted as a phase transition. There are several interpretations of penetration of the medium, leading to distinct critical probability definitions. However, the values agree for a broad class of undirected graphs.

For each vertex $v \in V(D)$, let $P(v, p)$ denote the probability that there exists an infinite open directed path in D beginning at v. $P(v, \cdot)$ is called the *percolation probability function* at v. The percolation probability function is identical for all vertices in a periodic class V_i, but may differ for vertices in distinct periodic classes. If D is a translation-invariant periodically-connected directed graph, then for any $v, w \in V(D)$, $P(v, p) > 0$ if and only if $P(w, p) > 0$. Thus, we may define the critical probability for the directed percolation model on D by

$$P_{DP}(D) = \inf\{p \in [0, 1] : P(v, p) > 0\},$$

which is independent of the choice of site v. (This version of the critical probability definition corresponds to that known as p_H for undirected percolation models.)

The exact critical probability is not known for any non-trivial directed percolation model. Much study has focused on the square lattice bond model. Successive improvements in lower bounds have been obtained by Hammersley (1957) by enumeration of self-avoiding paths and circuits, Mauldon (1961) by Markov chain approximations, Holley and Liggett (1975) by consideration of an equivalent interacting particle system, and Gray, Smythe, and Wierman (1980) using random walk barriers. The best lower bound is .6298, obtained from the dual diode-resistor model of Dhar, Barma, and Phani (1981). This value may be compared to numerical estimates clustered between .632 and .645 due to Kertesz and Vicsek (1980), Kinzel and Yeomans (1981), Dhar and Barma (1981), and Blease (1977).

3. Filling model

Assumptions

For directed site percolation models on planar graphs, there may exist distinct "right-dual" and "left-dual" filling models. This is also the case for the broader class of graphs discussed in this paper. The theorem will state conditions which insure the existence of a "left-dual" filling model, with the understanding that the right-dual is obtained (if it exists) as the left-dual of the directed percolation model obtained by reversing the direction of every bond.

The most substantial extension of the Wierman (1985) result is to a class of nonplanar graphs. (Employing the nonplanar matching graphs, introduced by Sykes and Essam (1964), a theory of dual models has previously been developed for undirected site percolation models.) A *mosaic* M is a planar graph imbedded in R^2 such that any two edges of M are either disjoint or have only one endpoint in common, and such that each component of $R^2 \setminus M$ is bounded by a Jordan curve consisting of a finite number of edges of M. To close-pack a face F of a mosaic M, we insert an edge between each pair of vertices on the perimeter of F which are not yet adjacent. If M is a mosaic and $\mathscr{F}$ is a subset of faces of M, the *matchable graph based* on $(M, \mathscr{F})$ is obtained from M by close-packing all faces in $\mathscr{F}$. A directed graph D for which $G(D)$ is a matchable graph is called a *directed matchable graph*. We will consider percolation models on periodic directed matchable graphs, imposing conditions on the directions of edges to insure the existence of a left-dual filling model.

To describe other conditions on the directedness of the graph D, we need to describe an imbedding of the diagonal edges into the plane also. Consider a close-packed face F of $G(D)$. Each of the edges $\{v_1, v_2\}$ which are not in $E(M)$ may be represented by a piecewise linear Jordan curve between v_1 and v_2 which lies entirely in the interior of F except for the endpoints v_1 and v_2. Such curves may be constructed so that:

(a) At most 2 edges pass through each interior point of F.

(b) If $v_1, v_2 \in V(D)$ on the perimeter of F, dividing the perimeter of F into two segments, then the edges $\{v_1, v_2\}$ and $\{w_1, w_2\}$ intersect at exactly one interior point of F if w_1 and w_2 are in different segments of the perimeter of F, and do not intersect if w_1 and w_2 are in the same segment.

An additional extension allows directed graphs which contain some undirected edges. A weaker condition on the directedness at each vertex is required than in Wierman (1985). For each vertex $v \in D$, we define $O(v)$ as the minimal sector containing all exits from v and containing no entrances to v in its interior, and define $I(v)$ as the minimal sector containing all entrances to v and containing no exits from v in its interior. These sectors $O(v)$ and $I(v)$ are uniquely determined unless only two edges are incident to v, and these are both undirected and lie on a common straight line. We assume that for each vertex $v \in D$ the interiors of $O(v)$ and $I(v)$ are disjoint. Thus, we allow up to two undirected edges incident at a vertex, which must then necessarily lie on the boundaries of $I(v)$ and $O(v)$. This permits consideration of the "stiff" model discussed in Smythe and Wierman (1978), p. 167–168, for example.

The sectors $O(v)$ and $I(v)$ may be described as

$$O(v) = \{x \in \mathbf{R}^2 : x = v + (r\cos\theta, r\sin\theta), r \geq 0, \theta_1 \leq \theta \leq \theta_2\}$$

$$I(v) = \{x \in \mathbf{R}^2 : x = v + (r\cos\theta, r\sin\theta), r \geq 0, \gamma_1 \leq \theta \leq \gamma_2\}$$

where $0 \leq \theta_1 < 2\pi$, $\theta_2 \leq \gamma_1$, and $\gamma_2 \leq \theta_1 + 2\pi$. For each $v \in D$, let $e_O(v)$ denote the edge incident to v which in part lies on the boundary ray $v + (r\cos\theta_2, r\sin\theta_2)$ of $O(v)$ near v, and let $e_I(v)$ denote the edge incident to v which in part lies on the boundary ray $v + (r\cos\gamma_1, r\sin\gamma_1)$ of $I(v)$ near v. Unless $e_I(v) = e_O(v)$ is an edge of the underlying mosaic M, the edges $e_I(v)$ and $e_O(v)$ are diagonals or boundary edges of a common face, which we denote by $F(v)$ for discussion of the filling process later. An additional restriction on the directedness, within each close-packed face, is required which has no counterpart in the Wierman (1985) result. To prevent the filling process from crossing an open directed path in the percolation model, we assume that no edge $e_I(v)$ is crossed from left to right by a diagonal edge, and no $e_O(v)$ is crossed from right to left.

A path or doubly-infinite path is a *self-avoiding path* if all its vertices are distinct. A self-avoiding path in a graph imbedded in $\mathbf{R}^2$ is a *simple* self-avoiding path if no pair of its edges intersect at an interior point of an edge. A doubly-infinite directed simple self-avoiding path r divides the plane into two connected components of which one, denoted $C(r)$, lies on the left side of the directed path. If $C(r)$

is contained in a half-plane, then r is called a *counterclockwise* path. As in Wierman (1985), it is assumed that each edge of D is an edge of a counterclockwise simple self-avoiding directed path.

A *self-avoiding circuit* is a sequence $v_0, e_1, v_1, \ldots, e_n, v_n = v_0$ such that all vertices except the endpoints v_0 and v_n are distinct. If no pair of edges intersect at an interior point of an edge, it will be called a *simple self-avoiding circuit*. After this adaptation to the non-planarity of D, we require as in Wierman (1985), that there exist no counterclockwise simple self-avoiding circuits in D.

Formulation

For a directed matchable graph D based on a mosaic M, we let M^* denote the graph obtained by close-packing every face of M. The filling model for D will be defined on M^*.

For a directed matchable graph D, and $0 \leq p \leq 1$, let $\{X_v : v \in V(D)\}$ be independent random variables with $p = P(X_v = 0) = 1 - P(X_v = 1)$. If $X_v = 0$ ($X_v = 1$), the vertex v is called restricted (unrestricted, respectively).

To describe the filling process from a restricted vertex, for each vertex v define the sector

$$R(v) = \{x \in \mathbf{R}^2 : x = v + (r\cos\theta, r\sin\theta), r \geq 0, \theta_2 \leq \theta \leq \gamma_1\}.$$

$R(v)$ may consist of a single ray along an edge of M^* near v, in which case the corresponding edge of D is undirected. If $R(v)$ is a ray, then the edge is an exit from v in the filling model. If $R(v)$ is not a ray and $F(v)$ is not close-packed, then $\{v, w\}$ is an exit from v in the filling model, for every w on the perimeter of $F(v)$. If $R(v)$ is not a ray and $F(v)$ is close-packed, then $e_I(v)$ and $e_O(v)$ are exits from v in the filling model.

If v is an unrestricted vertex, then every edge of M incident to v is an exit from v in the filling model, and every edge that is an exit if v were restricted is also an exit when v is unrestricted.

Note that it is possible for an edge $\{v, w\} \in E(M^*)$ to be both an exit from v and an exit from w in the filling model. In this case, $\{v, w\}$ is called undirected.

A *filling path* is an alternating sequence of vertices and edges $v_0, e_1, v_1, \ldots, e_n, v_n$, where e_i is an exit from v_{i-1} and an entrance to v_i for each $i = 1, 2, \ldots, n$. An *automatic filling path* is a filling path in which e_i is an exit from v_{i-1} which lies in $R(v_{i-1})$ near v_{i-1} for each $i = 1, 2, \ldots, n$, in which case e_i is an exit from v_{i-1} whether v_{i-1} is restricted or not.

The *wedge filled from* v, denoted $W(v)$, is the set of all vertices w such that there is a filling path from v to w. The *wedge automatically filled from* v, denoted $A(v)$, is the set of all vertices w such that there is an automatic filling path from v to w.

If sufficiently many vertices are unrestricted vertices in the filling model, $W(v)$ will contain a complete periodic class of vertices. The critical probability p_F of the filling model is defined by

$$p_F = \inf\{p \in [0, 1] : P_p[W(v) \supseteq V_i \text{ for some } i] > 0\},$$

where P_p denotes the probability measure on configurations under which each vertex is an unrestricted vertex with probability p independently. If no periodic class of vertices is completely filled, we say that there is *incomplete filling*.

It is useful to visualize the filling process in terms of regions of the plane rather than vertices in the graph M^*, using the imbedding of the diagonals of close-packed faces discussed previously. If the vertex v is restricted and $F(v)$ is not close-packed, the entire face $F(v)$ is filled from v. If the vertex v is restricted and $F(v)$ is close-packed, the region bounded by $e_I(v)$, $e_O(v)$ and the edge between their endpoints is filled from v. If v is unrestricted and a face F incident to v is not close-packed, all of F is filled from v. If v is unrestricted and a face F incident to v is close-packed, then the region bounded by the edges between v and the two vertices of F adjacent to v is filled from v. In each case, when a region is filled, all the edges and vertices bordering the region are filled.

A directed percolation model on D induces a filling model on the corresponding M^*. If v is open in the directed percolation model, it is a restricted site in the filling model. If v is closed in the directed percolation model, it is an unrestricted site in the filling model.

Statement

For a directed percolation model on D, by a left-dual filling model, we mean a filling model such that the filling process is incomplete if and only if there is a doubly-infinite simple self-avoiding open directed path in D, such that the filled region lies entirely on the left of the path. If this is the case, for any probability p, either the critical probability $p_F > p$ or the critical probability $p_{DP} \geq 1 - p$, so the critical probabilities of the models sum to one: $p_F + p_{DP} = 1$.

The principal result may now be stated.

Theorem: *Let D be a periodic directed matchable graph. A left-dual filling model exists for the directed site percolation model on D if the following hold:*

(a) *For each $v \in V(D)$, the sectors $O(v)$ and $I(v)$ are uniquely determined and their interiors are disjoint.*

(b) *D contains no counterclockwise simple self-avoiding circuits.*

(c) *Each edge of D is in a doubly-infinite counterclockwise simple self-avoiding directed path.*

(d) *No edge $e_I(v)$ is crossed from left to right, and no edge $e_0(v)$ is crossed from right to left.*

Condition (a) was used to define a filling model previously in this section. Condition (b) insures that the filled region is always unbounded, and thus the

Figure 1.

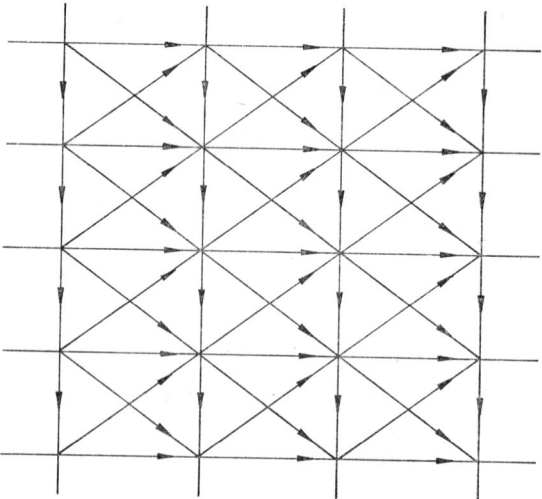

(a) The stiff percolation model on the square lattice.

(b) A directed percolation model on the square lattice mosaic with all faces close-packed.

filling model is a candidate for the role of dual model. Condition (c) insures that the boundary of the filled region is a directed path in the percolation model on D. Condition (d) prevents an open doubly-infinite directed path in D from being crossed from left to right by the filling process.

Examples of directed graphs which satisfy the hypotheses of the Theorem, but not the result of Wierman (1985), are illustrated in Figure 1 (see p. 346).

Figure 2. Two configurations of the filling model corresponding to the directed percolation model in Figure 1(b), showing the filled wedge in each case. Solid circles represent restricted vertices, and open circles represent unrestricted vertices. The source vertex is denoted by s.

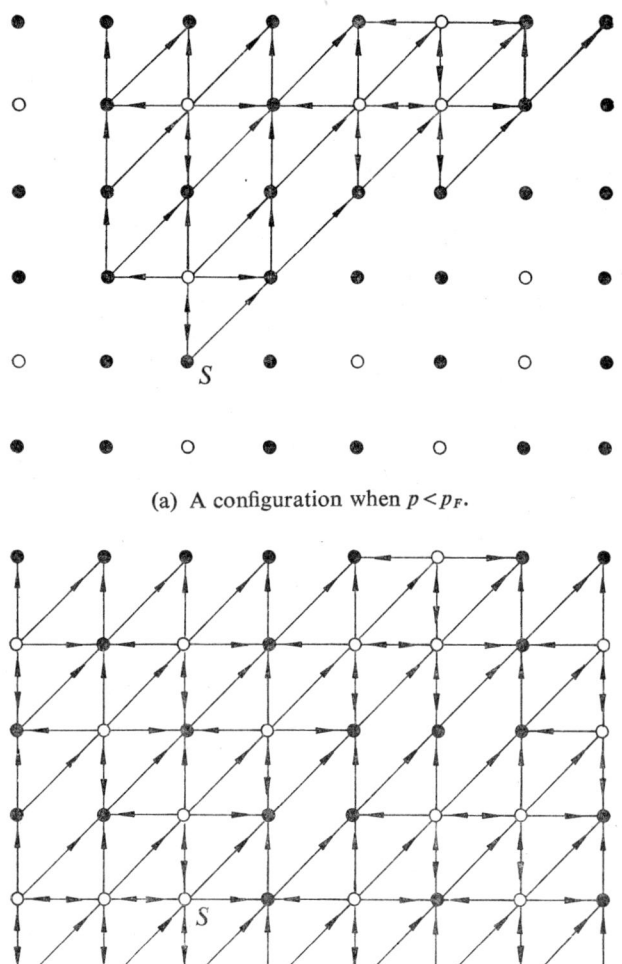

(a) A configuration when $p < p_F$.

(b) A configuration when $p > p_F$.

4. Proof of the Theorem

The simplest part of the proof is to show that a doubly-infinite open path in the directed percolation model prevents complete filling in the filling model.

Suppose that p is a doubly-infinite self-avoiding open directed path in the directed percolation model. (If there is a doubly-infinite directed path, then, by eliminating circuits from it, a doubly-infinite self-avoiding path is constructed with a subset of its vertices and edges.) However, a self-avoiding path may not correspond to a simple self-avoiding path in the plane, since it may contain two diagonal edges which cross in a face of the graph M. The path then contains a closed curve in the plane. Suppose the curve is determined by the vertices v_0, $v_1, \ldots, v_n$, where the edges (v_0, v_1) and (v_{n-1}, v_n) cross. If the closed curve is directed clockwise, then $\{v_0, v_n\}$ must be directed from v_0 to v_n, or else the edge (v_0, v_1) crosses $e_I(v_n)$ from left to right. The closed curve cannot be directed counterclockwise, since if $\{v_1, v_{n-1}\}$ is directed from v_1 to v_{n-1} a counterclockwise circuit is formed, but if it is directed from v_{n-1} to v_1 then $e_I(v_1)$ is crossed from left to right by (v_{n-1}, v_n). Thus, for each self-avoiding path, there is a corresponding path which is a self-avoiding curve in the plane. Furthermore, if the original path p is open in the directed percolation model, the corresponding self-avoiding curve p' is also open, since it replaces a loop with an edge between two open vertices.

Note that p' divides the plane into two connected components, $L(p')$ and $R(p')$, where $L(p')$ lies on the left side of p' and $R(p')$ lies on the right side of p' as p' is travelled in its proper direction. Each vertex $v \in p'$ has an entrance in p' and an exit in p'. Therefore, $e_I(v)$ and $F(v)$ are all contained in $L(p')$.

Suppose that the source vertex v_0 for the filling process is in $L(p')$. If a vertex $z \in R(p')$ were to be filled, there must exist a filling path r from v_0 to z. Either there exists a vertex y which lies on both p' and r and fills a vertex of $R(p')$, or there exists a diagonal edge of p' which is crossed by a diagonal edge of r. The first alternative is impossible, since y is open in the directed percolation model and thus fills only vertices in $L(p')$ in the filling model. The second alternative implies that there exists a vertex $w \in L(p')$ such that either $e_I(w)$ or $e_O(w)$ crosses a diagonal edge of p'. Hence, either $e_I(w)$ is crossed from left to right, or $e_O(w)$ is crossed from right to left, contradicting condition (d). Therefore, the filled wedge $W(v_0)$ is contained in $L(p') \cup p'$.

The more involved argument is to show that incomplete filling implies the existence of a doubly-infinite open path in the directed percolation model on D. The doubly-infinite path will correspond to the boundary of the filled wedge, so the first step shows that this boundary is unbounded.

In the filling model, the *exit boundary* at v is the path $v = v_0, e_O(v_0), v_1, e_O(v_1), \ldots$, and the *entrance boundary* at v is the path $v = v_0, e_I(v_0), v_1, e_I(v_1), \ldots$. The union of the entrance and exit boundaries bounds the automatically filled wedge $A(v)$. Note that in D, the exit and entrance boundaries at v are directed paths, directed from v and toward v respectively.

If the directed model contains undirected edges, initial segments of the entrance and exit boundaries at a vertex v coincide when the initial segment consists of undirected edges. If the entrance and exit boundaries at v coincide completely, they and $A(v)$ are necessarily infinite. (If the number of distinct vertices were finite, then eventually the vertex sequence alternates between two vertices which are connected by an undirected edge. Using the assumption that the other edges incident to these vertices are each in a doubly-infinite counterclockwise self-avoiding path, one finds that either condition (b) or condition (d) is violated.) Once the entrance and exit boundaries differ, they cannot intersect at a vertex (or portions of each combine to form a counterclockwise circuit) and they cannot cross using diagonal edges (by conditions (c) and (d)). Therefore, they must remain disjoint.

To show that the automatically filled wedge $A(v)$ is infinite for every site v, it remains to show that the exit and entrance boundaries each contain infinitely many distinct vertices. The case when the entrance and exit boundaries coincide was treated in the previous paragraph. In the case when the entrance and exit boundaries differ (after an initial segment of undirected edges), if one of these boundaries contains only finitely many distinct vertices, then it either contains a counterclockwise circuit (violating (b)) or a clockwise circuit (violating (c)). Therefore, $A(v)$ is infinite for every v.

A periodic class V of vertices is *self-filling* if there exists a pair of vertices v and w in V such that there is an automatic filling path from v to w. Suppose that a periodic class $\mathcal{U}$ is completely filled in a realization of the filling process. Let $u \in \mathcal{U}$ and let $u = v_0, e_1, v_1, e_2, \ldots$ be the exit boundary at u. Since there are a finite number N of periodic classes, $v_0, \ldots, v_N$ cannot be from distinct periodic classes. Thus, there exist $i < j \leqslant N$ for which v_i and v_j are in the same self-filling periodic class V. Using translates of the exit boundary from u to v_i, each vertex of V is automatically filled from a vertex of $\mathcal{U}$. Thus, if any periodic class of vertices is completely filled, a self-filling periodic class is completely filled. Note that the filling path used was a portion of an exit boundary.

Suppose that such a self-filling class V is not completely filled. Let $v \in V$ be a vertex that is not filled. By the self-filling property, there are vertices x and y in V such that there is an automatic filling path from x to y, which is an exit boundary path. Thus, there is an infinite automatic filling path p ending at v, consisting of translates of the path from x to y. No vertex of p is filled, or v would be filled. An edge of p that is in M is not filled, and thus does not border on the filled region. A diagonal edge of p cannot border on the filled region, since its endpoints are unfilled, and it cannot border or intersect the region filled from an unrestricted vertex on its left in the face. A diagonal edge of p cannot be crossed by automatic filling through a diagonal edge from right to left, since that violates condition (d). It cannot be crossed by a diagonal edge from a filled vertex from left to right, since the crossing edge would be an exit boundary for which (d) is violated. It cannot be crossed by $e_I(w)$ from a vertex w on its right side, since either $e_0(w)$ fills the terminal endpoint or $e_0(w)$ crosses the edge, violating (d).

Therefore, the curve representing the path does not border or intersect the filled region, and there exists a connected unbounded component of the plane, containing the path p, which is not filled. Thus, the filled wedge $W(s)$, where s denotes the source vertex of the filling process, has a self-avoiding curve, which corresponds to a doubly-infinite self-avoiding path, as part of its boundary. The vertex sequence for this path in G is just the sequence of vertices encountered as one follows the self-avoiding curve, since if the curve follows portions of the curves representing diagonals in a face, the resulting diagonal is present in G. All vertices in this path are restricted in the filling model, and thus open in the directed percolation model. It remains to show that the path is a directed path in D.

Consider two adjacent vertices v and w on the path bounding $W(s)$. Either v automatically fills w, w automatically fills v, or neither automatically fills the other.

If v automatically fills w, w is on the boundary of $A(v)$, and the edge $\{v, w\}$ is directed so that $A(v)$ lies along its left side, as desired. Similar reasoning applies when w automatically fills v.

Consider the case when neither v nor w automatically fills the other. The edge $\{v, w\}$ separates two faces of M, in exactly one of which both v and w are adjacent to a filled region. If the face is not close-packed, $\{v, w\}$ is part of the boundary of the filled region. If the face is close-packed, a curve (consisting of portions of diagonal edges) from v to w, which passes through no other vertex, is part of the boundary of the filled region. We will show that $\{v, w\}$ is directed so this face is on the left side of $\{v, w\}$.

It is possible that the automatically filled wedges $A(v)$ and $A(w)$ do not intersect. However, if they do not intersect, the entrance and exit boundaries of v and w are all "parallel" in the following sense: Two exit or entrance boundary paths q and r are parallel if there exist vertices $q_1, q_2 \in q$ from a self-filling class and $r_1, r_2 \in r$ from a self-filling class for which the line segments from q_1 to q_2 and r_1 to r_2 are parallel. It is easily seen that the exit boundaries of all vertices have a common direction, as do all entrance boundaries. Thus, if $A(v)$ and $A(w)$ do not intersect, the entrance and exit boundaries are all parallel, and there is an infinite strip of bounded width between $A(v)$ and $A(w)$, bounded by $\{v, w\}$, one exit boundary and one entrance boundary. By considering translates of a path between a vertex of $A(v)$ and a vertex of $A(w)$, we see that if the probability $p > 0$, the infinite strip is crossed by infinitely many filling paths, by the Borel-Cantelli lemma. Thus, it does not contribute to the infinite boundary path of $W(s)$ in G.

The union of $A(v)$, $A(w)$ and the edge $\{v, w\}$ has an infinite boundary path (which is disjoint from the boundary of the infinite strip between $A(v)$ and $A(w)$ if they do not intersect). In order to have the filled region on the left side, $\{v, w\}$ must then be directed from v to w. We prove this by contradiction.

Assume that $\{v, w\}$ is directed from w to v. By condition (c) there is a counter-clockwise doubly-infinite self-avoiding path r through (w, v). The path r must contain an entrance to w (which necessarily lies between $e_I(w)$ and (w, v) counter-clockwise) and an exit from v (which lies between $e_0(v)$ and (w, v) counterclockwise). The portion of r preceding (w, v), denoted r_1, cannot cross the entrance boundary of w, or else condition (d) is violated. The portion of r following (w, v), denoted r_2, cannot cross the exit boundary of v, since that would violate condition (d). Since r is a simple self-avoiding path, r_1 lies between the entrance boundary of w and r_2, and r_2 lies between r_1 and the exit boundary of v and r_1.

If $A(v)$ and $A(w)$ intersect, the exit boundary of v and the entrance boundary of w intersect, so r is finite, contradicting condition (c). If $A(v)$ and $A(w)$ do not intersect, r is contained in the infinite strip between them, separating the plane into two components, but bounds the component contained in the strip clockwise rather than counterclockwise. Thus $\{v, w\}$ must be directed from v to w, so the filled region is on the left.

Since the reasoning above applies to every edge on the boundary of $W(s)$, the filled region is bounded by a doubly infinite open directed path in the directed percolation model on D.

References

Blease, J., "Directed bond percolation on hypercubic lattices", *J. Phys. C*, **10** (1977), 925–936.

Dhar, D., and M. Barma, "Monte Carlo simulation of directed percolation on a square lattice", *J. Phys. C.* **14** (1981), L1–L6.

Dhar, D., M. Barma, and M. Phani, "Duality transformations for two-dimensional directed percolation and resistance problems", *Phys. Rev. Letters*, **46** (1981), 1238–1241.

Gray, L., R. T. Smythe, and J. C. Wierman, "Lower bounds for the critical probability in percolation models with oriented bonds", *J. Appl. Prob.*, **17** (1980), 979–986.

Hammersley, J. M., "Bornes superieres de la probabilité critique dans un processus de filtration", *Le calcul des probabilités et ses applications.* CRNS Vol. 87 (1957), 17–37.

Holley, R. A., and T. M. Liggett, "Ergodic theorems for weakly interacting infinite systems and the voter model", *Ann. Probability*, **3** (1975), 643–663.

Kertesz, J., and T. Vicsek, "Oriented bond percolation", *J. Phys. C*, **13** (1980), L343–L348.

Kesten, H., *Percolation theory for mathematicians*, Birkhäuser, Boston, 1982.

Kinzel, W., and J. Yeomans, "Directed percolation: a finite size renormalization approach", *J. Phys. A*, **14** (1981), L163–L168.

Mauldon, J., "Asymmetric oriented percolation in the plane", *Proc. Fourth Berkeley Symposium*, **1** (1961), 337–345.

Smythe, R. T., and J. C. Wierman, *First-passage percolation on the square lattice*, Lecture Notes in Mathematics, Vol. 671, Springer-Verlag, Berlin, 1978.

Sykes, M. F., and J. W. Essam, "Exact critical percolation probabilities for site and bond problems in two dimensions", *J. Math. Phys.*, **5** (1964), 1117–1127.

Wierman, J. C., "Duality for directed site percolation", *Particle Systems, Random Media and Large Deviations*, Contemporary Mathematics series, American Mathematical Society, Providence, 1985.

RANDOM GRAPHS '85: OPEN PROBLEMS

Problem 1 (*A. Frieze*) Let $D_{n,p}$ be a random digraph obtained from a complete digraph $\vec{K}_n$ by deleting each of $n(n-1)$ arcs independently with probability $1-p$. Find the threshold function for the property that $D_{n,p}$ is hamiltonian.

Conjecture: If $p=p(n)=\dfrac{1}{n}(\log n + c_n)$ then

$$\lim_{n\to\infty} P(D_{n,p} \text{ is ham.}) = \begin{cases} 0 & \text{if } c_n \to -\infty \\ e^{-e^{-c}} & \text{if } c_n \to c \\ 1 & \text{if } c_n \to +\infty. \end{cases}$$

Problem 2 (*A. Frieze*) Let k be a fixed natural number and $D_{n,k}$ be a random digraph on $\{1, 2, ..., n\}$ constructed as follows: Each vertex v chooses independently two disjoint sets of k vertices: $V_{\text{out}}(v)$ and $V_{\text{in}}(v)$. The arc (v, w) is in $D_{n,k}$ iff either $v \in V_{\text{in}}(w)$ or $w \in V_{\text{out}}(v)$. Does there exist such k_0 that for all $k \geq k_0$ $D_{n,k}$ is a.s. hamiltonian?

Conjecture: $k_0 = 2$.

Problem 3 (*A. Frieze*) Are almost all bipartite 4-regular graphs hamiltonian? (Cubic are!)

Problem 4 (*A. Frieze*) How many edges are needed so that a graph on n vertices a.s. contains a spanning maximal planar subgraph (i.e. with $3n-6$ edges)?

Problem 5 (*A. Frieze*) Let $P_{n,M}$ be a graph chosen at random from the family of all planar graphs on n vertices and M edges. What can be said about the properties of $P_{n,M}$? The simplest question about $P_{n,M}$ is to find the expected number of isolated vertices of $P_{n,M}$.

Problem 6 (*A. Frieze*) What is the chromatic index of a random r-regular graph?

Problem 7 (*D. Matula*) The graph $G=(V, E)$ is edge-path-regular if for some $m, k \geqslant 1$, the $m|E|$ edges of the multigraph $G^{[m]}$, formed by taking m distinct copies of each edge of E, can be partitioned into $k\binom{|V|}{2}$ paths, where k of the paths are shortest paths between i and j for each $i \neq j \in V$. Are almost all graphs edge-path-regular?

Problem 8 (*J. Cohen*) Let T_n be the number of vertices of out-degree zero in a random acyclic digraph with an arc probability $p = c/n$ $(c>0)$. Compute

$$\lim_{n \to \infty} \frac{1}{n} E(T_n).$$

Conjecture: The answer is e^{-c} (proved for $c<1$).

Problem 9 (*K. Weber*) Determine the threshold for planarity for random spanning (induced) subgraphs of the n-cube Q_n obtained from Q_n by independent deletion of its edges (vertices together with incident edges) with probability $1-p$.